The Use of High-purity Oxygen in the Activated Sludge Process

Volume I

Editor

J. R. McWhirter

Vice-president and General Manager
Environmental Systems Department
Linde Division
Union Carbide Corporation
New York, New York

Editor-in-Chief
Water Pollution Control Technology Series

Richard Prober

Adjunct Professor
Department of Chemical Engineering
Case Western Reserve University
Cleveland, Ohio

T0138846

CRC Press
Taylor & Francis Group
Boca Raton London New York

CRC Press is an imprint of the
Taylor & Francis Group, an **informa** business

CRC Press
Taylor & Francis Group
6000 Broken Sound Parkway NW, Suite 300
Boca Raton, FL 33487-2742

Reissued 2019 by CRC Press

© 1978 by Taylor & Francis Group, LLC
CRC Press is an imprint of Taylor & Francis Group, an Informa business

No claim to original U.S. Government works

A Library of Congress record exists under LC control number:

Publisher's Note
The publisher has gone to great lengths to ensure the quality of this reprint but points out that some imperfections in the original copies may be apparent.

Disclaimer
The publisher has made every effort to trace copyright holders and welcomes correspondence from those they have been unable to contact.

ISBN 13: 978-0-367-25950-1 (hbk)
ISBN 13: 978-0-367-25951-8 (pbk)
ISBN 13: 978-0-429-29071-8 (ebk)

Visit the Taylor & Francis Web site at http://www.taylorandfrancis.com and the CRC Press Web site at http://www.crcpress.com

FOREWORD

This second volume in the Uniscience Series on Water Pollution Control Technology is a comprehensive state-of-the-art report on the use of purified oxygen in the activated sludge wastewater treatment process. Economically feasible technology for use of oxygen rather than air in the activated sludge process is probably the most significant single development in wastewater treatment within the last decade. Dr. McWhirter and his colleagues, pioneers in the development and leaders in the commercialization of this process, have culled and critically reviewed data from numerous technical reports, journal articles, and conference presentations. Their objective was to provide a comprehensive and self-contained reference manual covering the waste treatment process itself as well as the necessary auxiliaries.

This book and other volumes in the Uniscience Series on Water Pollution Control Technology are addressed to design engineers, planners, and managers in industry and government. Our objective is to provide reference manuals that can be used by these diverse groups in this present critical period for implementation of water pollution control.

Richard Prober
Editor-in-Chief

PREFACE

The use of oxygen in the activated sludge process has grown over the past decade from a position of relative obscurity and limited academic interest to a position today of occupying a prominant role as a secondary wastewater treatment process. The economical substitution of oxygen for air in the activated sludge process was referred to by former Secretary of the Interior, Walter G. Hickel, as "the most significant technological advance in the secondary treatment of wastewater since the development of the activated sludge process in the early 1900s." The process has, indeed, achieved rapid and widespread use throughout the U.S. and is now receiving considerable attention throughout the world. Oxygen-activated sludge systems are already in operation in Japan, Mexico, Canada, the U.K., and Western Europe, and are being designed for use in the Philippines, South Africa, Finland, and Eastern Europe.

The advent of oxygen use in secondary treatment has brought with it a wealth of new technological information concerning the design of biochemical oxidation processes and, in particular, the activated sludge process. The use of oxygen in the activated sludge process has enabled many basic and substantial changes to be made in the fundamental design parameters of the process. Oxygen use and operation at high mixed-liquor dissolved oxygen levels removes many of the traditional constraints and limitations that have long plagued the standard air activated sludge process. This has resulted in the practical attainment of a high-rate, high-efficiency process which, in most cases, proves to be more economically attractive than the conventional air-activated sludge process as well as other secondary treatment alternatives.

The purpose of this two-volume series is to present a consolidated and comprehensive reference on oxygen-activated sludge technology. The subject matter is treated in considerable breadth and depth. The fundamental advantages of oxygen use and operation of the activated sludge process at high mixed-liquor dissolved oxygen levels is covered in detail. Many basic concepts of activated sludge system process design, equally applicable to both air and oxygen systems, are also included. The series is primarily aimed at wastewater treatment experts and, in particular, the practicing design engineer. There is also a substantial amount of material, however, that should be of interest and use in advanced undergraduate or graduate courses in environmental engineering and chemical engineering.

The present status of the oxygen-activated sludge process is due in large measure to the development and commercialization of the UNOX® System by the Linde Division of the Union Carbide Corporation. Consequently, this series is devoted almost exclusively to the UNOX oxygen-activated sludge system. Volume I of the series is divided into two parts which includes, (1) historical and background material relating to the use of oxygen and the development of the UNOX system, and (2) the basic process design considerations involved in oxygen-activated sludge system design. Volume II of the series is also divided into two parts consisting of (3) overall oxygenation system design considerations and additional applications, and (4) oxygen supply. As such, the two volumes effectively complement each other and provide a logical sequence for the presentation of the material which flows quite naturally from Volume I into Volume II. The individual parts of both volumes as well as the individual chapters of each part, however, develop in detail a particular aspect of the subject matter which enables it to virtually stand alone. Consequently, the reader is able to quickly focus on a subject area of particular interest and delve into this topic in depth without making extensive reference to the other parts of the series.

My sincere gratitude and admiration are extended to all of my colleagues at Union Carbide too numerous to name who were involved in the formulation and implementation of this work, many of whom also contributed greatly to the development of the UNOX System from its very early days in the laboratory to its current status of extensive world-wide application. Special thanks in this regard are particularly extended to Dr. L. C. Matsch, Mr. E. H. Zander, Mr. J. L. Steele, Mr. J. C. LeFever, Mr. R. H. Harris, and to Marie Costanzo for her tireless efforts in coordinating the preparation of the manuscript material by the numerous authors. Special appreciation is extended to Dr. Richard Prober, Editor-in-Chief of the CRC Press Uniscience Series on Water Pollution Control Technology, for his many helpful comments and suggestions, to the editorial staff of CRC Press, to Terri Weintraub, Gerald A. Becker, and Margaret Saulino for their cooperation and patience in the preparation of this book.

J.R. McWhirter
Westport, Connecticut
1977
November

EDITOR-IN-CHIEF

Richard Prober is a principal engineer with GMP Associates, and Adjunct Professor of Chemical Engineering at Case Western Reserve University, Cleveland, Ohio.

Dr. Prober received his B.S. in chemical engineering in 1957 from the Illinois Institute of Technology. In 1958 he received his M.S. degree and in 1962 his Ph.D. degree in chemical engineering from University of Wisconsin.

Dr. Prober's accomplishments include curriculum development for wastewater treatment plant operator training; development of low-flow dissolved oxygen models for the Cuyahoga River and Tinkers Creek, including direction of stream surveys to calibrate the models; development of process-design oriented B.S. and graduate level programs in wastewater engineering; and extensive research into process development of activated carbon treatment and treatment for industrial wastes containing cyanides.

His professional associations include the Water Pollution Control Federation, American Institute of Chemical Engineers, and the American Chemical Society. Dr. Prober has also served as Symposium Chairman for national meetings of the U.S. Environmental Protection Agency, American Institute of Chemical Engineers, and Wastewater equipment Manufacturer's Association.

THE EDITOR

John R. McWhirter is Vice-President and General Manager, Environmental Systems Department, Linde Division of the Union Carbide Corporation, New York, New York.

Dr. McWhirter was graduated from the University of Illinois in 1959 with a B.S. in chemical engineering. In 1961 he received his M.S. degree and in 1962 his Ph.D. in chemical engineering from The Pennsylvania State University.

Dr. McWhirter is the inventor of Union Carbides UNOX® System and has been issued a number of patents related to the use of oxygen and ozone in the field of wastewater treatment. Among the recent citations Dr. McWhirter has received is the 1970 National Chemical Engineering Personal Achievement Award for his pioneering efforts in the development of the UNOX System for wastewater treatment. The UNOX development also led directly to the 1971 Kirkpatrick Award for Chemical Engineering Achievement given to the Union Carbide Corporation. Dr. McWhirter is also the recipient of the 1976 Jacob F. Schoellkopf Award of the Western New York Section of the American Chemical Society.

Dr. McWhirter is a member of the American Institute of Chemical Engineers, American Chemical Society, Water Pollution Control Federation, and Water and Wastewater Equipment Manufacturers Association. He has also authored numerous articles and given many presentations before technical society meetings over the past 10 years in the area of wastewater treatment.

CONTRIBUTORS

J. G. Albertsson, M.S.
Manager
Equipment Engineering
Wastewater Treatment Engineering
  Department
Linde Division
Union Carbide Corporation
Tonawanda, New York

J. F. Andrews, Ph.D.
Professor
Department of Civil and Environmental
  Engineering
University of Houston
Houston, Texas

C. R. Baker, M.S.
Consultant
Industrial Gases Distribution and Process
  Technology
Process and Product Development Department
Linde Division
Union Carbide Corporation
Tonawanda, New York

R. C. Brenner, M.S.
Sanitary Engineer
Biological Treatment Section
Wastewater Research Division
Municipal Environmental Research
  Laboratory
U.S. Environmental Protection Agency
Cincinnati, Ohio

G. P. Breitbach, M.S.
Assistant Staff Engineer
Proposal and Pilot Plant Engineering
Wastewater Treatment Engineering
  Department
Linde Division
Union Carbide Corporation
Tonawanda, New York

R. V. Carlson, M.B.A.
Manager
Safety Engineering
Engineering Services Department
Linde Division
Union Carbide Corporation
Tonawanda, New York

J. J. Collins, M.S.
Product Manager
Process Systems
Engineering Products and Processes
  Department
Linde Division
Union Carbide Corporation
Tarrytown, New York

G. L. Culp, M.S.
President
Culp/Wesner/Culp
Clean Water Consultants
El Dorado Hills, California

R. F. Drnevich, M.S.
Section Engineer
Process Development
Wastewater Treatment Engineering
  Department
Linde Division
Union Carbide Corporation
Tonawanda, New York

P. J. Gareis, B.A.
Associate Director
Cost Engineering and Scheduling
Project Management
Linde Division
Union Carbide Corporation
Tonawanda, New York

D. W. Gay, M.S.
Region Sales Manager
Municipal Systems
Southeast Region
Environmental Systems Department
Linde Division
Union Carbide Corporation
Atlanta, Georgia

R. J. Grader, M.S.
Product Manager
Municipal Ozonation Systems
Environmental Systems Department
Linde Division
Union Carbide Corporation
Tonawanda, New York

W. E. Grunert, M. S.
Manager
Process Engineering
Wastewater Treatment Engineering
  Department
Linde Division
Union Carbide Corporation
Tonawanda, New York

R. F. Gyger, B.S.
Supervisor
Pilot Plants
Proposal and Pilot Plant Engineering
Wastewater Treatment Engineering
  Department
Linde Division
Union Carbide Corporation
Tonawanda, New York

D. W. Kaltrider, B.S.
Associate Director
Engineering Services Department
Linde Division
Union Carbide Corporation
Tonawanda, New York

S. E. King, M.S.
Marketing Specialist
Southeast Region
Environmental Systems Department
Linde Division
Union Carbide Corporation
Atlanta, Georgia

R. J. Kulperger, B.S.
General Manager
Gas Products Department
Union Carbide Canada Limited
Toronto, Canada

L. M. LaClair., M.B.A.
Product Manager
Nutrient Removal and Sludge Disposal
Environmental Systems Department
Linde Division
Union Carbide Corporation
Tonawanda, New York

M. S. Lipman, B.Ch.E.
Associate Director
PUROX Systems Engineering Department
Linde Division
Union Carbide Corporation
Tonawanda, New York

M. R. Lutz, B.Ch.E.
Manager
Field Engineering
Wastewater Treatment Engineering
  Department
Linde Division
Union Carbide Corporation
Tonawanda, New York

L. C. Matsch, Ph.D.
Senior Engineering Fellow
Process and Systems Development
Wastewater Treatment Engineering
  Department
Linde Division
Union Carbide Corporation
Tonawanda, New York

J. R. McWhirter, Ph.D.
Vice President and General Manager
Environmental Systems Department
Linde Division
Union Carbide Corporation
New York, New York

M. A. Miller, B.S.
Manager
Proposal and Pilot Plant Engineering
Wastewater Treatment Engineering
  Department
Linde Division
Union Carbide Corporation
Tonawanda, New York

N. Nenov, M.S.
Senior Engineer
New Product Development
Heat Transfer Technology Department
Linde Division
Union Carbide Corporation
Tonawanda, New York

G. J. Novak, B.S.
Manager
Operations Engineering
Air Separation Engineering Department
Linde Division
Union Carbide Corporation
Tonawanda, New York

D. A. Okun, Sc.D.
Kenan Professor
Environmental Engineering
The School of Public Health
Department of Environmental Sciences and
 Engineering
The University of North Carolina at Chapel
 Hill
Chapel Hill, North Carolina

H. M. Rosen, Ph.D.
Sales Manager
Ozonation Systems
Environmental Systems Department
Linde Division
Union Carbide Corporation
Tonawanda, New York

C. Scaccia, Ph.D.
Systems Development Supervisor
Process and Systems Development
Wastewater Treatment Engineering
 Department
Linde Division
Union Carbide Corporation
Tonawanda, New York

R. L. Shaner, B.Ch.E.
Senior Engineering Associate
Plant Systems Optimization
Air Separation Engineering Department
Linde Division
Union Carbide Corporation
Tonawanda, New York

M. J. Stankewich, Jr. M.S.
Program Manager
PUROX Systems Engineering
Linde Division
Union Carbide Corporation
Tonawanda, New York

N. P. Vahldieck, M.S.
Engineering Associate
Process and Systems Development
Wastewater Treatment Engineering
 Department
Linde Division
Union Carbide Corporation
Tonawanda, New York

E. A. Wilcox, M.S.
General Manager
Linde Products
Union Carbide Eastern, Inc.
Tokyo, Japan

K. W. Young, M.S.
Manager
Development
PUROX Systems Engineering
Linde Division
Union Carbide Corporation
Tonawanda, New York

E. H. Zander, M.S.
Associate Director
Wastewater Treatment Engineering
 Department
Linde Division
Union Carbide Corporation
Tonawanda, New York

DEDICATION

This book is dedicated to my colleagues at Union Carbide
whose skill, energy, and devotion to the UNOX effort over the past decade
has resulted in the many technical advancements embodied in,
and the commercial success of, the Unox system.

TABLE OF CONTENTS

VOLUME I

VOLUME II

OVERALL OXYGENATION SYSTEM DESIGN CONSIDERATIONS AND ADDITIONAL APPLICATIONS

OXYGEN SUPPLY

Background and Introduction

Chapter 1

INTRODUCTION

J. R. McWhirter

TABLE OF CONTENTS

The use of relatively high-purity oxygen gas in the activated sludge process has evolved since the mid-1960s from a position of primarily academic interest to a point today where it enjoys broad commercial application. A large and rapidly increasing number of oxygen-activated sludge wastewater treatment plants are already in operation and under construction, and the process continues to realize vigorous growth in the U.S., Japan, Europe, and Canada. As of July 1977, 90 oxygen-activated sludge plants were in operation worldwide with a total combined treatment capacity of approximately 1.5 billion gallons per day. An additional 61 plants were under construction throughout the world, with a cumulative capacity of over 2.8 billion gallons per day. Furthermore, another 97 oxygen plants are in various stages of design which represent an additional 3.1 billion gallons per day of total treatment plant capacity. Collectively, these 248 plants represent an overall combined treatment capacity in excess of 7.4 billion gallons per day which is committed to the use of the oxygen-activated sludge process. By 1985, it is projected that the total secondary treatment capacity which will be using or committed to the oxygen process will exceed 11 billion gallons per day. This represents a highly significant portion of the anticipated total secondary treatment capacity and is ample testimony to the importance and widespread application of the

oxygen-activated sludge process. It is also evidence of the fact that the development, commercialization, and utilization of the oxygen-activated sludge process has been considerably more rapid and extensive than normally associated with new wastewater treatment processes.

I. BRIEF HISTORY OF OXYGEN USE IN WASTEWATER TREATMENT

A. Early Studies

Perhaps the earliest known reference regarding commercial consideration of the use of pure oxygen in wastewater treatment occurred in 1934 in an internal memorandum of the Linde Division of Union Carbide Corporation.[1] This was long before the development of the modern industrial gas industry as we know it today, and concerned the desirable market characteristics of potential oxygen use in wastewater treatment. It was thought that this application would provide a steady, base-load oxygen requirement which would stimulate the use and supply of industrial gases. In his December 4, 1934 letter to Dr. L. I. Dana, then Director of the Linde Research Laboratory, Mr. J. J. Murphy wrote:

During a conversation on one of your recent visits to New York we both agreed that new and larger uses for oxygen would be well worth finding. In addition, I personally believe that we could find uses which would be steady and not greatly affected by the normal depression in business cycles. With a very steady load we could probably be in a position to sell oxygen rather cheaply. The two uses I have in mind are water purification and sewage treatment.

The subject was again revived after World War II when the work of Pirnie[2] and Okun[3-5] on the bio-precipitation process was stimulated by the development of "Cheap Tonnage Oxygen" through the use of relatively low-cost, simple oxygen plants for the production of "low-purity" (95%) oxygen. The bio-precipitation process was further studied in the early 1950s by Budd and Lambeth,[6] but little subsequent interest developed. This process, as discussed in Chapters 2 and 3, Volume I, involved not only the use of oxygen but also a new activated sludge process configuration employing a combined reactor-clarifier concept which received preoxygenated feed.

In the late 1950s and early 1960s, interest in the use of oxygen diminished considerably as attention was primarily focused on the effects of the dissolved oxygen concentration on the rate of bio-oxidation or oxygen uptake rate for conventional activated sludge operation conditions.[7,8] These studies revealed that under the typical operating conditions employed in the conventional air-activated sludge process (low MLVSS levels and relatively low oxygen uptake rates), increased dissolved oxygen levels in the mixed-liquor did not increase the rate of biochemical oxidation or the rate of substrate removal. These apparently negative results regarding the potential benefits of higher dissolved oxygen concentrations in the conventional activated sludge process, when coupled with the assumed relative inefficiency and high cost of oxygen use, resulted in a general lack of interest in the concept. As will be discussed further in subsequent chapters, these studies did not reveal many of the advantages of oxygen use and operation at high dissolved oxygen levels for some very fundamental reasons. Earlier studies [9-11] that revealed little or no improvement in treatment efficiency with the use of oxygen were negative for essentially the same reasons.

McKinney and Pfeffer[12] reported on the use of oxygen-enriched air in biological wastewater treatment in 1965. They argued that it is not the level of dissolved oxygen concentration that can be obtained with the use of oxygen that is important, but rather the rate at which oxygen can be transferred to the mixed-liquor. They observed that at lower dissolved oxygen levels, the rate of transfer of oxygen would be significantly increased through the use of oxygen-enriched air. They reasoned that higher oxygen gas concentrations could only be advantageous under oxygen-limiting conditions. If higher oxygenation rates could be obtained, it would be possible to substantially increase the organic loading on the system. By simultaneously increasing the organic loading and the mixed-liquor volatile suspended solids levels, a high degree of treatment efficiency could be obtained at the higher loading and higher oxygen transfer rate conditions. Such, of course, is the case and represents one of the important process and economic benefits of oxygen use in the activated sludge process.

McKinney and Pfeffer[12] further pointed out that there must be some economic gain or process improvement associated with the use of oxygen in comparison to conventional air aeration techniques. They referred to the following areas where such improvements should be obtained:

1. The avoidance of excessively high aeration rates and a reduction in the power required per unit of oxygen transferred, particularly at high oxygen uptake rate conditions

2. Increased rate of stabilization of organic material through elimination of oxygen transfer limitations

3. Reduction in, or elimination of, periods of zero dissolved oxygen concentration to improve overall process operating conditions

4. Reduction in plant size and capital investment

5. Increased capacity of organically overloaded plants without the need for increasing the plant hydraulic capacity

6. Operation of high rate, high organically loaded systems at high removal efficiencies by removing oxygen transfer limitations

As will be amply illustrated and documented throughout these two volumes, these and other process and economic advantages can be practically and effectively realized through the efficient use of oxygen in the activated sludge process.

The problem remained, however, as to how to achieve the inherent potential process advantages of oxygen use while simultaneously obtaining

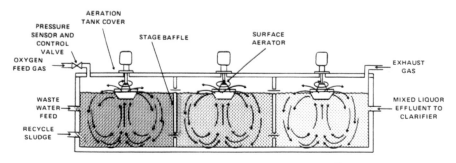

FIGURE 1. Simplified, schematic diagram of three-stage UNOX® System using surface aerators. (Courtesy of Union Carbide Corporation.)

efficient and economical oxygen utilization and retaining the desirable features of the conventional activated sludge process flow configuration.

II. THE UNOX® SYSTEM

The major impetus for the use of oxygen in wastewater treatment came with the development of the UNOX System by the Linde Division of Union Carbide Corporation. A study to evaluate the use of oxygen in the secondary treatment of wastewater was begun by Union Carbide in late 1966, followed by the invention of the UNOX System. The UNOX System is illustrated schematically in its usual form in Figure 1. The system uses a covered, multistage oxygenation and reaction system in which the wastewater and activated sludge are contacted in a series of cocurrent, gas-liquid mixing stages. The system is operated at essentially atmospheric pressure with relatively high-purity oxygen gas, wastewater, and recycle sludge being introduced into the first stage and flowing cocurrently through the successive contacting stages. The exhaust gas is vented from the last stage and the mixed-liquor is settled in a conventional gravity clarifier. The oxygen feed gas rate can be automatically controlled in direct proportion to the actual oxygen uptake rate in the system by means of a simple pressure controller in the first stage. As the oxygen uptake rate diminishes, the pressure in the first stage tends to increase and the oxygen feed rate is reduced. Alternatively, as the oxygen demand increases, the pressure in the first stage tends to decrease and the oxygen feed rate is increased. The number of stages employed, the overall reactor geometrical configuration, tank depth, and the specific type of gas-liquid contacting devices or aerators used in the stages can be economically optimized for each specific installation and process requirement. The system is extremely flexible in design and construction and employs standard mechanical equipment used in conventional air aeration practice. Standard surface aerators are typically used in most installations. The relatively simple construction features and the use of standard equipment components are two of the strong attributes of the system.

Although quite simple and straightforward in its physical embodiment, the underlying mass transfer and reaction processes involved in the UNOX System are considerably more complex than those involved in a conventional air aeration system. The enclosed, multistage recycle oxygenation system enables a high oxygen utilization efficiency to be achieved and maintains a high average oxygen partial pressure in the aerating gas. This results in efficient and economical oxygen utilization while simultaneously achieving a substantial reduction in the oxygen dissolution energy requirement per unit of oxygen dissolved. Also, economical operation at relatively high dissolved oxygen levels (4 to 6 mg/l) relative to standard aeration practice (1 to 2 mg/l) can be achieved. As will be discussed in subsequent chapters, this results in significant additional process and economic advantages with respect to mixed-liquor settling rates, sludge dewatering characteristics, and sludge production rates. It is this unique combination of effects that results in the UNOX System's overall process advantages and the accompanying economic benefits.

The first experimental tests of the UNOX System were accomplished in late 1967 and early 1968 in a simulated bio-oxidation system pilot plant. This unit (shown in Figure 2) employed a surface aeration system with four gas stages and a 1500-gal aeration tank, 14 ft in length by 3 ft in

FIGURE 2. Simulated Bio-Oxidation system pilot plant unit used to conduct first experimental tests of the UNOX® System. (Courtesy of Union Carbide Corporation.)

FIGURE 3.   Pilot plant unit used to conduct the first activated sludge tests of the UNOX® System at Batavia, New York. (Courtesy of Union Carbide Corporation.)

width with a water depth of approximately 3 ft. These tests provided basic verification of the capability of the multistage oxygenation system to achieve a high oxygen utilization efficiency while simultaneously maintaining a high average oxygen partial pressure in the aeration gas under the multicomponent mass transfer conditions encountered in the biochemical oxidation of wastewater. Following these simulated bio-oxidation tests, pilot plant tests involving actual activated sludge operation on municipal wastewater were conducted at Batavia, New York, during 1968 (Figure 3). These tests confirmed the activated sludge process advantages achievable with oxygen aeration as well as the economical utilization and dissolution of the oxygen gas.

The next phase of the UNOX program involved a full-scale plant test at the Batavia, New York municipal wastewater treatment plant under the sponsorship of the Federal Water Quality Administration (FWQA)[13,14] the predecessor of the current Environmental Protection Agency

(EPA). This program was begun in late 1968 and spanned a period of approximately 2½ years. It involved a direct side-by-side comparison of the UNOX System with a conventional air-activated sludge system. Half of the Batavia treatment plant was converted to the use of the UNOX System while the other half was operated using the conventionally designed diffused air aeration system (Figure 4*). This comprehensive study provided full-scale verification of the process and economic advantages of oxygen vs. air aeration in the activated sludge process.

Following the successful conclusion of the Batavia demonstration program, Union Carbide announced its commercial entry into the wastewater treatment business in May of 1970. Shortly thereafter, the city of Detroit formally announced its decision to redesign one half of the first segment of its new secondary wastewater treatment plant for the use of the UNOX System. This marked the first commercial commitment to the use of the UNOX System. The half of the

* For Figure 4 see color insert, following page 50.

FIGURE 5. Overall view of 300 MGD UNOX® System at Detroit, Michigan with 180 ton/day cryogenic oxygen plant and 900 ton liquid oxygen storage tank in the background. (Courtesy of Union Carbide Corporation.)

plant designed for UNOX had a treatment capacity of 300 MGD while using the same basic tank design as the other half of the plant which was designed for 150 MGD of capacity using a diffused air aeration system. The UNOX System was designed for a retention time of 1.14 hr and the air system was designed for a retention time of 2.28 hr (each design based on total flow through the aeration tanks including both influent waste and sludge recycle). Each system was designed to produce an effluent total BOD$_5$ of less than 25 mg/l from a total BOD$_5$ in the feed to the secondary system of 140 mg/l. The two systems were then to be tested and directly compared in performance and operation prior to a future decision regarding the design of the remainder of the facility. However, long before these first two plant segments were completed, the city of Detroit decided to base its remaining plant capacity on the oxygenation system design. In

mid-1974, construction of an additional 600 MGD oxygen-activated sludge system was begun. The initial 300 MGD UNOX System in Detroit was placed into operation in 1975 and has more than confirmed its ability to meet or exceed the design performance requirements. Figure 5 shows an overall view of the 300 MGD UNOX System in Detroit with the 180 ton/day cryogenic oxygen plant and 900 ton liquid oxygen storage tank in the background.

In the early 1970s, Union Carbide initiated an extensive and comprehensive pilot plant program to demonstrate the advantages and capability of the UNOX System on a wide variety of wastewaters. Since that time, over 200 pilot plant or treatability studies have been conducted on virtually every type of municipal and industrial wastewater. Most of these tests were run using on-site mobile pilot plants mounted on trailer vans, such as the one shown in Figure 6. These

FIGURE 6. Typical mobile UNOX® pilot plant unit used for conducting on-site pilot plant tests. (Courtesy of Union Carbide Corporation.)

tests were conducted over an extended time period to enable evaluation of various climatic and wastewater conditions as well as operation over a wide range of process parameters. This pilot plant program is believed by many in the field to be the most exhaustive and comprehensive test program ever conducted on a new wastewater treatment process. The results of this program were aptly summarized in a news article that appeared in the November 1974 issue of the *Journal of the Water Pollution Control Federation:*[15]

The oxygen activated sludge process is no longer an interesting concept of dubious practicality. Careful investigation and pilot plant testing have largely confirmed its publicized potential. As a process, oxygen activated sludge has come into its own.

III. CURRENT STATUS OF OXYGEN-ACTIVATED SLUDGE SYSTEM USE

During the last several years, numerous oxygen-activated sludge plants have come on-stream with an additional 20 to 25 plants going into operation every year. Tables 1 to 4 summarize the current status of oxygen-activated sludge system applications in the U.S. and world-wide and provide a breakdown into municipal and industrial wastewater usage. These tables are updated and slightly revised versions of those originally published by Brenner[16] in June 1976. As seen in these tables, application of the process is broadly based both geographically and by type of wastewater.

Brenner also published extensive operating and performance data case histories for 11 operating oxygen-activated sludge plants.[16] These plants cover a wide variety of process applications, system component configurations, and plant sizes. Figures 7 and 9 show pictures of several of the plants reviewed. With only one exception, these plants have met or substantially exceeded the design performance requirements. The lone case of subdesign performance resulted from excessive fat

TABLE 1

World-wide Oxygen-activated Sludge Plant Status, July 1977

Parameter		Operating plants	Plants under construction	Plants being designed	Totals
No. of plants					
Municipal		45	56	92	193
Industrial		45	5	5	55
	Totals	90	61	97	248
Design flow (MGD)					
Municipal		1261.0	2615.7	2958.3	6835.0
Industrial		283.2	191.4	150.1	624.7
	Totals	1544.2	2807.1	3108.4	7459.7
O_2 supply capacity (tons/day)					
Municipal		1749.4	4201.0	4257.0	10207.4
Industrial		982.7	204.5	340.3	1527.5
	Totals	2732.1	4405.5	4597.3	11734.9

Modifed from Brenner, R. C., Updated Status of Oxygen Activated Sludge Wastewater Treatment, U.S. Environmental Protection Agency, Municipal Environmental Research Laboratory, Cincinnati, Ohio, June 1976.

TABLE 2

U.S. Oxygen-activated Sludge Plant Status, July 1977

Parameter		Operating plants	Plants under construction	Plants being designed	Totals
No. of Plants					
Municipal		39	44	74	157
Industrial		24	4	2	30
	Totals	63	48	76	187
Design flow (MGD)					
Municipal		1245.7	2326.4	2504.2	6076.3
Industrial		204.2	165.0	45.0	414.2
	Totals	1449.9	2491.4	2549.2	6490.5
O_2 supply capacity (tons/day)					
Municipal		1734.4	3805.5	3602.0	9141.9
Industrial		709.7	188.0	180.0	1077.7
	Totals	2444.1	3993.5	3782.0	10219.6

Modified from Brenner, R. C., Updated Status of Oxygen Activated Sludge Wastewater Treatment, U.S. Environmental Protection Agency, Municipal Environmental Research Laboratory, Cincinnati, Ohio, June 1976.

and grease loadings in the influent (often in excess of 100 mg/l) from a local poultry processor. This excessive amount of grease, which escapes the secondary clarifiers as a consequence of inadequate skimming and grease removal capability, has prevented consistent attainment of the effluent quality objectives.

The overall process and mechanical reliability of these plants has been excellent. A few minor mechanical problems with the equipment were encountered with some of the early Pressure Swing Adsorption (PSA) oxygen generators and with some of the aeration system components; however, these were readily corrected. The plants reported on by Brenner,[16] as well as all of the 90 oxygen-activated sludge plants in operation world-wide by July 1977 have confirmed the process and economic advantages of oxygen use.

TABLE 3

Breakdown of Oxygen-activated Sludge Plants by Country, July 1977

Country	Operating plants	Plants under construction	Plants being designed	Totals
U.S.	63	48	76	187
Japan	22	3	5	30
Canada	1	1	1	3
Mexico	1	–	–	1
United Kingdom	1	–	3	4
Germany	1	2	3	6
Denmark	–	1	–	1
Switzerland	–	1	–	1
Belgium	1	–	–	1
Italy	–	3	2	5
France	–	2	2	4
Poland	–	–	1	1
Austria	–	–	2	2
Finland	–	–	1	1
Taiwan	–	–	1	1
Totals	90	61	97	248

Modified from Brenner, R. C., Updated Status of Oxygen Activated Sludge Wastewater Treatment, U.S. Environmental Protection Agency, Municipal Environmental Research Laboratory, Cincinnati, Ohio, June 1976.

TABLE 4

Breakdown of Oxygen-activated Sludge Plants by Industrial Application, July 1977

Industrial application	Operating plants		Plants under construction		Totals	
	No. of plants	Design flow (MGD)	No. of plants	Design flow (MGD)	No. of plants	Design flow (MGD)
Chemicals	10	38.9	0	0	10	38.9
Dyestuffs	1	3.1	0	0	1	3.1
Food processing	2	2.5	0	0	2	2.5
Petrochemical	9	27.4	0	0	9	27.4
Pharmaceutical	2	1.7	0	0	2	1.7
Pulp and paper	18	194.3	5	165.0	23	359.3
Steel	1	13.7	0	0	1	13.7
Synthetic rubber	1	0.8	0	0	1	0.8
Brewing	1	0.8	0	0	1	0.8
Totals	45	283.2	5	165.0	50	448.2

Modified from Brenner, R. C., Updated Status of Oxygen Activated Sludge Wastewater Treatment, U.S. Environmental Protection Agency, Municipal Environmental Research Laboratory, Cincinnati, Ohio, June 1976.

FIGURE 7.   12 MGD UNOX® System at Winnipeg, Canada. (Courtesy of Union Carbide Corporation.)

FIGURE 8.   7.5 MGD UNOX® System at Speedway, Indiana. (Courtesy of Union Carbide Corporation.)

FIGURE 9. 8 MGD UNOX® System at Morganton, North Carolina. (Courtesy of Union Carbide Corporation.)

IV. ALTERNATIVE OXYGENATION SYSTEM DESIGNS

The vast majority of the development and commercial application of the oxygen-activated sludge process to date has been with the UNOX System. The attention drawn to the use of pure oxygen has, however, stimulated other developments. Among these are the OASES® System by Air Products and Chemicals, Incorporated, Allentown, Pennsylvania; the MAROX® System by FMC, Chicago; and the F³O System by Air Reduction Company, Murray Hill, New Jersey.

The OASES System for wastewater treatment was first introduced on the market in 1971 by Air Products and Chemicals, Incorporated. This system uses covered oxygenation basins that are typically divided into a number of gas and liquid contacting stages with gas and liquid flowing cocurrently through the reactor. Oxygen feed control is normally accomplished by means of reactor pressure control. Air Products also uses a

design modification referred to as the Controlled Back Mix OASES System. This design employs a covered oxygenation basin with multi-stage gas contacting. The mixed-liquor is contacted cocurrently with the gas in an elongated, "serpentine," liquid flow circuit defined by partitions in the reactor.

The MAROX System, marketed by FMC Corporation, utilizes an uncovered oxygenation basin design in which the oxygen feed gas is directly diffused into the mixed-liquor. This sytem uses rotating active diffusers for oxygen dissolution. Oxygen feed gas is pressurized and fed through a hollow shaft mixer assembly to the diffusers which disperse the oxygen gas into small bubbles. Oxygen feed gas control is accomplished by the use of mixed-liquor D.O. probes that maintain a preset dissolved oxygen concentration in the mixed-liquor.

The most recent development in the design of oxygen-activated sludge systems is Airco, Incorporated's forced free-fall oxygenation (F^3O) sys-oxygen dissolution is accomplished in the process. Physically, the F^3O System is a self-contained module of monolithic cast concrete construction designed to be fully submerged within an aeration basin. Mixed-liquor is pumped into the top center of the module by means of an axial flow impeller. The flow enters a central draft tube, is forced up an outer annular region and falls through an enclosed region where a high purity oxygen atmosphere is maintained. Oxygen transfer is effected in this region. The pumped mixed-liquor then leaves the module through exit ports in the base and returns to the external aeration basin.

Mixed-liquor D.O. levels are used to control the height of the liquid fall and the oxygen feed rate.

This two-volume series is principally devoted to the technology and application of the UNOX System. The editor and many of the chapter authors have been intimately involved with, and responsible for, the development of the UNOX System since its very beginnings. The concentration on the UNOX System is due to its widespread commercial acceptance and their familiarity, experience, and access to the extensive body of design and performance information related to this method of oxygen-activated sludge wastewater treatment.

REFERENCES

1. Union Carbide Corporation, Internal Memorandum, Tonawanda, New York, December 4, 1934.
2. Pirnie, M., Cheap Oxygen Possible New Tool in Sanitary Engineering, Office Memorandum, Malcolm Pirnie Engineers, White Plains, New York, 1946.
3. Okun, D. A., Pure oxygen in bio-precipitation process may reduce sewage treatment costs, *Civ. Eng.* New York, 18, 283, 1948.
4. Okun, D. A., A System of Bio-Precipitation of Organic Matter From Sewage, Dissertation, Harvard University, Cambridge, Mass., 1948.
5. Okun, D. A., System of bio-precipitation of organic matter from sewage, *Sewage Works J.*, 21, 253, 1949.
6. Budd, W. E. and Lambeth, G. F., High purity oxygen in biological waste treatment, *Sewage Ind. Wastes*, 29, 237, 1957.
7. Okun, D. A. and Lynn, W. R., Preliminary investigations into the effect of oxygen tension on biological sewage treatment, in *Biological Treatment of Sewage and Industrial Wastes*, Rheinhold, New York, 1956, chap. 2–4.
8. Okun, D. A., Discussion – high purity oxygen on biological sewage treatment, *Sewage Ind. Wastes*, 29, 253, 1957.
9. Calvert, A. T., Report of Water Pollution Board for Year Ending June 30, 1937, British Dept. of Scientific and Industrial Research, 1937.
10. Grant, S., Hurwitz, E., and Mohlman, F. W., Oxygen requirements of the activated sludge process, *Sewage Works J.*, 2, 228, 1930.
11. Williams, E., Effect of bubbles of gas on mixtures of sewage and activated sludge, *J. Soc. Chem. Ind. London*, 59, 39, 1940.
12. McKinney, R. E. and Pfeffer, J. T., Oxygen-enriched air for biological waste treatment, *Water Sewage Works*, 112, 381, 1965.
13. Albertsson, J. G., McWhirter, J. R., Robinson, E. K., and Vahldieck, N. P., Investigation of the Use of High Purity Oxygen Aeration in the Conventional Activated Sludge Process, *Water Pollut. Control Res. Ser.*, Publ. 17050 DNW, Federal Water Quality Administration, May 1970.
14. Linde Division, Union Carbide Corporation, Continued Evaluation of Oxygen Use in the Conventional Activated Sludge Process, Water Quality Office, Environmental Protection Agency, Project No. 17050 DNW, September, 1971.
15. Ward, P. S., Interest in oxygen process growing, *J. Water Pollut. Control Fed.*, 46(11), 2451, November 1974.
16. Brenner, R. C., Updated Status of Oxygen Activated Sludge Wastewater Treatment, U.S. Environmental Protection Agency, Municipal Environmental Research Laboratory, Cincinnati, Ohio, June 1976.

Chapter 2
OXYGEN AND THE BIO-PRECIPITATION PROCESS

D. A. Okun

TABLE OF CONTENTS

I. INTRODUCTION

On October 26, 1946, the late Malcolm Pirnie, a consulting engineer of New York City, sent the late Gordon M. Fair, then Dean of the Graduate School of Engineering of Harvard University, an office memorandum entitled "Cheap Oxygen Possible New Tool in Sanitary Engineering."[1] Pirnie had been intrigued with the development of relatively low-cost and simpler plants for the production of oxygen by air liquefaction and rectification during World War II. The low cost was possible because the percent purity was somewhat lower than oxygen produced conventionally, and because it might be used directly without compression or liquefaction. Such "tonnage" oxygen appeared to have promise for industry, particularly in increasing the productivity of steel mills. It seemed to Mr. Pirnie that such oxygen might find a place in biological wastewater treatment, where increasing the availability of oxygen might increase the rate of treatment.

With his memorandum, Pirnie attached a proposed plant design that made use of an upflow precipitator that he had developed for the removal of coagulated floc particles in water treatment.

The modification was that the influent to this upflow precipitator was oxygenated with pure oxygen, thereby carrying dissolved oxygen into the floc blanket (Figure 1).

This writer was at that time a Teaching Fellow and a doctoral candidate at Harvard University and Dean Fair suggested that he might undertake a study of the potential of pure oxygen for wastewater treatment. Fill-and-draw activated sludge treatment studies with pure oxygen were undertaken to establish whether or not the microbial life associated with biological wastewater treatment was benefited or deleteriously affected. A search of the literature revealed that others had earlier made studies of the use of pure oxygen in activated sludge and, if the results were not conclusive, they provided no evidence that high dissolved oxygen concentrations interfered with microbial activity.[2-5] It was noted that nitrification may be inhibited in wastewaters and diluted wastewater samples, particularly in biochemical oxygen demand (BOD) tests incubated at high dissolved oxygen concentrations.[6]

When the fill-and-draw activated sludge studies proved promising, parallel continuous-flow pilot plants were installed in the laboratory to demon-

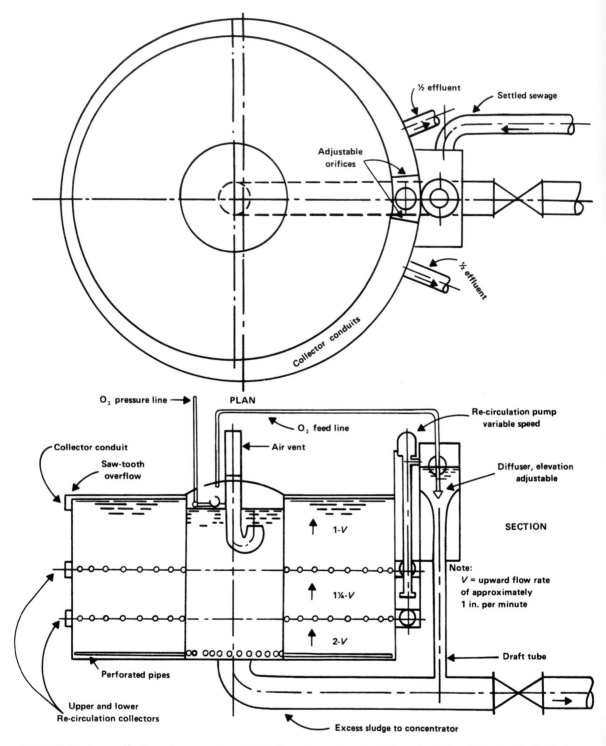

FIGURE 1. An application of oxygen for stabilization of organic material in water. (From Malcolm Pirnie Engineers, New York, 1946.)

strate the feasibility of using pure oxygen in a system involving preoxygenation and subsequent treatment in an upflow blanket clarifier. The first report of this work was made in an address delivered by Okun at the Centennial Celebration of the founding of the Lawrence Scientific School at Harvard University on February 13, 1948 and published under the title "Pure Oxygen in Bio-Precipitation Process May Reduce Sewage Treatment Costs."[7]

The process differed substantially from the conventional activated sludge process, with all of the activity occurring in a single biological precipitation chamber. The separate aeration and sedimentation units of conventional activated sludge plants were not required.

Accordingly, a name for the process was sought, and bio-precipitation was selected from its early employment by Buswell, Shive, and Neave,[8] who wrote:

Colloids .... may be said to be precipitated if they are taken up as food by micro-organisms of such growth habits that they form large compact flocs which will settle out. This process for the removal of sewage colloids we have called "bio-precipitation."

A complete description of the research, including an extensive bibliography of 132 items, the raw data and their analyses, numerous photomicrographs, and an appendix describing studies into a polarographic method for the determination of dissolved oxygen (which was to become important in the control of the process) is included in a 290-page dissertation[9] which is summarized in a paper published in 1949.[10]

II. CHARACTERISTICS OF PURE OXYGEN

Pure oxygen is attractive as a consequence of Henry's law which states that the saturation concentration of a gas in a liquid is directly proportional to the partial pressure of the gas in the atmosphere in contact with the liquid:

$$c_s = k_s p$$

where c_s is the saturation concentration of the gas in the liquid in milligrams per liter, p is the partial pressure of the gas in the gas phase in atmospheres, and k_s is the coefficient of absorption or Henry's law constant in milligrams per liter per atmosphere.

The absorption coefficient or Henry's law constant for oxygen in water at $0°C$ is 70.5 mg/l/atm. At a concentration of oxygen in air of 21% (p = 0.21 atm), the equilibrium saturation concentration of oxygen in water at $0°C$ is the product of these, or 14.8 mg/l. At $0°C$ and 760 mm of mercury pressure, the weight of oxygen is 1.43 mg/ml, so that the saturation concentration is 10.4 ml/l.

If the proportion of oxygen in the gas increases from 21 to 100%, the equilibrium saturation concentration increases some fivefold, raising the saturation concentration from 14.8 to approximately 70.5 mg/l. (The above calculations neglect the partial pressure of water vapor which does not change the order of magnitude of the figures significantly.)

Additionally, the saturation concentration is proportional to the absolute pressure. The saturation concentration of oxygen increases approximately 50% for both air and pure oxygen when the gas is introduced in a submerged injector in 17 ft of water and 100% with 34 ft of submergence. The rate of gas absorption is proportional to its degree of undersaturation in the absorbing liquid:

$$\frac{dc}{dt} = k_g (c_s - c_t)$$

where dc/dt is the rate of change of concentration or rate of absorption of the gas at time t; c_s is the saturation concentration; c_t is the concentration in the liquid at time t; and k_g is a proportionality factor (mass transfer coefficient) for the conditions of exposure. From this relationship, it is clear that, if c_t is low, the rate of absorption of oxygen is directly proportional to the saturation concentration, or therefore approximately fivefold greater with oxygen than air.

It is also known that wastewaters are likely to be saturated with nitrogen with respect to the nitrogen in air since nitrogen is essentially unreactive. Therefore, when wastewater is exposed to pure oxygen, the rate of transfer of nitrogen into the oxygen gas would be very high. The longer a bubble of oxygen is exposed to wastewater that had been in equilibrium with air, the more it will tend to become a mixture of oxygen and nitrogen, although always richer in oxygen than is air. Accordingly, even when pure oxygen is used in aerating water or wastewater, gas bubbles that escape are no longer pure oxygen but are mixtures of oxygen, nitrogen, and carbon dioxide. Continuous recirculation of escaping gas with the expectation that all of the oxygen might be used is, therefore, not feasible.

III. THE DEVELOPMENT OF THE BIO-PRECIPITATION PROCESS

A. Laboratory Fill-and-draw Studies

Two one-liter fill-and-draw activated sludge units, one oxygenated with air and the other with pure oxygen were operated in battery jars during a period of 3 months to establish whether activated sludge, stabilized in an atmosphere of pure oxygen, would behave significantly differently from sludge developed in an air atmosphere. The experiments were begun with wastewater taken from a sewer in Cambridge, Massachusetts, after which the sludges were built up and supported with dextrose nutrient broth as a synthetic wastewater.

The degree of treatment attained with pure oxygen was found to be consistently slightly better than with air. With approximately 33 mg/l of dissolved oxygen maintained in the oxygen-fed unit, nitrification was not inhibited. The so-called sewage fungus, *sphaerotilus natans,* did not develop in the oxygen-fed unit, although it did appear in the air-fed unit. Many of the protozoa reported for normal activated sludge were found in the oxygen-supported sludge. This sludge was easier to filter, had a lower sludge volume index, and was less subject to deterioration than the sludge developed with air. Accordingly, there was every reason to move ahead with continuous-flow units using pure oxygen.

B. Laboratory Bio-precipitation Process Studies

Because pure oxygen permits substantially higher concentrations of dissolved oxygen to be maintained than is possible with air, a flow process similar to that suggested by Pirnie is feasible with oxygen where it would be difficult to operate with air. Research was simultaneously undertaken on the use of pure oxygen and the bio-precipitation process. Bio-precipitation differs from conventional activated sludge in that, instead of separate aeration and sedimentation tanks, the wastewater is preoxygenated and a precipitation unit used for: (a) carrying the oxygen to the microorganisms to sustain their microbial activity, (b) providing the nutrients for the biological floc, which simultaneously removes this nutrient from the wastewater, some of it being metabolized and some used to create new cell material, and (c) separation of the biological floc from the flowing liquid, permitting a clear effluent to pass from the unit. The laboratory apparatus is shown schematically in Figure 2. Two identical units were built and used primarily for comparing variables, but occasionally for operation in series.

Wastewater was pumped from a combined sewer into a large storage tank that provided primary sedimentation. The settled wastewater then flowed continuously to two 2-l glass influent reservoirs, one for each plant. From the reservoir, the wastewater descended through a tapered oxygenation tube, countercurrently to the upward movement of oxygen bubbles released at the bottom of the tube. A pump carried the oxygenated wastewater into the bottom of the precipitation unit (1¾-in. diameter by 52½-in. length) containing the suspended biological floc. The wastewater rose through the blanket of suspended floc, with the effluent being collected at the top. Part of the effluent was recirculated to the influent reservoir to increase the amount of oxygen delivered to the biological floc. A glass funnel was mounted on top of the tube, as a control section in which changes in the condition of the floc suspensions or hydraulics of the system could be neutralized by the decrease in upward velocity with increasing cross-sectional area. A stirrer was operated at 9 r/min to prevent agglomeration of the floc and to control short circuiting of flow in the precipitation unit. The transfer gradient was maintained at a relatively high and consistent level by operating the oxygenation tube in a countercurrent flow pattern. Pure oxygen was added at the point where the dissolved oxygen concentration was the highest, while at the influent end of the oxygenation tube, the diluted oxygen bubbles were exposed to influent wastewaters containing almost zero oxygen. With no particular attempt made to achieve efficient oxygen utilization, up to 50% utilization was achieved with an average utilization of approximately 30%.

The parallel bio-precipitation units were operated continuously with floc formed in the unit itself. Oxygen utilization of the biological floc is readily determined within the bio-precipitation unit as the product of the difference between the oxygen concentrations at the influent and effluent of the precipitation unit and the flow rate. A polarographic method for dissolved oxygen determinations *in situ* was developed to facilitate the control of the bio-precipitation process and to permit the unit to be used for oxygen utilization studies.[11] The results of the experimental investigations are summarized below:

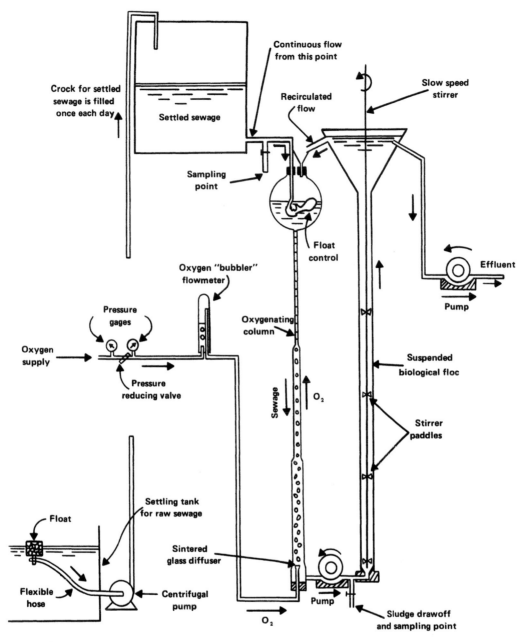

FIGURE 2. Schematic diagram of experimental laboratory bio-precipitation apparatus. (After Okun, D. A., *Civ. Eng.* (New York), 18, 32, 1948.)

1. The efficiency of BOD removal was a function of the weight of the biological floc volatile matter in the system. Where the percent volatile matter is relatively constant, the efficiency is a function of the total weight of solids.

2. Provision of the oxygen required for the biological floc by preoxygenation made possible the maintenance of a concentration of biological floc in the system more than twice as great as that possible in the aeration tanks of activated sludge plants.

3. A smaller volume of tank is required because each unit volume in the precipitation unit can carry a greater weight of biological floc than a conventional aeration tank. In order to attain a given high degree of treatment, the volume of the bio-precipitation unit was only approximately one fourth of the volume of aeration and sedimenta-

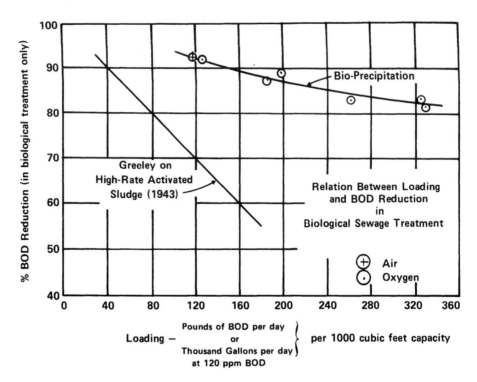

FIGURE 3. Comparison of bio-precipitation with high-rate activated sludge treatment. (After Okun, D. A., *Civ. Eng.* (New York), 18, 32, 1948.)

tion tanks necessary for the activated sludge process. Figure 3 compares the results achieved in this experimental unit with high-rate activated sludge treatment experience at that time.

4. Higher loadings produce higher rates of growth of biological floc per unit of influent, a greater volatile content in the biological floc, and a lesser degree of nitrification.

5. Blanket rising of sludge in final settling tanks of activated sludge plants, resulting from reduction of nitrites and nitrates to gaseous nitrogen in the absence of dissolved oxygen, cannot occur in the bio-precipitation unit because of the continuous presence of dissolved oxygen.

6. High loading did not result in any deterioration of the biological floc when sufficient oxygen was continuously available to the microorganisms. This is assured when there is a positive dissolved oxygen concentration in the effluent. When air was used in the same flow pattern, it was virtually impossible to maintain dissolved oxygen in the effluent, and there was greater difficulty in controlling the sludge settleability. In particular, extensive growths of *sphaerotilus* occurred at low dissolved oxygen levels, although the organism was entirely absent when the same floc was exposed to

high concentrations of dissolved oxygen under equal conditions of loading. The *sphaerotilus* reappeared when the air replaced the oxygen.

7. The presence of abnormal amounts of sugar in the wastewater also encouraged the development of *sphaerotilus* with low density or bulking sludges even when the dissolved oxygen concentrations were high. Because of the low level of agitation in the precipitation unit, the presence of *sphaerotilus* did not impair the degree of treatment and, in fact, assured a highly polished effluent.

8. The system of preoxygenation does not facilitate washing out of carbon dioxide produced by the microbial metabolism. Consequently, the pH dropped from the influent to the effluent. This appeared to have no noticeable effect in the operation of the process.

9. Protozoa were present in greater numbers and appeared to be more active in the biological floc at the bottom of the precipitation unit than at the top. This difference was more marked between the biological flocs in the first and second stages when the two units were operated in series. The operation of the two units in series seemed to offer promise of greater efficiency than was possible by operation of the same units in parallel.

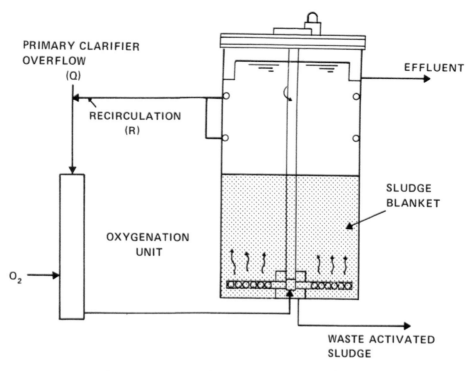

PRIMARY CLARIFIER
OVERFLOW
(Q)

RECIRCULATION
(R)

EFFLUENT

SLUDGE
BLANKET

OXYGENATION
UNIT

O₂

WASTE ACTIVATED
SLUDGE

FIGURE 4. Experimental bio-precipitation plant flow diagram. (Reprinted with permission from Budd, W. E. and Lambeth, C. F., *J. Water Pollut. Control Fed.*, 29, 238, 1957.)

10. Since the total oxygen requirements of the process are met by preoxygenation of the influent, excessive rates of oxygen utilization that occur at the influent of the precipitation unit did not result in disappearance of oxygen at that point. In fact, the highest oxygen gradients are maintained where the rates of oxygen utilization are the greatest.

Subsequent studies at the University of North Carolina at Chapel Hill revealed conclusively that, so long as a minimum of dissolved oxygen on the order of 0.2 mg/l is maintained in the immediate presence of the microorganisms responsible for biological wastewater treatment, the level of D.O. has no effect on the metabolic rate of the organisms.[12,13] The value of pure oxygen is in increasing the saturation concentration and, thereby, the oxygen transfer rate, thus permitting greater penetration of dissolved oxygen into the zoogleal masses of the biological floc and helping to maintain aerobic conditions even in the center of the floc particle. The anticipated savings in plant construction appeared to be great enough to justify further development.

C. Pilot and Large-scale Studies

The nonprofit Research Corporation of New York, to whom the patent for the process was assigned, licensed Dorr-Oliver, Incorporated, Stamford, Connecticut, to develop the process. Investigations were initiated in 1953 at the Back River Sewage Treatment Plant in Baltimore, Maryland. A pilot plant was designed for constant flows of 10 to 30 gal/min and consisted of a 4-ft diameter by 12-ft side water depth upflow unit.

Because of the upflow principle, control of the sludge volume index at a low level was found to be necessary in order to attain reasonable overflow rates. This was accomplished by keeping the BOD loading between 0.2 and 0.4 lb of BOD per day per pound of suspended solids in the system.

The results of this pilot plant were sufficiently promising to lead to the construction and operation of a full-scale experimental plant at the Stamford, Connecticut treatment plant early in 1955. The plant was a 12-ft diameter steel tank, 22.5 ft deep, equipped with a mechanism having four submerged rotating arms located at the bottom of the tank which served as influent inlet to the tank (Figure 4). Settled wastewater was pumped from the main plant primary clarifier at a variable rate of from 25 to 120 gal/min which corresponded to the flow variation at the treatment plant. Oxygen absorption was accomplished in a

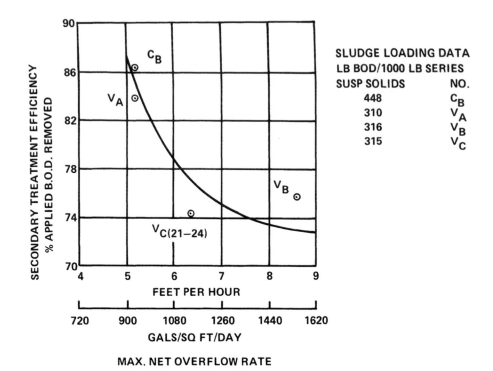

FIGURE 5.   BOD removed as a function of overflow rate. (Reprinted with permission from Budd, W. E. and Lambeth, C. F., *J. Water Pollut. Control Fed.*, 29, 246, 1957.)

20-in. countercurrent column, 9.5 ft high. Mechanical mixing and baffles were used to provide reasonable oxygen utilization efficiencies. The results of these investigations were reported by Budd and Lambeth.[14]

For any given oxygen input to the column, it was possible to vary the absorption efficiency by varying the horsepower applied to the oxygenation column. The efficiencies obtained ranged from 30 to 80%, although no serious effort was made to attain maximum efficiency of oxygen absorption. The major effort in the research was to assess the treatment efficiency of the bio-precipitation unit.

For a relatively constant load, the treatment efficiency of the bio-precipitation system was primarily a function of the overflow rate. Figure 5 illustrates the impact on the BOD removal of various overflow rates in the experimental plant.

In comparing the bio-precipitation system with conventional activated sludge treatement, a useful comparison is the tank area required for the precipitation unit in bio-precipitation and for aeration tanks and final sedimentation in conventional activated sludge treatment. Figure 6 illustrates that bio-precipitation for relatively high degrees of treatment results in approximately a

50% reduction in area requirements. For higher rate treatment with somewhat lower efficiency, the savings would be greater. Oxygen requirements were on the order of 0.75 ton per million gallons of average strength domestic wastewater. The bio-precipitation process using pure oxygen appeared to overcome some of the less desirable features of activated sludge treatment even where pure oxygen is used in the conventional activated sludge process:

1. By preoxygenation and maintaining a positive concentration of dissolved oxygen in the effluent, oxygen is present at all points in the biological floc blanket. This situation is not possible with conventional activated sludge where dissolved oxygen disappears rapidly in the sludge in the final sedimentation tank, even when oxygen is used.

2. Separate final sedimentation tanks are not required.

3. The high shear velocities customary in activated sludge aeration tanks are not experienced, and floc particles are allowed to build up to larger size. This results in much higher settling rates than is possible with the conventional activated sludge process.

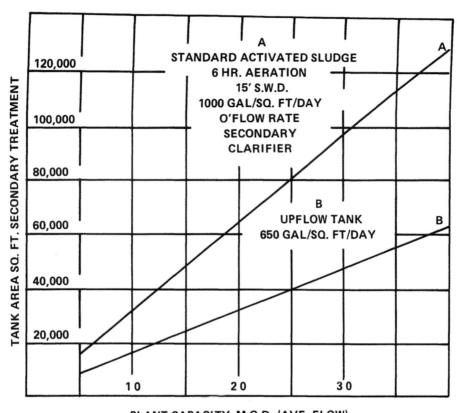

FIGURE 6. Comparison of plant area requirements: activated sludge vs. bio-precipitation. (Reprinted with permission from Budd, W. E. and Lambeth, C. F., *J. Water Pollut. Control Fed.*, 29, 249, 1957.)

Because of this more sympathetic treatment of the biological floc, and because more of it can be maintained in a given space than in the activated sludge process, higher treatment efficiencies per unit of tank volume are achieved. Economy in the use of pure oxygen may be achieved by improved oxygenation units and even possibly by pressurizing the gas and liquid to several atmospheres, thereby attaining saturation values of hundreds of milligrams per liter and eliminating the need for recirculation.

IV. THE HIATUS IN INTEREST IN OXYGEN

Despite these promising results, there was little interest in adopting bio-precipitation because the process was novel and the problems of pure oxygen use uncertain. Except for a study by Robbins of the application of the bio-precipitation process to the treatment of neutral sulfite semichemical wastes from kraft mills by the National

Council for Stream Improvement,[15] and a modest study and review of the use of pure oxygen by Pfeffer and McKinney,[16] nothing of note was done until the Linde Division of the Union Carbide Corporation actively undertook demonstration of the UNOX® System with a plant at Batavia, New York.[17] The Union Carbide Corporation made a considerable commitment in studying and developing the use of pure oxygen in wastewater treatment. With the development of the UNOX System, the use of pure oxygen in biological waste treatment achieved a considerable measure of acceptance. The UNOX System appeared to be very attractive and more ready for adoption because it used the conventional activated sludge-flow configuration of a separate aeration tank and clarifier and would permit the upgrading of existing facilities.

V. PROSPECTS FOR THE FUTURE

The interest in a two-volume series such as this,

and a symposium devoted entirely to the subject "Applications of Commercial Oxygen to Water and Wastewater Systems,"[18] is testimony to the fact that the use of pure oxygen in wastewater treatment has come of age. Now that the fear associated with the use of pure oxygen has been dissipated, many additional applications of pure oxygen in wastewater treatment can be expected.

REFERENCES

1. **Pirnie, M.** Cheap Oxygen Possible New Tool in Sanitary Engineering, Office Memorandum, Malcolm Pirnie Engineers, New York, 1946.
2. **Grant, S., Hurwitz, E., and Mohlman, F. W.,** Oxygen requirements of the activated sludge process, *Sewage Works J.*, 2, 223, 1930.
3. **Calvert, H. T.,** Report of the Water Pollution Board, British Department of Scientific and Industrial Research, 1937.
4. **Williams, E.,** Effect of bubbles of gas on mixtures of sewage and activated sludge, *J. Soc. Chem. Ind.* (London), 59, 39, 1940.
5. **Viehl, K.,** Influence of content of oxygen on biological treatment of sewage, *Zantralbl. Bakteriol. Parasitenkd. Infektionskr.*, 104, 161, 1941.
6. **Gurbaxani, M.,** Influence of Dissolved Oxygen on Biochemical Oxygen Demand of Sewage, Report to Department of Sanitary Engineering, Graduate School of Engineering, Harvard University, 1947.
7. **Okun, D. A.,** Pure oxygen in bio-precipitation process may reduce sewage treatment costs, *Civ. Eng.* (New York), 18, 32, May 1948.
8. **Buswell, A. M., Shive, R. A., and Neave, S. L.,** Bio-Precipitation Studies (1921-1927), Illinois State Water Survey, Bulletin No. 25, 1928.
9. **Okun, D. A.,** A System of Bio-Precipitation of Organic Matter from Sewage, Dissertation, Harvard University, 1948.
10. **Okun, D. A.,** System of bio-precipitation of organic matter from sewage, *Sewage Works J.*, 21, 763, 1949.
11. **Moore, E. W., Morris, J. C., and Okun, D. A.,** Polarographic determination of dissolved oxygen in water and sewage, *Sewage Works J.*, 20, 1041, 1948.
12. **Okun, D. A. and Lynn, W. R.,** Preliminary investigations into the effect of oxygen tension on biological sewage treatment, *Biological Treatment of Sewage and Industrial Wastes*, Reinhold, New York, 1956, chaps. 2-4.
13. **Okun, D. A.,** Discussion – high purity oxygen in biological sewage treatment, *Sewage Ind. Wastes*, 29, 253, 1957.
14. **Budd, W. E. and Lambeth, C. F.,** High purity oxygen in biological sewage treatment, *Sewage Ind. Wastes*, 29, 237, 1957.
15. **Robbins, M. H., Jr.,** Use of molecular oxygen in treating semi-chemical pulp mill wastes, *Proceedings, 16th Industrial Waste Conference*, Purdue University, Lafayette, Ind., 1961.
16. **Pfeffer, J. T. and McKinney, R. E.,** Oxygen-enriched air for biological waste treatment, *Water Sewage Works*, 112, 381, 1965.
17. Federal Water Quality Administration, Investigation of the Use of High Purity Oxygen Aeration in the Conventional Activated Sludge Process, Report No. 17050-DNW 05/70, Washington, D.C., 1970.
18. **Speece, R. E. and Malina, J. F.,** *Applications of Commercial Oxygen to Water and Wastewater Systems*, Water Resources Symp. No. 6, University of Texas, Austin, 1973.

Chapter 3

OXYGEN AND THE ACTIVATED SLUDGE PROCESS

J. R. McWhirter

TABLE OF CONTENTS

I. INTRODUCTION

The potential use of high-purity oxygen in biochemical oxidation processes for secondary treatment was a reasonably popular topic of discussion in the wastewater treatment field prior to the mid-1960s. As discussed in Chapters 1 and 2, there are many such references in the literature, some of which go back to the 1930s. The two most extensively discussed aspects of the subject in the early literature revolve around the use of oxygen in the bio-precipitation process and the effect of dissolved oxygen level on the rate of bio-oxidation in the activated sludge process. These two very different considerations will be discussed separately.

II. OXYGEN AND THE BIO-PRECIPITATION PROCESS

Malcolm Pirnie, Sr. was the original inventor of the bio-precipitation process which was first discussed in the literature in 1948.[1] The system was studied in the laboratory over a period of several years by Okun[2] and later was further evaluated in a comprehensive pilot plant and full-scale program by Budd and Lambeth[3] in the 1950s. The Budd and Lambeth study encompassed two evaluations including an initial 10 to 30 gal/min unit tested at the Baltimore, Maryland Back River Sewage Treatment Plant in 1953 and a 25 to 120 gal/min unit tested at the Stamford, Connecticut Municipal Sewage Treatment Plant during the period from May 1955 to March 1957. A detailed description and discussion of the process and these tests are contained in Chapter 2.

One of the most interesting aspects of the bio-precipitation system has been the emphasis associated with the use of oxygen in this process. In fact, the main thrust of the bio-precipitation process is a new type of activated sludge system flow configuration employing a combined reactor-clarifier unit. Oxygen is supplied by predissolving it in the wastewater feed and recycle stream being fed into the reactor-clarifier unit. The system can employ either air or oxygen preaeration of the feed stream to the reactor-clarifier; both were evaluated by Okun.[4] In order to obtain adequate dissolved oxygen and minimize the wastewater recycle flow through the reactor-clarifier unit, however, the use of oxygen becomes particularly attractive and virtually essential in attempting to achieve an economically viable system. The use of pure oxygen in the counter-current oxygenation column (as shown in Figure 4 of Chapter 2) can increase the equilibrium dissolved oxygen concentration to approximately 90 mg/l at the base of a 34-ft deep oxygenation column operating at atmospheric pressure. A 90% approach to equilibrium then results in an oxygenation column effluent dissolved oxygen concentration of approximately 81 mg/l. The inherent process advantages and economic attractiveness of the use of high-purity oxygen in the bio-precipitation process thus becomes readily apparent from the standpoint of substantially reducing the recycle flow, the size, and cost of the reactor-clarifier unit.

Although the bio-precipitation process has some potentially desirable features as pointed out by Okun,[5] it also has some significant constraints and disadvantages that limited its economic attractiveness and discouraged its use. This is particularly true in comparison to the use of oxygen in the conventional activated sludge process flow configuration.

III. OXYGEN SUPPLY IN THE ACTIVATED SLUDGE PROCESS

An adequate supply of dissolved oxygen has long been recognized as one of the key factors in obtaining good operating performance in the activated sludge process. Oxygen supply is also one of the major operating cost elements in the overall process; it is, therefore, entirely logical that a great deal of attention has been devoted to maximizing the efficiency and minimizing the cost of aeration systems. In fact, many of the modifications made to the conventional plug flow air-activated sludge process design, such as step aeration and complete mixing, were to a large extent motivated by the need to prevent oxygen transfer limitations at the head end of the plug flow aeration tanks. It was observed for many years that the high oxygen uptake rate conditions that existed at the feed end of plug flow aeration tanks often resulted in a lack of adequate dissolved oxygen, which adversely affected the operation of the process. The step aeration and complete-mix-activated sludge process modifications help to alleviate these oxygen transfer limitations by feeding the wastewater at a multiple number of locations along the length of the aeration tanks, in the case of step aeration, and by completely mixing the total contents of the aeration basin, in the case of the complete-mix process. Both of these techniques tend to smooth out or average the oxygen uptake rate over the entire aeration basin. They achieve this benefit, however, at the expense of increasing the residence time required to achieve the desired effluent quality. Thus, these process modifications are undesirable from a reaction kinetic and effluent quality standpoint for a given size aeration basin. The optimum situation, of course, would be the use of a plug flow or multi-stage aeration system design without concern for oxygen transfer limitation problems.

The basic requirements of the activated sludge process place many criteria, and often seemingly conflicting demands, on the performance of the oxygenation system, whether it be an air or oxygen system design. Several of the more important oxygenation system requirements are summarized as follows:

1. High oxygen transfer rate capability
2. Operation at low dissolution power inputs and low mixed-liquor shear levels to minimize floc breakup and enhance the settling properties of the biomass
3. Good bulk liquid mixing to ensure intimate contact of the biomass, substrate, and dissolved oxygen at all points throughout the aeration tank as well as suspension of the biomass
4. Maintenance of a high dissolved oxygen level throughout the aeration tank to ensure a healthy biomass and operation at peak process efficiency
5. Low capital investment and power requirements to minimize overall costs
6. Good process control characteristics to enable rapid response to changes in oxygen uptake rate in order to prevent oxygen transfer limitations under highly loaded conditions and conserve operating power costs under low load conditions
7. Ease of operation and mechanical simplicity

The achievement of all of these performance criteria is further complicated by the fact that oxygen gas is only sparingly soluble in water. This results in the overall interphase mass transfer process being liquid-phase controlled, which leads to an inherently high energy requirement in the oxygen dissolution process. This is a characteristic of all processes involving the dissolution of sparingly soluble gases in liquids. The presence of the large concentration of nitrogen in air (approximately 80%) further aggravates the oxygen transfer process by reducing the available oxygen partial pressure driving force for the interphase mass transfer process. As a result, substantial amounts of energy are expended in creating a large gas-liquid interfacial area and a high degree of liquid phase interfacial turbulence to enhance the interphase mass transfer rate. In spite of substantial efforts and many technological advances over the years, the dissolution of oxygen in water remains an inherently costly process in terms of the quantity of oxygen dissolved per unit of energy expended. This will always be the case because of the low solubility of oxygen in water, which results in the mass transfer process being liquid-phase controlled.

Table 1 summarizes the oxygen dissolution energy requirements for both diffused air and surface aeration systems commonly employed in air-activated sludge plants. The energy requirements are tabulated as a function of the mixed-liquor dissolved oxygen concentration for typical operating conditions encountered in the aeration of activated sludge for the treatment of municipal wastewater. It is readily seen and well known that the energy required is strongly influenced by the mixed-liquor dissolved oxygen (D.O.) concentration. The energy requirement at a mixed-liquor D.O. level of 2.0 mg/l is, for example, about 31% higher than at a mixed-liquor D.O. of 0.0 mg/l. The energy requirement is approximately 55% higher at a mixed-liquor D.O. of 3.0 mg/l compared to that at a D.O. of 0.0 mg/l. Thus, the strong tendency to design air aeration systems for operation at low dissolved oxygen levels is readily understood as a consequence of the rapid increase in power requirement and cost with the mixed-liquor D.O. level.

In contrast to the oxygen dissolution process, the separation of oxygen from air is a highly efficient process in terms of the energy required per unit of oxygen generated. Using modern air separation techniques, the energy required to separate oxygen from air is only about 1.1 to 1.6 kWh/100 ft^3 of oxygen generated. This corresponds to approximately one third to one fourth the amount of energy required to dissolve oxygen from air into activated sludge mixed-liquor at a

TABLE 1

Typical Energy Requirements for Oxygen Dissolution into Mixed-liquor with Air

Mixed-liquor D.O. (mg/l)	Diffused air oxygen dissolution energy requirement		Surface aerator oxygen dissolution energy requirement	
	kWh/100 CF O_2	lb O_2/H.P. Hr	kWh/100 CF O_2	lbs O_2/hp hr
0.0	3.7	1.6	2.6	2.3
1.0	4.2	1.4	3.0	2.0
2.0	4.9	1.2	3.4	1.8
3.0	5.7	1.1	4.0	1.5
4.0	6.8	0.9	4.9	1.2
5.0	8.6	0.7	6.2	1.0
6.0	11.7	0.5	8.4	0.7

Note: Aeration conditions: Mixed-liquor temperature, 20°C; Alpha factor, 0.80; Beta factor, 0.95; Blower compressor efficiency, 0.70; Motor efficiency, 0.90; Compressor discharge pressure, 8.1 psig; Aerator gear efficiency, 0.95.

Data courtesy of Union Carbide Corporation.

D.O. level of 2.0 mg/l. Therefore, if high-purity oxygen can be effectively and efficiently dissolved into mixed-liquor at a significantly lower power requirement per unit of oxygen dissolved than is required for dissolution of oxygen from air, then high-purity oxygen aeration would be more energy efficient than air aeration. In other words, it would take less power to first separate oxygen from air and efficiently dissolve it into mixed-liquor than to transfer oxygen directly from atmospheric air into mixed-liquor.

IV. HIGH-PURITY OXYGENATION SYSTEM REQUIREMENTS AND CHARACTERISTICS

As shown above, high-purity oxygenation offers potential energy savings in comparison to air aeration, if high-purity oxygen gas can be practically and efficiently dissolved into mixed-liquor. Therefore, in addition to the general oxygenation system requirements listed previously, a high-purity oxygen aeration system has some additional constraints and requirements that must be satisfied in order for the system to be economically attractive. These are summarized as follows:

1. An economical on-site oxygen gas genera-tion system must be employed.
2. A high percentage oxygen absorption and

utilization efficiency must be achieved at all times (e.g., > 90%).
3. A high overall oxygen dissolution energy transfer efficiency must be obtained as compared with air aeration (e.g., 2 to 3 times higher energy transfer efficiency).

All of these performance criteria must be simul-taneously met for the system to be economically attractive in comparison to air aeration.

These further performance requirements result in some inherent characteristics of oxygenation systems which are considerably different from those encountered in normal air aeration practice. First, a very large volume of mixed-liquor must be effectively and efficiently contacted with a very small amount of relatively high-purity oxygen gas. In comparison to a typical diffused air system, for example, approximately 1/50 as much aeration feed gas must be contacted with the mixed-liquor. Second, the interphase mass transfer process is a more complex, multicomponent process involving the transfer of significant quantities of gases other than oxygen. This occurs because wastewater is typically saturated with nitrogen in equilibrium with air and because significant amounts of carbon dioxide are generated in the bio-oxidation process. Therefore, the interphase transfer of other gaseous components, including nitrogen, carbon dioxide, argon, and water vapor must be properly taken into account. Third, efficient mechanical or gas-mixing energy input is required to achieve the required level of bulk fluid mixing to ensure intimate contact of the biomass with the wastewater and to provide uniform distribution of the dissolved oxygen throughout the mixed-liquor.

These unique requirements and characteristics of high-purity oxygenation systems lead to the conclusion that system designs which are effective or optimal for air aeration will probably not be effective for direct oxygenation. Thus, a new oxygenation system design is required to economically adapt the use of high-purity oxygen to the activated sludge process while simultaneously obtaining the additional process and performance advantages of aeration with a high oxygen partial pressure aeration gas.

V. OXYGEN AND THE UNOX® SYSTEM

A. Basic UNOX System Design

In its most common form, the UNOX System accomplishes the objectives and requirements of an activated sludge high-purity oxygenation system as discussed above through the use of an enclosed, multi-stage gas-liquid contacting system. Figure 1 shows a simplified, schematic, cross-sectional view of such a multi-stage contacting system. Surface aerators are used as the means of achieving gas-liquid contacting and mixing in the individual stages. The aeration tank is divided into a number of discrete sections or stages (three stages are shown in Figure 1) by means of baffles

or tank walls and is completely covered to provide a relatively gas-tight enclosure. The system is operated at essentially atmospheric pressure, so that effective gas sealing with minimal leakage is readily achievable. The feed wastewater, recycle sludge, and oxygen feed gas are introduced into the first stage and flow cocurrently through the multi-stage system. The number of stages employed and the method of gas-liquid contacting are quite flexible and can be optimized for any particular application or plant size. The overall aeration tank dimensions and stage configuration can also be varied over quite broad ranges to facilitate any specific design limitations or site requirements. The mixed-liquor depth, for example, can be varied from a low of about 8 ft to as high as about 30 ft to enable optimization for particular plant design restrictions or economic conditions. The design of the specific aeration devices, of course, must be matched to the particular stage configuration and liquid depth to ensure both an adequate rate of oxygen transfer and bulk fluid mixing within the individual stages as well as a high oxygen dissolution energy transfer efficiency.

The oxygen feed gas is introduced into the first stage and maintained at a slight positive pressure of approximately 1 to 4 in. of water column. The successive oxygenation stages are connected to each other in a manner that allows gas to flow freely from stage to stage with only a very slight pressure drop sufficient to prevent gas backmixing or interstage mixing of the aeration gas. This is accomplished by appropriate sizing of the interstage gas passages. These design considerations are

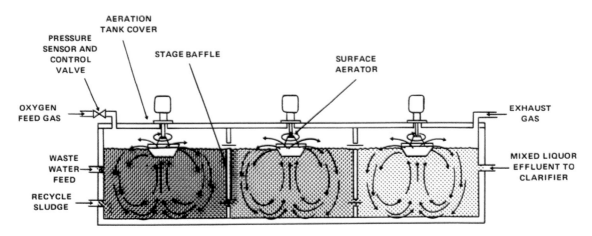

FIGURE 1. Simplified, schematic diagram of three-stage UNOX® system using surface aerators. (Courtesy of Union Carbide Corporation.)

discussed in considerably more detail in Chapter 10.

B. UNOX System Operation and Control

In its simplest form, the UNOX System can be operated at its full rated capacity by operating the oxygen generator at its normal design output and feeding all of the oxygen gas produced directly to the aeration tanks. With this mode of operation, wastewater feed fluctuations will cause the oxygen absorption efficiency and the mixed-liquor D.O. levels to vary above and below the average design values. At low loads, the mixed-liquor and effluent D.O. concentrations would be above their normal design values and the oxygen absorption efficiency would be below the design value. Conversely, at high loads, the mixed-liquor and effluent D.O. levels would be below their normal design average values and the oxygen absorption efficiency would be higher than the design average. This method of operation of a UNOX System is completely analogous to the normal operation of an air-activated sludge system with the aerators operated at their full load rated capacity, independent of the wastewater feed variations. The UNOX System has an important additional feature, however, which is the ability to cope with peak oxygen demands that exceed the capacity of the oxygen generator. This is accomplished by automatic liquid oxygen vaporization from the liquid oxygen storage tank when the demands of the system exceed the capacity of the on-site oxygen generator. This provides an additional degree of flexibility for operation at high load conditions which cannot be obtained with an air system. The oxygen absorption efficiency of the system under these peak load conditions is lower than the design average value, but the process does not become oxygen supply limited, as would be the case with an air system. A liquid oxygen storage tank and automatic liquid oxygen vaporization system are incorporated into the design of every UNOX System for peak load shaving and oxygen generator backup in the case of both scheduled and unscheduled outages.

In addition to the above mode of operation, the UNOX System can be operated with a simple pressure regulator system to enable the oxygen generator output to be automatically controlled to respond to changes in system oxygen demand as a consequence of wastewater feed fluctuations. The basis for this method of control is constant pressure operation of the gas space within the enclosed aeration tanks. A pressure controller connected to the first stage gas space detects changes in gas pressure which are caused by increases or decreases in the system oxygen demand. The oxygen generator production rate is then automatically adjusted to maintain a constant gas pressure in the first stage. Thus, the oxygen generator production rate and power consumption can be effectively and easily controlled by a simple, reliable pressure regulating system. This can result in substantial operating cost power savings by maintaining the oxygen generator power consumption proportional to system demand. This type of pressure control system cannot be used in air aeration systems. With air aeration, one must resort to mixed-liquor D.O. control to adjust the aerator power input to conserve power under low loaded conditions. Such D.O. control systems are much less reliable and require considerably more maintenance than pressure control systems.

C. UNOX System Gas-Liquid Contactor Design

The type of gas-liquid contactor or aerator used in a UNOX System can be specifically tailored to meet the particular process requirements and geometric considerations involved in the overall reactor or aeration tank design. The only important criterion is that the aerator be capable of efficiently contacting and recirculating one of the fluids within the stage (either gas or mixed-liquor) against the other fluid while achieving adequate bulk fluid mixing within the stage and maintaining a reasonable overall oxygen dissolution energy transfer efficiency. Most, if not all, of the basic types of gas-liquid contacting devices used in air aeration systems could be employed in a UNOX System. This includes surface aerators, submerged turbine devices, and diffusers.

In most cases, surface aerators, such as shown in Figure 1, are the most economically attractive and simplest gas-liquid contacting devices to employ in a UNOX System. Surface aerators recirculate the mixed-liquor through the gas space to accomplish oxygen transfer and provide the necessary bulk liquid mixing. If necessary, supplemental bulk fluid mixing can be accomplished by attaching a bottom mixing impeller to the surface aerator to ensure a uniform composition of the mixed-liquor and maintain the activated sludge solids in suspension.

Submerged turbines or diffusers can also be employed in a UNOX System to achieve gas-liquid contacting. Submerged turbine systems, such as that schematically diagramed in Figure 2, are particularly attractive for very large plants which use deep aeration tanks to conserve space. The gas within each stage is recirculated and contacted with the mixed-liquor by means of small gas blowers. These recirculating gas blowers pump the oxygen gas to a submerged sparging device and/or impeller combination at a rate sufficient to provide the needed oxygen transfer and maintain a uniform mixed-liquor suspension. The particular system design shown in Figure 2 employs a rotating sparger integrally connected to a submerged impeller. Oxygen gas recirculating blowers pump the gas through a hollow shaft to the rotating sparger at the end of the mixer shaft. The downward pumping action of the mixing impeller located on the same shaft as the sparger disperses the gas bubbles throughout the mixed-liquor and provides effective bulk fluid mixing. This flow configuration effectively disperses the gas bubbles throughout the stage and achieves a high degree of oxygen absorption per pass by promoting relatively long residence times for the dispersed oxygen gas bubbles. This type of gas-liquid contacting system is discussed in considerably more detail in Chapter 1, Volume II.

A gas diffusion-type of gas-liquid contacting system could also be employed in a UNOX System. Recirculating gas blowers would pump the gas to the diffusers instead of to the submerged turbine, sparger system shown in Figure 2. Again, the only important performance criteria are that the diffusers be capable of a reasonable oxygen dissolution energy transfer efficiency under the gas flow rate conditions employed and that adequate bulk fluid mixing is maintained to ensure a uniform mixed-liquor suspension.

D. UNOX System Reactor or Aeration Tank Design

The UNOX System is normally designed with a discrete number of mixed-liquor stages to enhance the overall rate of reaction and BOD_5 removal within the system and, thereby, produce the highest quality effluent with a given size aeration tank. A relatively small number of completely mixed stages in series very closely approximates the performance of an ideal plug flow system from an overall reactor design standpoint. This subject is discussed in more detail in Chapter 1, Volume II.

The use of plug flow or multipass aeration tanks in conventional air activated sludge systems, however, has traditionally led to oxygen transfer limitation problems at the head end of such plug flow systems as discussed previously. A very desirable feature of the multi-stage gas contacting system used in UNOX is that it lends itself very

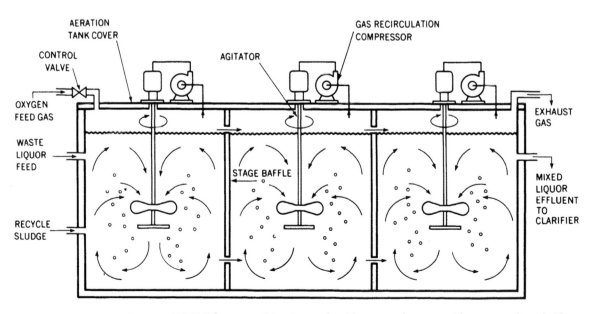

FIGURE 2. Schematic diagram of UNOX® system with submerged turbine contacting system (three stages shown). (Courtesy of Union Carbide Corporation.)

well to simultaneous staging of the mixed-liquor while eliminating the oxygen transfer limitation problem. Staging increases the oxygen demand at the feed end of the system as compared to the effluent end. The oxygen transfer capacity of a multi-stage, cocurrent oxygenation system naturally varies from stage to stage with the inherent transfer rate tending to decrease from the feed stage to the final or exhaust stage. This is a result of the decreasing gas phase oxygen composition as the oxygen gas is utilized in the system. Thus, a multi-stage, cocurrent mixed-liquor and oxygen gas contacting system tends to yield an oxygenation system design that naturally tends to match the process oxygen demand variation of the mixed-liquor with the inherent oxygen mass transfer capability of the staged oxygen gas contacting system. This enables the size of the individual stages and the gas-liquid contacting power input per stage to be more nearly equal throughout the system without encountering oxygen transfer limitations.

The number of gas and mixed-liquor contacting stages in a UNOX System need not be the same and can be specifically designed to meet any system requirement. In fact, a multi-stage gas contacting system can be readily employed with a completely mixed reactor basin and still achieve the basic performance advantages and characteristics of multi-stage gas contacting. Thus, if for any particular process operating conditions or wastewater characteristics it is deemed advisable to employ a complete mix reactor design, the multi-stage gas contacting system can be readily adapted to this type of overall system configuration. Similarly, the sizes of the individual gas stages or the individual mixed-liquor stages need not be identical. In some industrial applications with extremely wide variations in waste strength, it is desirable to make the feed mixed-liquor stage larger than the other stages in the system. This design approach provides greater capability for dampening feed fluctuations to mitigate against toxicity or other shock load conditions which may adversely affect overall process performance.

The use of a multi-stage reactor design employing a series of completely-mixed liquid stages in essence allows the designer to specifically tailor the system to accommodate any conceivable type of process requirement. The number and size of the stages in the system allows the degree of divergence from true plug flow conditions to be controlled precisely and independently of variables such as feed rate, power input, and fluid properties. In other words, staging permits control of the degree of backmixing within the system. At the one extreme, a single stage, complete mix system provides an infinite degree of liquid backmixing to provide a uniform liquid phase composition throughout and the maximum possible degree of buffering against feed fluctuations. On the other hand, a relatively modest number of completely mixed stages in series (4 to 6) provides most of the advantages of an ideal plug flow system by minimizing the required reactor volume to achieve a particular level of effluent quality.

Both liquid and gas staging in a UNOX System are typically most readily and economically obtained by the use of discrete staging walls or baffles in the aeration tank to provide the desired number of gas and liquid stages. This approach also results in maximum control of the degree of staging under all ranges of wastewater flow and other operating conditions. It is also possible, however, to design the reactor or aeration tank so that the length of the mixed-liquor flow path is very large relative to its width and depth. This type of reactor design has been used for years in the conventional multipass, plug flow design employed in diffused air aeration systems. For a given overall aeration tank volume, such a geometry increases the velocity of the mixed-liquor flow from the feed end to the discharge end. This suppresses backmixing of the mixed-liquor from the downstream zones into the upstream zones. Thus, in such a system, the mixed-liquor is "effectively" staged or contacted in a series of "zones" from the feed to the discharge end of the aeration tank even though the stages are not physically partitioned from each other. As discussed in more detail in Chapter 1, Volume II, this design approach is economically feasible only in very large plants and has undesirable characteristics which limit its overall attractiveness.

VI. UNOX SYSTEM OXYGEN MASS TRANSFER PERFORMANCE

The use of relatively high-purity oxygen in the activated sludge process provides an entirely new and important degree of freedom in the design of the aeration system. This additional design parameter is, of course, the oxygen composition of the feed gas. In contrast to air aeration systems, which are by definition constrained to a feed gas oxygen composition of 21% by volume, high-purity oxygenation systems can be operated over a wide range of feed gas oxygen concentrations. Relatively high-purity feed gas compositions (90 to 98% oxygen) are economically more attractive; however, this is primarily a matter of overall system optimization as opposed to a design constraint. This additional degree of freedom results in added oxygenation system performance requirements and unique gas-liquid contacting and mass transfer characteristics when compared to air aeration systems.

The design and operation of a high-purity oxygenation system is considerably more complex than that of an air aeration system from the standpoint of the interphase mass transfer process. Whereas in air aeration systems only the transfer of oxygen from air to the mixed-liquor need be considered in the practical design of the aeration system, in high-purity oxygen aeration systems the mass transfer process involves the simultaneous transfer of several important gaseous components between the gas and liquid phases. This multicomponent mass transfer process results from wastewater being saturated with nitrogen in equilibrium with atmospheric air and from the substantial quantities of carbon dioxide generated in the bio-oxidation process. Thus, when relatively high-purity oxygen gas is brought into contact with activated sludge mixed-liquor, nitrogen and carbon dioxide are desorbed from the mixed-liquor into the oxygen gas simultaneously with the transfer of oxygen from the gas phase to the mixed-liquor. An accurate prediction of the mass transfer process must also properly take into account the argon and water vapor content of the aeration gas.

The design and performance of the oxygenation system is further influenced by the rather complex water chemistry and related equilibria involving the major ionic species present in the mixed-liquor. Chief among these constituents are the various dissociation products of carbonic acid (H_2CO_3) and the other components which determine the overall alkalinity of the mixed-liquor. A detailed presentation of these design considerations is contained in Chapter 9, Volume I, The present discussion will be limited to the overall oxygenation system design and performance parameters as they relate to oxygen mass transfer.

A. General Oxygen Mass Transfer Performance Characteristics of a Multistage, Cocurrent Gas-Liquid Contacting System

The general oxygen mass transfer performance characteristics of a multi-stage, cocurrent gas-liquid contacting system can best be obtained by reference to Table 2 which summarizes the overall performance parameters for a six-stage, surface aeration system design treating typical municipal sewage. This table contains the individual stage aeration gas compositions and flow rates, volumetric oxygen uptake rates, and oxygen absorption efficiencies on a stage-wise and cumulative basis. As seen, the oxygen gas composition decreases from 99.5% in the feed gas to 40.0% in the sixth stage, and vent gas as nitrogen and carbon dioxide are progressively desorbed into the oxygen feed gas. An overall oxygen gas absorption efficiency of 95% is obtained, even though the vent or exhaust gas oxygen concentration is still 40.0%. Thus, a high average oxygen partial pressure is maintained in the aeration gas while simultaneously achieving a high oxygen utilization efficiency.

This very important characteristic is a consequence of the substantially lower flow rate of the vent gas in comparison to that of the feed gas (0.6 lb mol/hr of vent gas as compared to a feed flow rate of 4.6 lb mol/hr). The shrinkage in volumetric flow rate of the aeration gas as it flows through the multi-stage system results from the relatively small quantity of nitrogen gas dissolved in the incoming wastewater in relation to the quantity of oxygen utilized in the biochemical oxidation process and the very high solubility of carbon dioxide as compared with the solubility of oxygen and nitrogen. The quantity of oxygen used in the bio-oxidation process is generally about 10 to 20 times the amount of dissolved nitrogen in the feed wastewater. The solubility of carbon dioxide in water is about 35 times greater than that of oxygen. Therefore, the quantity of nitrogen in the untreated sewage is not great enough to significantly dilute the oxygen feed gas until a high percentage oxygen utilization is achieved. Simultaneously, most of the carbon dioxide generated remains dissolved in the mixed-liquor with only a very small amount being present in the gas phase. The relatively small amount of carbon dioxide desorbed into the gas phase is a result of its very high water solubility compared to oxygen and nitrogen and, hence, a relatively large amount of

TABLE 2

Oxygen Mass Transfer Performance Summary for Six-stage Cocurrent Oxygenation System Treating Typical Municipal Sewage

	Stage number					
Performance parameter	1	2	3	4	5	6
Aeration gas composition (% by volume)						
Oxygen	86.8	78.9	72.3	65.2	55.8	40.0
Nitrogen	8.5	14.5	20.1	26.4	35.1	50.3
CO_2	2.9	4.4	5.4	6.1	6.8	7.4
Argon and H_2O	1.8	2.2	2.2	2.3	2.3	2.3
Feed gas rate to stage (lb mol/hr)	4.6	3.5	2.6	2.0	1.6	1.1
Exit gas rate from stage (lb mol/hr)	3.5	2.6	2.0	1.6	1.1	0.6
Oxygen uptake rate (mg/l/hr)	120.0	69.0	45.0	34.0	31.0	28.0
Feed oxygen absorbed per stage (%)	35.0	20.0	13.0	10.0	9.0	8.0
Overall cumulative feed oxygen absorbed (%)	35.0	55.0	68.0	78.0	87.0	95.0

Note: Feed wastewater BOD_5, 250 mg/l; Feed gas oxygen composition, 99.5% by volume; Total aeration residence time (based on Q), 2 hr (20 min per stage); MLVSS concentration, 4000 mg/l; Mixed liquid D.O. level, 2.0 mg/l.

Data courtesy of Union Carbide Corporation.

carbon dioxide in the liquid phase generates only a modest gas phase partial pressure. Thus, the relatively small volumetric flow rate of vent gas is still high in oxygen content even though a high overall oxygen utilization efficiency has been achieved.

Table 2 also shows the variation of oxygen uptake rate and per cent of the feed oxygen absorbed per stage throughout the system. The oxygen uptake rate varies from a high of 120 mg/l/hr in the first stage to a low of 28 mg/l/hr in the sixth stage. Most of the variation in the oxygen uptake rate occurs in the first three stages with little variation in the later three stages. This is to be expected since the BOD_5 concentration in the mixed-liquor and, hence, the bio-oxidation rate, is considerably higher at the feed end of the system. The cocurrent contacting system is uniquely suited to this variation in oxygen demand since the oxygen composition of the aeration gas and the inherent oxygen transfer capability of the system is higher in the first stages. Since the oxygen gas concentration in Stage 1 is about 2.2 times as great as that in Stage 6, nearly 2.2 times as much oxygen per unit of dissolution power input can be dissolved in Stage 1 as compared with Stage 6. Thus, although the oxygen uptake rate in stage

one is about four times greater than that in stage six, the aerator dissolution power used in Stage 1 for oxygen transfer purposes is only about 1.8 times as great as that used in Stage 6. This is the naturally tapered characteristic of the cocurrent, multi-stage contacting system. In many cases, sufficient bulk fluid mixing power input to maintain adequate solids suspension becomes controlling in the final stages and fluid mixing power rather than oxygen transfer requirements determines the dissolution power input.

The per cent of the feed oxygen absorbed per stage throughout the system varies in the same manner as the oxygen uptake rate, as would be expected. In the example in Table 2, 55% of the feed oxygen is absorbed in the first two stages and 78% is absorbed in the first four stages.

A very important performance requirement for a high-purity oxygenation system, as noted earlier in Section IV, is the maintenance of a high average gas phase oxygen partial pressure in order to minimize the dissolution power requirement. The oxygen dissolution energy transfer efficiency should be at least three times that of an air aeration system, while maintaining a high oxygen utilization efficiency, if the total system is to be economically attractive from a power requirement standpoint. Figure 3 compares the typical average oxygen transfer driving force for a UNOX System operating at a mixed-liquor D.O. level of 6 mg/l with that for an air system operating at a D.O. level of 2 mg/l. The average liquid phase oxygen mass transfer driving force for the UNOX System is approximately 25 mg/l as compared with 7 mg/l for an air system. This means on an overall average basis the UNOX System will transfer about 3.6 times as much oxygen per unit of dissolution power input as an air system. Therefore, the requirement of achieving at least a threefold increase in the oxygen dissolution energy transfer efficiency is more than met even while operating at a substantially higher mixed-liquor D.O. level.

B. Optimum Oxygen Utilization Efficiency

A high oxygen utilization efficiency in a high-purity oxygenation system is basic and essential to obtaining an economically attractive overall system design. The higher the overall oxygen absorption efficiency, the less oxygen needs to be generated to satisfy the system oxygen demand. Intuitively, it might be expected that the highest possible oxygen utilization efficiency obtainable

would be the most economically attractive. This is not the case, however, since as the oxygen absorption efficiency increases, the overall average oxygen concentration in the aeration gas decreases and the dissolution power requirement increases. Thus, there is an optimum economic trade-off between reduced oxygen generation power and increased oxygen dissolution power. This relationship is shown in Figure 4 for a typical UNOX System design. The oxygen generation power requirement decreases linearly with increased oxygen utilization as would be expected. The oxygen dissolution power requirement increases very rapidly, however, as the oxygen absorption efficiency increases above approximately 90%. The sharp increase in the dissolution power requirement is caused by the extremely low oxygen concentration in the later stages at oxygen utilization efficiencies above 90%. This relationship is depicted in Figure 5 where the effect of oxygen utilization on gas stage oxygen purity is shown as a function of oxygen utilization efficiency. The last stage or vent gas purity rapidly decreases above an oxygen utilization efficiency of about 90%. In fact, the oxygen concentration in the vent gas

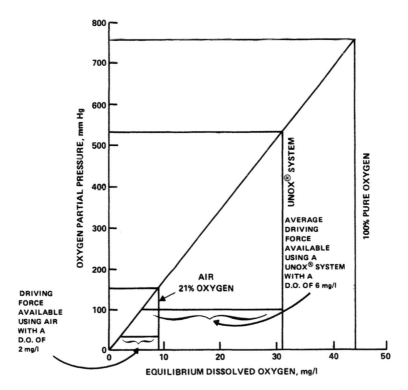

FIGURE 3. Comparison of average oxygen dissolution driving forces for oxygen and air aerated systems. (Courtesy of Union Carbide Corporation.)

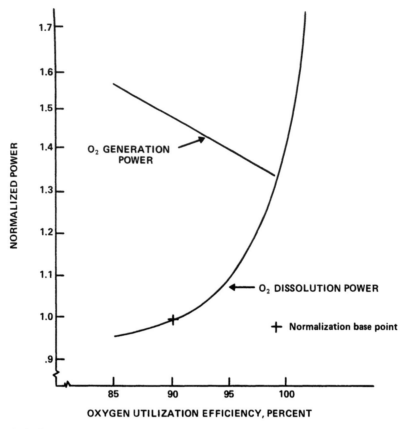

FIGURE 4. General oxygen dissolution and generation power requirement as a function of oxygen utilization efficiency (98% O_2 feed purity). (Courtesy of Union Carbide Corporation.)

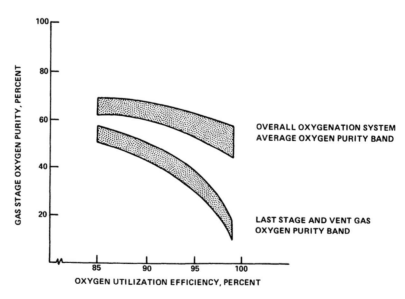

FIGURE 5. General effect of oxygen utilization efficiency on gas stage oxygen purity. (Courtesy of Union Carbide Corporation.)

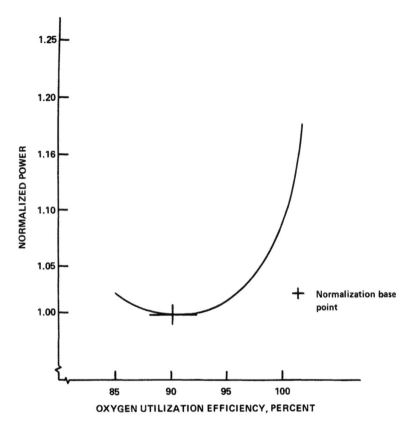

FIGURE 6. General overall oxygenation power requirement (generation plus dissolution power) as a function of oxygen utilization efficiency (98% O_2 feed purity). (Courtesy of Union Carbide Corporation.)

decreases to below 21%, or that of atmospheric air, at an oxygen utilization efficiency of 97 to 98%. The overall oxygenation system average oxygen gas purity also decreases with increased utilization efficiency, but not as rapidly as the vent gas purity, as shown in Figure 5.

Figure 6 depicts the overall oxygenation power requirement (generation plus dissolution power) as a function of the oxygen utilization efficiency. A distinct minimum power requirement or optimum oxygen utilization efficiency exists. Typically, the optimum oxygen utilization efficiency is approximately 90%. There is also little variation in the total power requirement in the range of 85 to 95% oxygen utilization. Above 95% absorption efficiency, however, the overall power required increases very rapidly as shown in Figure 6. Thus, an average design load oxygen utilization efficiency of about 90% provides a highly stable operating condition with considerable flexibility to accommodate changes in absorption efficiency with changing load demands on the system.

C. Effect of Number of Stages on Oxygenation System Performance

Figures 7 through 11 summarize the effect of the number of stages in the system on the oxygen utilization efficiency and optimum total power requirement. Figures 7 and 8 show that the dissolution power requirement to obtain a given degree of overall oxygen absorption efficiency decreases with an increase in the number of gas-liquid contacting stages. The decrease in dissolution power requirement is quite significant for an increase from one to three or four stages but becomes progressively less above five stages. The dissolution power required to obtain 90% oxygen utilization efficiency is approximately 55% higher in a single-stage system than in a four-stage system. An increase from four to twelve stages, however, results in only about an 8% further reduction in dissolution power as shown in Figure 8. Therefore, most of the benefit of staging is achieved with relatively few stages (three to six). Only a marginal further improvement is achieved with a signifi-

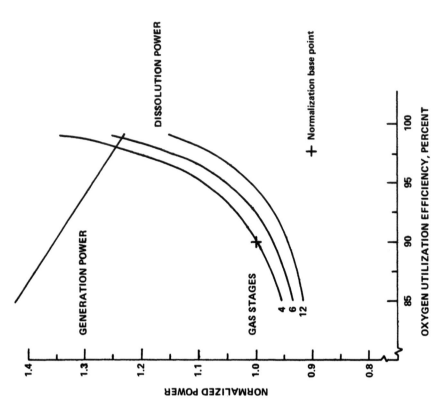

FIGURE 8. Oxygen dissolution and generation power requirement as a function of oxygen utilization efficiency and number of stages (98% O_2 feed purity). (Courtesy of Union Carbide Corporation.)

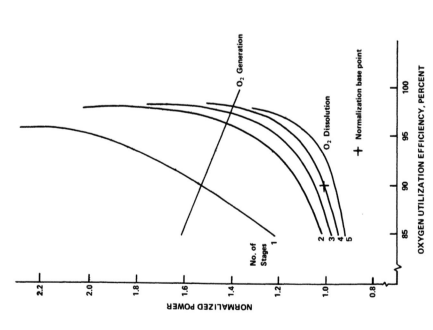

FIGURE 7. Oxygen dissolution and generation power requirement as a function of oxygen utilization efficiency and number of stages (98% O_2 feed purity). (Courtesy of Union Carbide Corporation.)

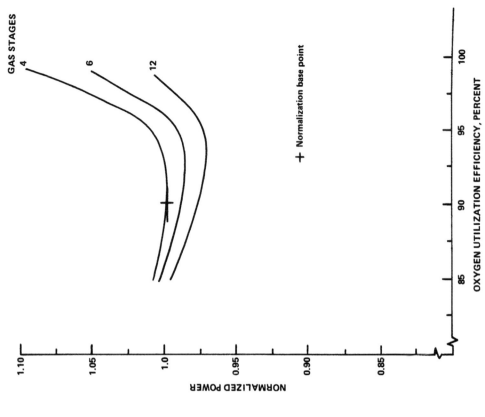

FIGURE 10. Overall oxygenation power requirement (generation plus dissolution power) as a function of oxygen utilization efficiency and number of stages (98% O₂ feed purity). (Courtesy of Union Carbide Corporation.)

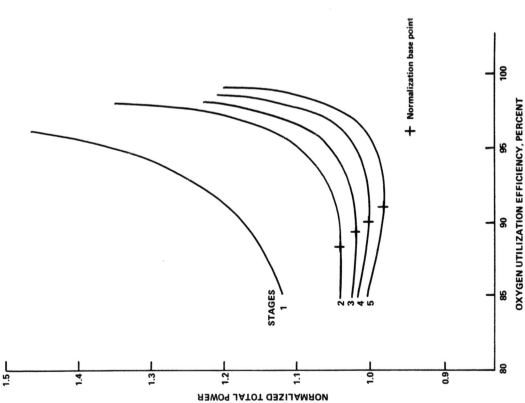

FIGURE 9. Overall oxygenation power requirement (generation plus dissolution power) as a function of oxygen utilization efficiency and number of stages (98% O₂ feed purity). (Courtesy of Union Carbide Corporation.)

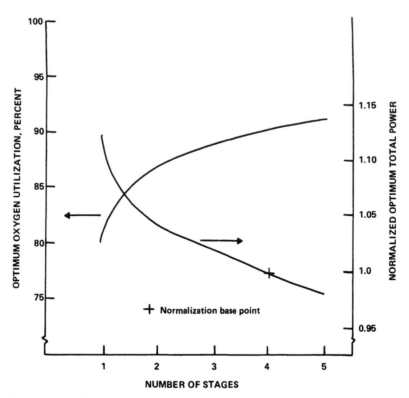

FIGURE 11. Optimum oxygen utilization efficiency and total power requirement as a function of the number of stages (98% O₂ feed purity). (Courtesy of Union Carbide Corporation.)

cantly greater number of stages. This is a highly desirable characteristic, since the capital cost of the gas-liquid contacting system obviously escalates with an increase in the number of stages employed.

Figures 9, 10, and 11 summarize the effect of the number of stages on the optimum total oxygenation power requirement and the optimum oxygen utilization efficiency. The optimum total power requirement significantly decreases with an increase in the number of stages up to about four with only a slight further decrease being obtained with an increase in the number of stages up to twelve. The optimum oxygen utilization efficiency also increases with an increase in the number of stages in the system. The optimum oxygen utilization efficiency increases from about 80 to 91% with an increase from one to five stages, while the optimum total power requirement decreases by approximately 15%.

D. Effect of Mixed-liquor D.O. Level on Oxygenation System Performance

A very important characteristic and advantage

of the UNOX System is the ability to operate at relatively high mixed-liquor D.O. levels at a low overall oxygenation system power requirement. As will be discussed subsequently, operation of the activated sludge process at sustained high mixed-liquor D.O. levels results in several very important and basic process improvements which enable significant changes in the key system design parameters. These changes lead to a substantially more economical system design.

Figures 12 to 14 summarize the effect of the mixed-liquor D.O. level on the dissolution power requirement, oxygen utilization efficiency, and optimum overall system power level. Figure 12 shows the relationship of the mixed-liquor D.O. level to the dissolution power requirement as a function of the oxygen utilization efficiency. As would be anticipated, the dissolution power requirement generally increases with an increase in the mixed-liquor D.O. level for any degree of oxygen absorption efficiency. At a 90% oxygen utilization efficiency, for example, the dissolution power requirement is approximately one third higher for a D.O. level of 8 mg/l than for a D.O.

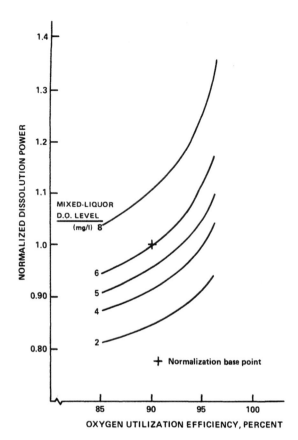

FIGURE 12. Oxygen dissolution power as a function of oxygen utilization efficiency and mixed-liquor D.O. level (four stages, 98% O₂ feed purity). (Courtesy of Union Carbide Corporation.)

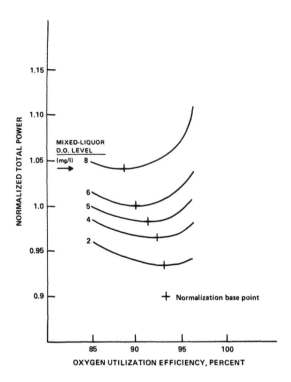

FIGURE 13. Overall total power requirement (generation plus dissolution power) as a function of oxygen utilization efficiency and mixed-liquor D. O. level (four stages, 98% O₂ feed purity). (Courtesy of Union Carbide Corporation.)

level of 2 mg/l for the four-stage system shown in Figure 12. The rise in dissolution power requirement also rapidly increases with the degree of oxygen utilization efficiency. The increase in dissolution power requirement between a D.O. level of 2 and 8 mg/l at a 95% oxygen utilization efficiency is substantially greater than that at 85% utilization efficiency (51% increase at 95% utilization compared to 27% increase at 85% utilization efficiency). Therefore, it is more economical to operate at lower oxygen utilization efficiencies at higher mixed-liquor D.O. levels. Figures 13 and 14 illustrate this effect quite clearly. The optimum oxygen utilization efficiency decreases from approximately 93% at a D.O. level of 2 mg/l to about 88% at a D.O. level of 8 mg/l. Of even greater significance is the fact that the total oxygenation system power requirement (including generation plus dissolution power) increases by only about 11% with an increase in the mixed-liquor D.O. level from 2 to 8 mg/l for a four-stage

system. This increase would be even less for a system with a greater number of stages. Therefore, the UNOX System can operate over a wide range of mixed-liquor D.O. levels (up to relatively high D.O. levels of 6 to 8 mg/l) with only a very modest increase in the total oxygenation system power requirement. This degree of flexibility in performance and operating conditions is totally impossible with an air aeration system.

E. Effect of Feed Gas Oxygen Purity on Oxygenation System Performance

The general effect of feed gas oxygen purity on the overall oxygenation system cost is shown in Figure 15. The relative total annual aeration cost (capital plus operating cost) is plotted as a function of the feed gas oxygen purity at mixed-liquor D.O. levels of 2 and 8 mg/l. The cost of the feed gas oxygen at concentrations below 99.8% was determined by taking the cost of 99.8% oxygen produced by the cryogenic method and blending it with atmospheric air to obtain the desired overall average feed gas oxygen concentration. The lowest overall system cost is obtained

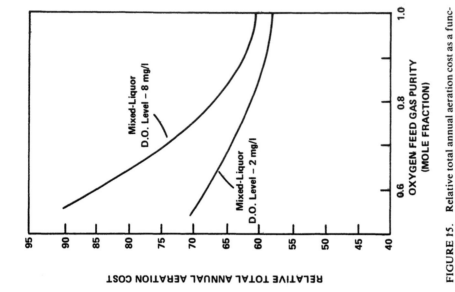

FIGURE 15. Relative total annual aeration cost as a function of feed gas oxygen purity (mixed-liquor D.O. levels of 2 and 8 mg/l). (Courtesy of Union Carbide Corporation.)

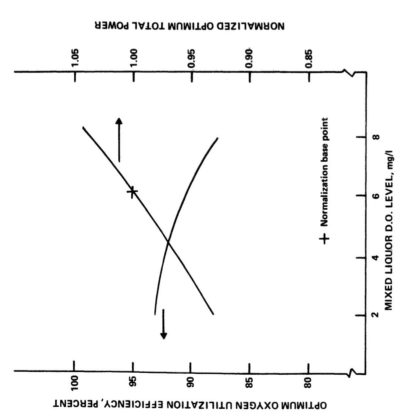

FIGURE 14. Optimum oxygen utilization efficiency and total power requirement (generation plus dissolution power) as a function of mixed-liquor D.O. level (four stages, 98% O₂ feed purity). (Courtesy of Union Carbide Corporation.)

with the highest oxygen purity, as shown in Figure 15. There is very little difference in cost for feed gas oxygen purities above 90%; however, below 90% purity, the overall cost increases rapidly, particularly at the higher mixed-liquor D.O. level. This effect of D.O. level is expected, since it becomes significantly more difficult to operate at high D.O. levels as the oxygen content of the feed gas decreases.

Figure 16 compares the total oxygenation system power requirement for Pressure Swing Adsorption (PSA) oxygen production at 90% purity with cryogenic oxygen production at 95 and 98% feed gas purities. There is about a 20% overall power increase (generation plus dissolution power) with PSA oxygen supply. This slight power increase is more than compensated for by the reduced capital cost and operational simplicity of the PSA oxygen generator. The difference in power requirement between a PSA oxygen generator and a cryogenic oxygen plant also decreases rapidly as the plant capacity decreases.

F. Cocurrent vs. Countercurrent Operation of Multi-stage Oxygenation Systems

Figure 17 compares the overall oxygen utilization efficiency and relative total aeration cost of countercurrent and cocurrent operation of multistage oxygenation systems. In contrast to the operation of most simple equilibrium mass transfer processes, such as distillation, liquid-liquid extraction, gas absorption, etc., cocurrent operation is considerably more efficient and economically attractive than countercurrent operation for the

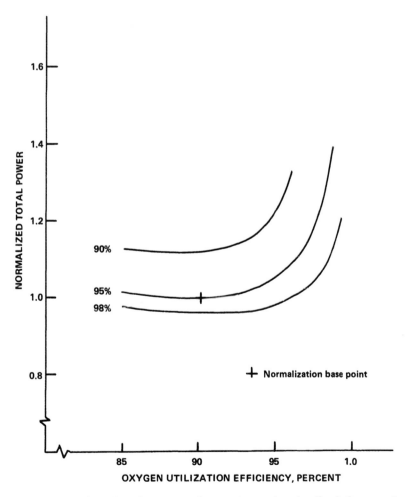

FIGURE 16.   Overall total power requirement (generation plus dissolution power) as a function of oxygen utilization efficiency and oxygen feed gas purity. (Courtesy of Union Carbide Corporation.)

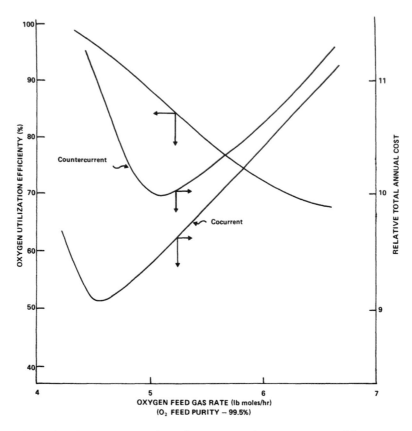

FIGURE 17. General performance comparison of cocurrent and countercurrent multi-stage oxygenation systems. (Courtesy of Union Carbide Corporation.)

oxygenation of activated sludge mixed-liquor. This is because the combined biochemical oxidation and mass transfer process occurring in the liquid phase takes place at a constant mixed-liquor dissolved oxygen level and the highest oxygen transfer rate demand occurs in the first or feed stage of the system. Thus, the highest oxygen concentration in the gas phase should be obtained where the oxygen demand is greatest. This leads to the rather unusual situation in which cocurrent operation is more efficient than countercurrent contacting of the gas and liquid phases.

G. Overall Comparison of UNOX System and Air-activated Sludge Oxygenation System Power Requirements

The oxygenation system power requirement to be considered when comparing a UNOX System with an air system must, of course, include both oxygen generation and dissolution power. Since a UNOX System typically is designed to dissolve about 90% of the oxygen generated, the generation power must be divided by 0.9 to convert it to

a per pound-dissolved basis prior to addition of the dissolution power requirement. The generation and dissolution power requirement for a PSA oxygen supply is 0.34 kW per pound of oxygen dissolved per hour; for a cryogenic oxygen supply, the power requirement is typically 0.28 kW per pound of oxygen dissolved per hour. These performance levels, translated into what may be more familiar units, correspond to 2.4 and 2.9 lb of oxygen transferred per hour per brake horsepower using PSA and cryogenic oxygen generators respectively.

Under typical air system process design conditions, a well-designed surface aerator will provide a standard transfer efficiency (STE) of approximately 3.5 lb of oxygen transferred per hour per shaft horsepower. When corrected for mixed-liquor design operating conditions and mechanical efficiencies, the actual operating transfer efficiency is about 1.9 lb of oxygen transferred per hour per brake horsepower. Allowing for normal electrical motor losses and converting to units of energy per unit rate of dissolution, the energy consumption is 0.43 kWh per pound of oxygen dissolved. A similar calculation results in a dissolution energy requirement of 0.63 kWh per pound of oxygen dissolved from air for coarse bubble diffusers. The range of 0.43 to 0.63 kWh per pound of oxygen dissolved represents the energy requirement that would be expected for dissolution equipment in air systems. These relative energy requirements are graphically summarized in Figure 18.

As shown in this figure, the UNOX System results in about a 35% reduction in energy requirement with cryogenic oxygen supply when compared to a well-designed, high-efficiency surface aeration system and about a 21% power reduction for the same comparison with a PSA oxygen supply. When compared to coarse bubble air diffusers, these UNOX System energy reductions correspond to approximately 56 and 46%, respectively. These reductions in oxygenation power requirement are also obtained with an average mixed-liquor D.O. level of 6.0 mg/l for the UNOX System as compared with an average D.O. level of 2.0 mg/l for the air systems. It is important to remember that the substantial power reductions associated with the UNOX System also apply to operation of the two systems at constant design point loads. The significant additional power savings which can be realized through automatic turndown of the oxygen generator at low loads is not included.

VII. UNOX SYSTEM PROCESS DESIGN AND OPERATING CONDITIONS

The activated sludge process involves the interaction of many biochemical and physical processes which determine its overall performance capability and limit its design parameters. A complete evaluation of the system must consider the following subprocesses and parameters and their interrelationships:

1. Diffusion characteristics within the biomass
2. Biokinetic substrate removal rate
3. Settling characteristics of the biomass
4. Active fraction of biomass in the system
5. Oxygen consumption requirement
6. Mass transfer requirements and limitations

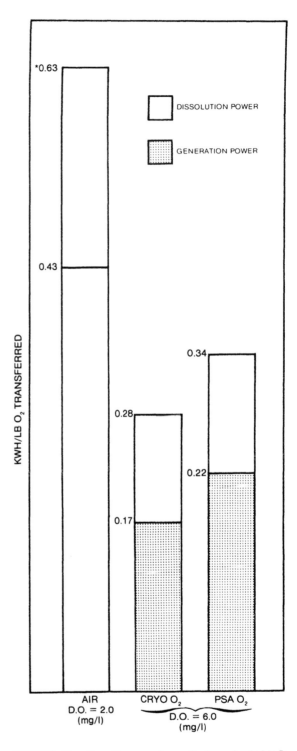

FIGURE 18. Overall comparison of air and UNOX® system oxygenation system energy requirements. (Courtesy of Union Carbide Corporation.)

7. Sludge production quantities and characteristics
8. System stability

The economical substitution of oxygen for air in the activated sludge process by means of the UNOX System removes many of the traditional design constraints and enables practical operation under process design conditions which were either impossible, impractical, or uneconomical with air as the source of oxygen. These distinguishing process performance and operating conditions are summarized as follows:

1. Economical operation at high mixed-liquor D.O. levels (4 to 8 mg/l)
2. Multi-stage (or plug flow) operation at high organic loadings and high MLSS levels without dissolved oxygen limitation
3. High volumetric oxygenation capacity per unit of gas-liquid contacting power input
4. Operation under high rate, high MLSS levels with good sludge settleability, compactability, and low sludge-recycle ratios
5. Low sludge production under low retention time, high organic loading conditions

An important and essential feature of the UNOX System and similar oxygen-activated sludge systems, which enables achievement of many of its basic process performance advantages, is its ability to operate at sustained high mixed-liquor dissolved oxygen levels. Such operation is a major factor resulting in:

1. Maximum substrate removal capability at all times

2. Ability to design for higher food to biomass ratio (F/M) than air systems at comparable overall BOD_5 removal efficiencies
3. Improved sludge settling characteristics
4. Improved sludge dewatering characteristics
5. Lower sludge production at high F/M loadings
6. Improved overall system stability

An in-depth discussion of the effect of high mixed-liquor D.O. levels on these parameters is contained in Section VIII.

Table 3 compares typical design parameters for the UNOX System with conventional air aeration systems for nominal municipal wastewater treatment conditions. The UNOX System can operate at MLSS levels several times greater and aeration detention periods several times less than those of air aeration systems while maintaining comparable or better overall levels of treatment. Among the important features of the UNOX System providing for practical and efficient operation under these high rate conditions are the inherently higher volumetric oxygenation capacity of the oxygenation system per unit of power input and the system's ability to economically operate at high mixed-liquor D.O. levels. The UNOX System is designed to operate at mixed-liquor D.O. levels of 4 to 8 mg/l as compared with 1 to 2 mg/l for air systems, as shown in Table 3.

A very important advantage of the high available oxygen driving force is the ability to

TABLE 3

Comparison of Process Design Conditions for the UNOX® System and for Conventional Air Aeration Systems for Typical Municipal Wastewater

Process design parameters	UNOX oxygenation system	Conventional air aeration systems
Mixed-liquor D.O. level (mg/l)	4—8	1—2
Aeration detention time (raw flow only) (hr)	1—3	3—8
MLSS concentration (mg/l)	4,500—8,000	1,000—3,000
	3,500—6,000	900—2,600
Volumetric organic loading (lb BOD_5/day/ 1000 ft³)	150—250	30—60
Food/biomass ratio lb BOD_5/lb MLVSS/day	0.4—1.0	0.2—0.6
Recycle sludge ratio lb recycle/lb feed	0.2—0.5	0.3—1.0
Recycle sludge concentration (mg/l)	15,000—35,000	5,000—15,000
Sludge volume index (Mohlman)	30—70	100—150
Sludge protection lb VSS/lb BOD_5 removed	0.4—0.55	0.5—0.75

Data courtesy of Union Carbide Corporation.

achieve a relatively high volumetric oxygen transfer rate per unit of power input. An oxygenation system can transfer an average of about 3.5 times as much oxygen per unit of interfacial area as a comparable air system. Hence, an oxygenation system can operate with about one third the mixed-liquor power input of an air system at comparable overall oxygen transfer rates. This characteristic is particularly important in high rate activated sludge systems. It results in minimum volumetric power input and therefore minimizes the turbulence and shear levels to which the biomass is exposed while permitting maintenance of high mixed-liquor D.O. levels. High turbulence levels tend to retard the flocculating tendency of the sludge, and this adversely affects the sludge's settling characteristics.

The need for and advantage of a high volumetric oxygenation capacity in the UNOX System can be best appreciated by consideration of the operating conditions in the first or feed stage of a typical four-stage system. With the system conditions outlined in Table 3, the feed stage is operating with an organic loading of from 600 to 800 lb of BOD_5 per day per 1000 ft^3 and a F/M of from 1.6 to 3.2 lb BOD_5 per pound of MLVSS per day. The retention time in this stage is from 15 to 30 min. Obviously, these conditions generate an extremely high oxygen demand that must be satisfied, in addition to maintaining high mixed-liquor D.O. levels in order to ensure maximum overall process effectiveness. The UNOX System is able to operate at mixed-liquor D.O. levels of 4 to 8 mg/l in the first stage under these operating conditions at relatively low volumetric power inputs.

The combination of operation at high mixed-liquor D.O. levels and at relatively low shear, low dissolution power levels produces superior activated sludge settling characteristics. Operation under these conditions produces a flocculant and rapid settling sludge with excellent compacting and dewatering characteristics. This enables UNOX System operation at high mixed-liquor solid levels (4500 to 8000 mg/l) with normal secondary clarifier overflow rates (600 to 800 gal/day/ft^2)

and relatively low activated sludge recycle ratios (low sludge volume index). Numerous studies of air aeration systems under comparable high solids and high loading conditions have shown poor sludge settling and compacting characteristics, necessitating low clarifier-overflow rates and high sludge recycle ratios. The resulting constraints in air system operation tend to significantly offset the economic incentive for operation under high rate conditions in the aeration tanks. These constraints do not exist for the UNOX System.

In addition to the factors mentioned above, operation at high mixed-liquor D.O. levels has the added benefit of providing a substantial buffer against sudden or shock organic loads which quickly increase the oxygen demand on the system. The high D.O. level provides a temporary "cushioning" effect to sustain process efficiency during short peak load periods. Any subsequent lowering of the D.O. level provides a greater oxygen partial pressure driving force, enabling a higher overall oxygenation capacity to be rapidly obtained. The automatic pressure control system immediately responds to this situation by increasing the oxygen supply rate to the first and later stages in the system to cope with the increasing bio-oxidation demand. Thus, an increased rate of oxygen transfer is achieved through both lowering of the D.O. level and increase in the gas phase oxygen composition.

Another very important feature of the UNOX System is its relatively low excess sludge production even under low retention time, high organic loading conditions. One of the traditional drawbacks to the operation of high rate air aeration systems has been the significantly greater sludge production of such systems compared to that of conventional, low rate systems. Since sludge disposal is a major cost factor in any secondary treatment bio-oxidation process, sludge production is an important consideration in the overall economics. The UNOX System produces considerably less excess sludge per pound of BOD_5 removed than comparable conventional processes operated under much lower organic loading conditions.

VIII. THE EFFECT OF HIGH MIXED-LIQUOR DISSOLVED OXYGEN CONCENTRATIONS ON THE ACTIVATED SLUDGE PROCESS

From the above discussion, it is apparent that the UNOX System offers a completely new degree

of operational and design freedom with respect to mixed-liquor dissolved oxygen concentrations.

Whereas air system designs are economically and practically constrained to operation at relatively low loadings and mixed-liquor D.O. levels of 1 to 2 mg/l, the UNOX System can economically operate at mixed-liquor D.O. levels of 4 to 8 mg/l at relatively low volumetric dissolution power inputs. In order to gain further insight into the process advantages attainable through operation at high mixed-liquor dissolved oxygen levels, it is necessary to consider the basic underlying biochemical and physical processes that control the operation of the activated sludge process.

A. Basic Concepts of the Activated Sludge Process

The activated sludge process basically involves the growth of a flocculated, aerobic biomass in a reactor or aeration tank, using the organic impurities in the influent as food, followed by gravity separation of the biomass in a secondary clarifier or settling tank. The biomass must flocculate and settle well in the secondary clarifier to remove and concentrate the biomass from the wastewater for recycle to the aeration tank and to minimize the suspended solids level in the treated effluent. The necessity for a flocculant biomass results in the need to consider diffusional resistances in the biofloc matrix. The individual biofloc particles contain numerous organisms, each of which requires nutrients, carbonaceous substrate, and dissolved oxygen to be aerobically active in the biochemical oxidation process. Diffusion of these components into the biofloc matrix is therefore of considerable importance.

Conceptualization of the diffusion process can be aided by considering each biofloc particle as a spherical globular mass as depicted schematically in Figure 19. The degree of penetration or diffusion of material into the biofloc particle depends on floc size, bulk liquid concentration of the diffusing components, the rate of uptake of the components by the organisms in the floc particle, and the rate of diffusion characterized by the diffusion coefficient of the components.

The molecular weight of the nutrients involved is substantially less than that of the organic substrates; thus, their diffusion coefficients are substantially higher. At normal nutrient concentrations, therefore, the nutrients would be expected to penetrate the floc particles at least as far or farther than the organic substrate and not be a limiting factor in the overall rate process.

The diffusion coefficient for oxygen is also

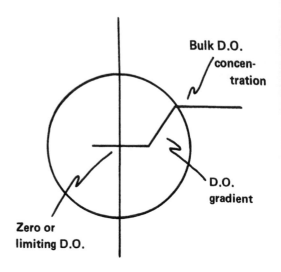

FIGURE 19. Schematic diagram of oxygen concentrations within biofloc particles. (Courtesy of Union Carbide Corporation.)

substantially greater than that for normal organic substrate materials. Since the uptake rate of oxygen is about the same as that of the organic substrate, the ratio of the bulk liquid concentrations of substrate and oxygen should be approximately inversely proportional to their diffusion coefficients in order that neither diffusion rate be controlling over the other. Matson et al.[6] have shown that for normal substrates contained in domestic wastewaters, this ratio is approximately 5. Thus, if substrate material is present in concentrations higher than approximately 5 times that of the dissolved oxygen concentration, then oxygen will be the diffusion controlling component. Matson et al.[6] have shown that for floc sizes and biokinetic oxidation rates typically encountered in activated sludge systems, an oxygen concentration of from 3 to 5 mg/l is required for complete floc penetration. Therefore, diffusional resistance should not be a factor in a system with this D.O. level until the substrate concentration is in the range of 15 to 25 mg/l.

The oxygen concentration gradient within the biofloc particles is schematically depicted in Figure 19. The gradient and depth of oxygen penetration are determined by the floc size, the bulk liquid D.O. level, the bulk liquid substrate level and diffusion rate, and the basic biokinetic oxidation rate. Under conditions where oxygen is the limiting nutrient, the kinetics of organic substrate removal are controlled by the oxygen reaction kinetics, which follow a Monod-type relationship shown in simplified form in Figure 20. When the D.O. level falls below a critical value, the uptake rate rapidly reduces to zero. This results in the situation schematically depicted in Figures 21A and 21B. The D.O. gradient and resultant biokinetic activity levels are plotted to the right and left, respectively, of the symmetrical axes drawn through the spherical biofloc particles. As shown, the rate of activity is essentially constant, and near the maximum level, until the critical D.O. level is reached; it then rapidly decreases to zero. This results in a central core of little or no activity in the biomass floc. The core is neither useful in BOD_5 oxidation nor auto-oxidation of the biomass. The higher the bulk D.O. level, the further oxygen will penetrate into the particle and the greater the

quantity of active biomass within the biofloc matrix. Thus, higher bulk liquid D.O. levels can maintain larger floc particles aerobic and active and/or maintain a given floc particle size aerobic at higher organic substrate levels and bio-oxidation rates.

Two of the most important factors affecting the diffusional resistance of the biofloc particles are the mean floc size and the distribution of floc sizes. The floc size distribution is important even though the number of particles substantially larger than the mean size are relatively few in number because the number of microorganisms in the larger floc particles is substantially greater than the number in the mean floc size particles. Thus, even if the diffusional resistance is overcome for the mean and smaller diameter particles, respiration in many of the microorganisms in the larger floc particles is diffusionally controlled.

It has been pointed out by Chapman et al.[7] and by many other investigators[8-10] that both micro- and macroscale mixing occurs in activated sludge aeration basins. Macroscale mixing is generally responsible for the overall bulk fluid mixing level and suspension of the biomass, whereas microscale mixing is basically responsible for mass transfer and particle shear. It can be shown using the theory of isotropic turbulence that the stable floc particle size is directly affected by the power density (power per unit volume) in the basin and the size of the basin. A higher power density will decrease the size of the smallest eddies for local isotropic turbulence in a given basin size. Since the smallest eddies are responsible for most of the

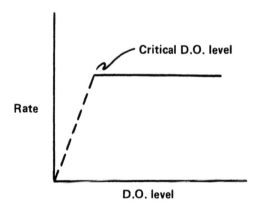

FIGURE 20. Monod-type kinetic relationship for rate of oxidation as a function of D.O. level for individual microorganisms. (Courtesy of Union Carbide Corporation.)

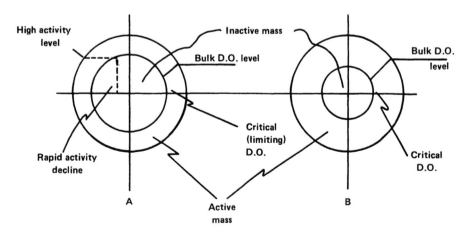

FIGURE 21. Schematic diagram of (A) active and (B) inactive portions of biofloc particles. (Courtesy of Union Carbide Corporation.)

energy dissipation and the stable floc size must be comparable to the size of the energy dissipating eddies, it follows that the floc size will be smaller at higher power densities. The floc size distribution will also be narrower because the higher power density causes more intense macromixing, which results in more frequent basin turnover and exposure of the floc particles to the areas of highest shear in the vicinity of the oxygen dissolution equipment. Furthermore, if one now compares large and small scale aeration basins at the same power density, it follows from the above theory that there is a strong effect of basin size on the average floc size. The mean floc size is smaller and the floc size distribution narrower in smaller basins because of the smaller energy dissipating eddies and the more intense and higher frequency of macromixing.

In addition to the above considerations, it is generally well known and accepted that a much higher power density is required in a laboratory bench-scale system than in a full-scale plant in order to maintain a given mixed-liquor D.O. level. Power densities in small-scale laboratory systems are typically an order of magnitude higher than in full-scale plants. Even pilot plant systems typically operate at 50 to 75% higher power densities than full-scale plants. Thus, the mean floc size will be smaller and the floc size distribution narrower for a small-scale laboratory or pilot plant system than for a full-scale plant. This large variation in power density and mean floc size must be taken into account when laboratory and pilot-scale tests are used to provide the basis for and predict the performance of full-scale plants.

In addition to the above considerations relating to the mean floc size and floc size distribution, the oxygen uptake rate must be properly considered in determining the importance of the diffusional resistance and the degree of oxygen penetration into the biofloc particles. The higher the volumetric oxygenation rate, the higher is the required bulk liquid D.O. level to obtain an equal degree of floc penetration. The importance of the substrate concentration must also be recognized. Higher substrate concentrations require a higher bulk liquid D.O. level for the same degree of penetration of oxygen into the floc.

These concepts will now be examined as they relate to the basic design parameters that control the operation of the activated sludge process and

the influence of mixed-liquor dissolved oxygen level on these parameters will also be discussed.

B. Effect of High Mixed-Liquor Dissolved Oxygen Level on Rate of Substrate Removal

A great deal of the early pioneering work on the use of oxygen in the activated sludge system focused on the effect of high mixed-liquor D.O. levels on the rate of substrate removal. Okun and Lynn,[11] Gaden,[12] and Wuhrmann,[13] as well as many others,[14-22] studied this effect. All of this work was carried out in laboratory bench-scale or small pilot plant complete-mix reactors bio-oxidizing domestic sewage. This means that the average power densities were necessarily relatively high and the average biofloc particle sizes relatively small. The studies were also conducted at the normal food to biomass ratios usually encountered in conventional air-activated sludge practice and operating at MLSS levels and residence times in a complete-mix reactor mode to give reasonably high quality effluent. Under these small floc size, low oxygen uptake rate conditions, the organic substrate is at relatively low bulk liquid concentration levels and the overall oxygen uptake rate is limited by the rate of organic substrate diffusion into the biomass particles. Hence, under these operating conditions, one would not expect to see any effect of higher D.O. levels, since the degree of penetration of oxygen into the biofloc particles cannot increase the substrate removal rate. Indeed, this is what all of these early investigators observed, and this seemingly negative conclusion regarding the effect of high mixed-liquor D.O. levels was responsible, to a considerable degree, for the early lack of interest in the use of oxygen.

In fact, most pilot plant and field studies show that the effluent BOD_5 is essentially the same for both oxygen- and air-activated sludge systems when treating domestic sewages at low overall system organic loadings. For most domestic wastewaters, the basic biokinetic oxidation rate is sufficiently high and the effluent organic substrate levels are sufficiently low that systems operate in the organic substrate diffusion limiting regime. These systems are not biokinetically controlled or oxygen diffusion controlled. Thus, effluent quality will be nearly the same for air and oxygen systems operated over a reasonable range of low organic loadings as long as the system flocculates well, and dispersed growth does not occur. However, there are many industrial or combination municipal-

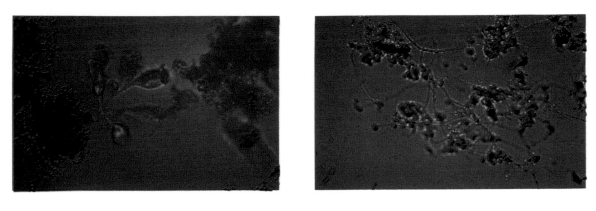

A   B

FIGURE 23.   A. Biomass obtained at Philadelphia, S.E., pilot plant. UNOX® System mixed-liquor D.O. = 5 to 8 mg/l. (Magnification × 2000). B. Biomass obttained at Philadelphia, S.E., pilot plant. Air system mixed-liquor D.O. 0.5 to 1.5 mg/l. (Magnification × 450.) (Courtesy of Union Carbide Corporation.)

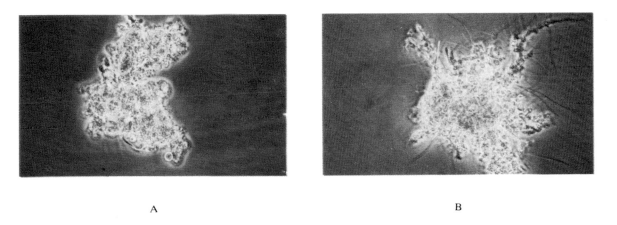

A   B

FIGURE 24.   A. Typical biofloc from oxygen system, D.O. = 6.7 to 8.0 mg/l. (Magnification × 200.) B. Typical biofloc from air system, D.O. 1.3 to 1.8 mg/l. (Magnification × 200.) (From Sezgin, Jenkins, and Parker, paper presented at Water Pollution Control Federation 49th Annual Conference, Minneapolis, Minn., 1976. With permission.)

FIGURE 4. First full-scale plant test of the UNOX® System at Batavia, New York. A conventional diffused air system is in the foreground and the OX System is in background. (Courtesy of Union Carbide Corporation.)

industrial wastewaters in which difficult to degrade materials cause the system to be biokinetically controlled at loadings where the system flocculates well. The effluent quality under these circumstances depends directly on the quantity of aerobically viable organisms in the system and, hence, directly on the degree of oxygen penetration into the biofloc particles. A high-purity oxygen system operating at higher dissolved oxygen levels would enable more of the floc particle to be maintained aerobically active. Therefore, the oxygen-activated sludge system would be expected to produce a better quality effluent. An example of such a case is shown in Table 4 where data for comparative air- and oxygen-activated sludge systems were obtained at comparable detention times and biomass loadings with the systems operating side by side in the same municipal-industrial wastewater in Wuppertal, Germany.[23] The significant improvement in effluent quality of the oxygen system when compared to the air system at the same loading is quite evident from Table 4. The COD, BOD$_5$, and SS levels in the effluent of the oxygen system were all substantially below those of the air system with the only difference in operating conditions for the two systems being the higher mixed-liquor D.O. level of the oxygen system.

Even for the treatment of normal domestic wastewaters, the conditions of oxygen uptake rate, substrate concentration, and biomass level in a multi-stage UNOX System are significantly different from those encountered in conventional air aeration systems. In the first stage of a UNOX System, the substrate concentration is at its maximum value and the system is operating at a very high loading rate (F/M). Consequently, a high mixed-liquor D.O. level is necessary to ensure maximum floc penetration and prevent oxygen diffusion into the biomass from becoming rate controlling. In the later stages of the system, the substrate concentration is low and its diffusion rate is controlling. In this portion of the system, a substantial portion of the inner floc is undergoing endogenous respiration as a result of the availability of adequate oxygen and the absence of substrate which was oxidized in the first stages.

C. Effect of High Mixed-liquor Dissolved Oxygen Level on Ability to Design for Higher Organic (F/M) Loadings

It has long been known that the flocculating nature of the biomass is a function of the ratio of the organic load on the system to the biomass in the system (F/M). Very high ratios lead to a dispersed floc and very low ratios lead to a poorly flocculated biomass characterized by large amounts of pin floc in the effluent. The flocculating nature of the biomass has been theorized[24, 25] to be due to the production of an exocellular biopolymer material during the endogenous respiration of the organisms. This biopolymer material acts as the "glue" between the individual organisms resulting in a flocculated matrix of biomass when the ratio of the biopolymer to biomass is in the correct range. When the loading is too high, the rate of growth of the organisms far exceeds the endogenous respiration rate and the amount of biopolymer generated is insufficient to glue the biomass together. The result is a dispersed growth system. On the other hand, when the loading is too low, food is scarce, and the organisms use the biopolymer material as food. Under these conditions, the result is an "ungluing"

TABLE 4

Pilot Plant Data: Wuppertal-Buchenhofen Wuppertal, Germany Municipal Plus Chemical Industry, Iron Processing, and Textile Waste

Performance Summary	Air	Oxygen
Retention time, (hr)	6.50	6.20
MLVSS (mg/l	3440.00	4270.00
D.O. (mg/l)	0.5—2.0	~15.00
Biomass loading, (lb COD/lb MLVSS/day)	0.81	0.70
Biomass loading (lb BOD$_5$/lb MLVSS/day)	0.35	0.32
Influent characteristics (mg/l)		
COD	751.00	767.00
BOD$_5$	324.00	348.00
SS	113.00	114.00
Effluent characteristics (mg/l)		
COD	281.00	176.00
BOD$_5$	97.00	18.00
SS	76.00	36.00
Removals (%)		
COD	63.00	77.00
BOD$_5$	70.00	95.00
SS	33.00	68.00
Sludge production (including effluent SS) lb ESS/lb BOD$_{5R}$	0.91	0.52
Clarifier-underflow concentration (%)	0.50	2.00

of the biofloc matrix and the appearance of pin floc in the effluent.

In view of the above considerations, the organic loading or F/M of an activated sludge system has rightfully become an important basic design parameter. The definition of F/M is the pounds of substrate (usually expressed as BOD_5) applied per day per pound of mixed-liquor volatile suspended solids. In this definition, it is assumed that the mixed-liquor volatile suspended solids level is a good indication of the quantity of active, viable organisms in the biomass in the system. The true loading parameter of importance is the mass of substrate per day per mass of active, viable organisms in the system.

When comparing high and low D.O. systems, as was done in Figures 21A and 21B, one can readily recognize that a high D.O. system has a greater active biomass fraction than a low D.O. system. The true ratio (mass of substrate per day per unit of active biomass) is lower in a system with high

D.O. than a system with low D.O. at the same measured value of F/M. Thus, a high D.O. system will hold a good flocculating and settling biomass at higher measured F/Ms than a low D.O. system. The mass of substrate per day per unit of true active biomass in high D.O. systems does not become equal to that of low D.O. systems until higher measured F/Ms are reached. Even in a biokinetically controlled system, the high D.O. system can operate at a higher F/M than a low D.O. system with the same effluent quality because more of the biomass is aerobic and active.

The ability of the UNOX System to effectively operate at higher measured F/Ms than conventional air systems while producing comparable or better effluent quality and improved sludge settling characteristics has been demonstrated many times. A typical comparison of this type is shown in Table 5 for a side by side air vs. oxygen study at Los Angeles County, California. Even though the UNOX System was operated at a measured F/M of

TABLE 5

Pilot Plant Data:
Los Angeles County municipal and refinery waste

	Air		Oxygen	
Performance Summary	Phase I	Phase II	Phase I	Phase II
Retention time, Q (hr)	7.60	6.40	2.00	2.00
Recycle ratio, R/Q (%)	47.00	75.00	30.00	25.00
Sewage temperature (°F)	76.00	80.00	76.00	80.00
MLVSS (mg/l)	1950.00	1980.00	2950.00	2500.00
D.O., (mg/l)	1.00	0.30	6.00	6.00
Biomass loading (lb COD/lb MLVSS/day)	0.79	0.93	2.23	2.11
Biomass loading, (lb BOD_s/lb MLVSS/day)	0.39	0.41	0.98	1.04
Influent characteristics (mg/l)				
COD	492.00	490.00	492.00	490.00
BOD_s	242.00	216.00	242.00	216.00
SS	160.00	168.00	160.00	168.00
Effluent characteristics (mg/l)				
COD	94.00	115.00	90.00	90.00
BOD_s	38.00	18.00	20.00	20.00
SS	36.00	45.00	23.00	20.00
Removals (%)				
COD	81.00	77.00	82.00	82.00
BOD_s	84.00	92.00	92.00	91.00
SS	78.00	73.00	86.00	88.00
Clarifier-overflow rate (gal/ft²/day)	268.00	320.00	580.00	580.00
Clarifier-underflow concentration (%)	0.89	0.63	1.60	1.80

Data courtesy of Union Carbide Corporation.

over twice that of the air system (an F/M of approximately 1.0 vs. 0.4), the UNOX System produced better quality effluent with respect to COD, BOD_5, and suspended solids. The sludge settling properties of the UNOX System were also superior, as evidenced by the substantially greater clarifier-underflow concentration even though the clarifier-overflow rates were significantly higher. In summary, the basic design selection of F/M should be determined by the following considerations:

1. The system loading must be within the range to produce a good flocculating biomass.
2. The F/M must be below the biokinetically limiting value in order to remove the substrate to a satisfactorily low effluent level.
3. The system must be capable of supplying adequate oxygen to maintain appropriately high mixed-liquor D.O. levels.
4. The energy or power input for oxygenation must not expose the biomass to excessively high shear levels, which would lead to floc destruction and smaller mean biofloc particle sizes.

The ability of oxygen-activated sludge systems to sustain high mixed-liquor D.O. levels enables Points 1 and 2 to be well satisfied while operating at relatively high F/M loadings. Points 3 and 4 are equally important. The use of oxygen provides additional advantages in these areas. As discussed earlier in Section V, the high average oxygen partial pressure maintained in the aerating gas in a UNOX System enables high oxygen transfer rates to be maintained at high mixed-liquor D.O. levels with relatively low dissolution power inputs for gas-liquid contacting. Volumetric oxygen transfer rates of 3 to 4 times those attainable with air aeration can be achieved at D.O. levels of 5 to 6 mg/l at comparable dissolution power inputs per unit volume of mixed-liquor. This means that high F/M operation can be sustained at high D.O. levels without excessive shear and consequent biofloc particle breakup. This combination of beneficial effects is uniquely provided by the multi-stage, cocurrent design of the UNOX System. In the head end or first stage of the system, the highest oxygen partial pressure driving force is matched to the highest oxygen demand requirement and F/M loading. This enables the oxygen demand to be satisfied with the lowest possible level of gas-liquid contacting power input. In the later stages of the system, where the oxygen demand is significantly reduced, the oxygen driving force is also the lowest. The oxygen demand in these later stages can usually be more than adequately supplied at the minimum power input levels required to sustain an appropriate degree of bulk fluid mixing and biomass suspension.

D. Effect of High Mixed-liquor Dissolved Oxygen Level on Sludge Settling Characteristics

The improved settling velocity of sludges grown at high D.O. levels is perhaps the most surprising effect of operation of activated sludge systems at high mixed-liquor D.O. levels. This probably accounts for it being one of the most controversial factors in comparisons of air and oxygen system performance. The effect has, nonetheless, been conclusively shown in more than a dozen comparative side by side air and oxygen pilot plant tests and in full-scale oxygen-activated sludge plants. Virtually all of the UNOX pilot plants and full-scale plants have shown that high MLSS concentrations of 4500 to 8000 mg/l and clarifier-underflow concentrations of from 1.5 to 3.0% are readily achieved at low recycle ratios and at conventional secondary clarifier-overflow rates of 600 to 800 GPD/ft.[2] This performance in itself is evidence of the improved settling rates of high D.O. sludges. Low D.O. air systems typically achieve MLSS concentrations of only 1500 to 2500 mg/l with recycle sludge concentrations of 0.5 to 1.5% at comparable recycle ratios and clarifier-overflow rates.

The settling velocity of activated sludges varies significantly depending primarily on the nature of the wastewater and the degree of inert material in the sludge. The degree of agitation and turbulence in the aeration tank can also influence the settling characteristics. Too high a shear level can adversely affect the mean biofloc particle sizes and reduce the settling rate.

The gathering of accurate sludge settling data also requires close control of the experimental procedure used. Great care must be exercised with respect to the settling column-filling procedure, the elimination of wall effects, and maintenance of an appropriate sludge depth. The best results can be obtained using at least a 6-in. diameter sludge settling column which is specifically built for this purpose.

One of the most important factors affecting measured sludge settling velocities, which has been largely ignored in the past, is the so-called preconditioning or preaeration of the sludge prior to running the settling tests. Most sludge settling tests have been run in locations remote from the aeration tank sampling point and some period of time after the sludge sample was taken. The standard test procedure has been to preaerate or reaerate the sludge to keep it fresh prior to running the settling tests. This preaeration, of course, raises the mixed-liquor D.O. level which, in turn, improves the sludge settling velocity. This effect demonstrates the beneficial effect of high mixed-liquor D.O. levels on sludge settling rates. Table 6 shows a comparison of the sludge settling rates for aerated and unaerated air-activated sludges. For the unaerated sludges, fresh sludge samples were taken directly from the aeration tank and the settling velocities were immediately determined at various solids concentrations at the aeration tank site. For the aerated sludges, the same sludge was aerated for a few hours and then the settling velocities again measured. Inspection of Table 6 clearly reveals that in all cases, except one, the settling velocities of the sludges after aeration were considerably higher than before aeration. It is most significant to note that the

only sludge for which settling velocity was not increased by reaeration was the one in which the plant was already maintained at a relatively high D.O. level of 3.6 mg/l prior to aeration.

Because of the many factors that influence sludge settling rates and the wide variation in sludge settling velocities for different wastewaters, meaningful direct comparison of initial settling velocities can only be obtained from direct side by side comparison on the same wastewater. Recognizing this fact, Union Carbide has made a special effort to obtain such data on as many air- and oxygen-activated sludge systems as possible operating in parallel on the same wastewater at the same time. Figure 22 shows some of these data together with additional air settling data reported in the literature and additional oxygen settling data. Inspection of Figure 22 clearly shows that the oxygen system settling velocities generally lie well above those of the air systems. The following conclusions can be drawn from these data:

1. The settling velocities of various sludges differ by a significant amount independent of whether the sludge is oxygen aerated or air aerated.

TABLE 6

Comparison of Aerated and Unaerated Air Settling Data

Plant location	D.O. (mg/l)	Unaerated ISR (ft/hr)	MLSS (mg/l)	D.O. (mg/l)	Aerated ISR (ft/hr)
Brockton, Mass.	1.5	4.90	2156	9.0	7.00
		5.00	2010		6.20
		2.10	4020		3.20
		0.95	6030		1.10
		0.49	8040		0.56
Hartford, Conn.	1.2	7.70	2482	7.0	11.90
		7.80	2224		12.40
		3.20	4448		6.70
		1.40	6672		2.80
		0.70	8896		1.05
Marlborough, Mass. East	1.2	7.70	1502	8.5	8.40
		1.75	3484		2.10
Marlborough, Mass. West	3.6	4.90	4070	6.0	4.00
		8.10	2213		8.40
		3.40	4426		5.20
		1.00	6639		1.30
		0.70	8852		0.80
Pensacola, Fla.	1.0	6.70	1650	6.0	8.10

Note: ISR, initial settling rate.

Data courtesy of Union Carbide Corporation.

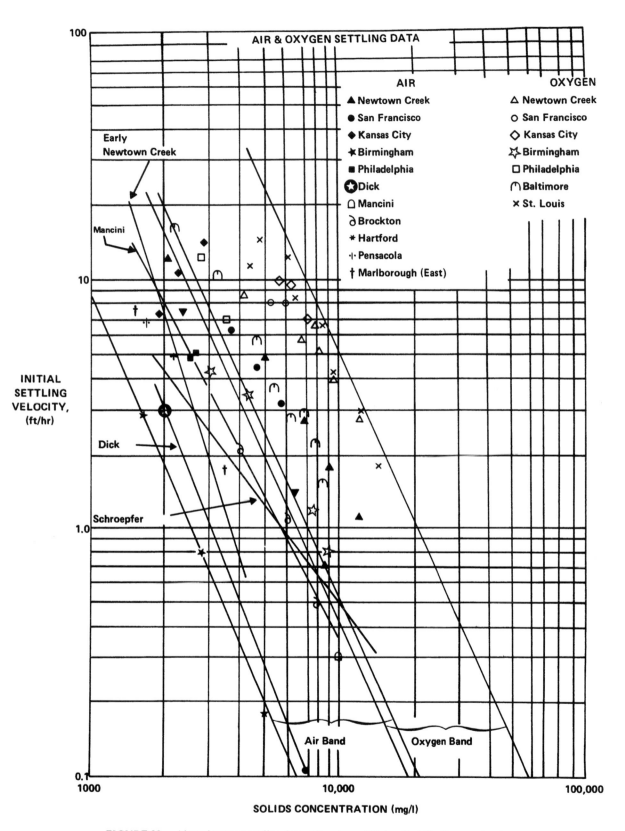

FIGURE 22.   Air and oxygen settling data. (Courtesy of Union Carbide Corporation.)

2. The "band" of oxygen-activated sludge settling velocities lies well above the band of air-activated sludge settling velocities.

3. The displacement between the air and oxygen system settling bands corresponds to a solids concentration increase of approximately threefold for the same settling velocity for the oxygen system sludges.

4. Settling velocities on the high or low side of the air band correspond to oxygen-activated sludge settling data on the high or low side of the oxygen band, with a significantly higher settling velocity being maintained for the oxygen system.

5. The highest air-activated sludge settling velocities are somewhat higher than the lowest oxygen-activated sludge settling velocities. These plants, however, were operated at unusually high D.O. levels of from 5 to 7 mg/l at Newtown Creek (because of half design load) and from 2 to 5 mg/l at San Francisco (because of an oversized pilot plant). Nevertheless, the oxygen sludge settling velocities at these two locations were still considerably higher than the respective air sludge settling velocities.

More recently, other investigators have also recognized the beneficial effect of high mixed-liquor D.O. levels on settling velocity. Jewell and Eckenfelder[26] reported on this effect in their air and oxygen system treatment studies on brewery wastewater. Ball et al.[27] have noted that an important benefit of oxygenation systems arises from "improved sludge settling and dewatering characteristics." The Water Pollution Research Laboratory of England[28] also studied the settling and filtration characteristics of air- and oxygen-activated sludges from parallel operating plants and stated that the oxygenated sludges had significantly higher settling velocity and lower specific resistance to filtration than the air sludges.

The diffusional resistance model depicted in Figures 19 through 21 can also provide considerable insight into the improved settling velocity of sludges developed under high mixed-liquor dissolved oxygen conditions. It has been proposed by Sezgin et al.[29] that the biofloc particles encountered in activated sludge systems consist of a mixture of filamentous and spherical organisms. They have proposed that the biofloc particles need some filamentous organisms to provide strength to the particle. It is postulated that the filaments act as a "backbone" structure to which the spherical

organisms attach themselves, much like "flesh on a bone." The overall character of the biofloc is then dependent on the relative amount of filamentous- and spherical-type organisms in the biofloc. Too few filaments and the biofloc is fragile and easily torn apart. Too many filaments and the biofloc settles poorly because the filaments grow not only within the biofloc particles but well out into the mixed-liquor, thus preventing the individual biofloc particles from closely approaching each other. It is further postulated that oxygen limitation within the biofloc interior causes the filaments to grow and reach out towards the outer portions of the particles. At low dissolved oxygen levels, the growth of the filaments appears to be more rapid than that of the spherical organisms, while at high dissolved oxygen levels, the reverse appears to be true. Thus, if the dissolved oxygen level in the center of the biofloc particles is too low, the filaments will outgrow the spherical organisms and tend to become more prevalent. The filaments will then extend out from the biofloc into the mixed-liquor and retard the settling velocity.

Therefore, a biomass wherein the biofloc particles are totally penetrated by oxygen would be expected to have better settling and dewatering properties than one in which the biofloc centers are continually or periodically exposed to very low dissolved oxygen conditions.

Experimental observation of these effects has been obtained in two different studies involving direct side by side comparison of air- and oxygen-activated sludge system performance. Figures 23 and 24 (these figures appear following page 50) show photomicrographs of sludges that illustrate the differences between biomass maintained at high mixed-liquor dissolved oxygen levels and that maintained at low mixed-liquor dissolved oxygen levels. The Philadelphia pilot plant photomicrographs (Figure 23) were of biomass obtained from mixed-liquor operating on the same wastewater taken on the same day (June 11, 1972). The more compact and larger biofloc particles of the oxygen system biomass are readily apparent. The photomicrographs taken by Sezgin et al.[29] (Figure 24) were obtained under very carefully controlled laboratory conditions and the difference in the compactness and concentration of filamentous-type organisms is easily observable. The relative size of the biofloc particles in this study is about the same, which also attests to the carefully controlled conditions under which the experiments were carried out. The authors were only interested in the effect of dissolved oxygen level and the degree of oxygen penetration into the biofloc particles. They, therefore, took great care to maintain the same mixed-liquor power density and shear level into each system using identical equipment. This can be accomplished in a laboratory-scale reactor, but would be completely impractical in a full-scale plant from an economic standpoint.

E. Effect of High Mixed-liquor Dissolved Oxygen Level on Sludge Production

The dewatering and disposal of excess activated sludge represents a significant portion of total secondary treatment costs. Consequently, considerable attention must be devoted to this aspect of overall plant design in order to obtain the most cost-effective integrated system design. It is well known that the nature and type of wastewater greatly influences the amount of sludge produced during the bio-oxidation process. It is also well recognized that the design of the system affects the sludge production rate. This is generally true for any type of activated sludge system design (air or oxygen). The higher the organic loading rate (F/M) the greater will be the amount of excess sludge produced. This has led the English for decades to design relatively low rate air-activated sludge systems. They contend that with higher rate systems the savings in aeration tank cost do not offset the attendant higher sludge disposal costs. Although this opinion is not shared by many experts around the world, the objective of minimizing excess sludge disposal costs is common to all. How high D.O. level operation minimizes sludge production while at the same time enabling operation at high F/M will now be examined.

The mechanism of reduced sludge production with high D.O. operation as compared with low D.O. operation is straightforward considering the prior discussion and reference to Figures 21A and 21B. The greater depth of oxygen penetration into the biofloc particles with high D.O. operation produces a greater fraction of active biomass which enables a higher degree of endogenous respiration than with a comparable low D.O. system at the same organic substrate loading. Alternatively, as in Section VII.A. it can be seen that the high D.O. level produces a higher level of active biomass and, hence, a lower true F/M than would be measured relative to a low D.O. system.

The lower true F/M results in a relatively lower sludge production rate at the same apparent measured value of F/M. The multi-stage UNOX System mixed-liquor reactor design further enhances this effect by enabling very low F/M operation at high D.O. levels in the later stages of the system which maximizes the endogenous respiration rate opportunity.

In spite of the above arguments and a wealth of substantiating data obtained over the past several years from many sources, there has been considerable debate as to whether there is a significant difference in sludge production rates between air and oxygen systems operating at the same measured overall F/M loadings. A large part of the problem is the high degree of variability of sludge production rates from different wastewaters and the difficulty of accurately measuring and comparing relative sludge production rates from the same wastewaters in side by side test runs. Such tests must be carefully controlled and run for relatively long periods of time to ensure that truly steady state data are being obtained and that the results are not influenced by shifts in sludge inventory in the system during the test period. Another reason for the debate is that this effect had not been previously suspected or considered in the literature and had gone undetected for such a long period of time.

Because of the measuring difficulties referred to above, the most meaningful and accurate data can be obtained from full-scale plant operations. Comparative data must be obtained from simultaneous operation of side by side air and oxygen systems operating on the same wastewater for a relatively long period of time. The two most comprehensive test comparisons of this type made to date are the Batavia, New York tests conducted in the late 1960s and the Newtown Creek, New York tests conducted in the early 1970s. Figure 25 shows a summary of the Batavia, New York relative sludge production data for all three phases of operation over a two-year period. It is clearly seen that the excess sludge produced per pound of biomass removed was substantially less for the oxygen system over a wide range of organic loadings (F/M from 0.3 to 0.9). During all of these tests, very careful overall material balance measurements were made for both the air and oxygen systems. Table 7 summarizes the overall performance and sludge production data for the Newtown Creek comparative tests. As in the

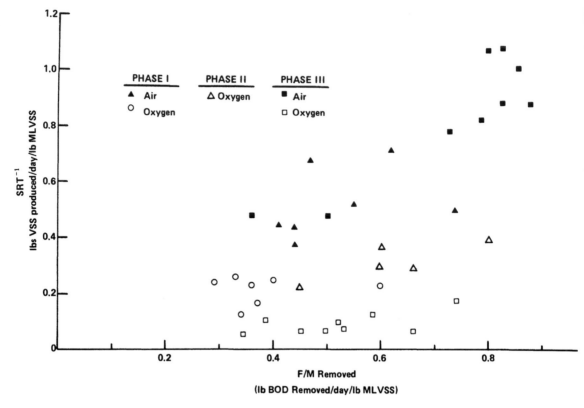

FIGURE 25.   Steady state air and oxygen data, Batavia, New York. (Courtesy of Union Carbide Corporation.)

Batavia tests, the oxygen system produced substantially less excess sludge at comparable, as well as at significantly higher, loadings than the air system.

In addition to these full-scale tests, a number of other smaller-scale tests conducted by a variety of investigators have revealed the same general conclusions. The U.S. Environmental Protection Agency (EPA) pilot plant tests conducted at Washington, D.C.[30] compared the sludge production rates between side by side air and oxygen units. Their data showed significantly less sludge production for the oxygen system at sludge retention times of over 6 days. Their data also appeared to show a diminished difference in sludge production rates between the two systems at retention times of less than 6 days. This apparent reduction in the relative sludge production rates, however, was merely due to the manner in which the data were reported. Sludge production data were reported on the basis of BOD_5 applied and not BOD_5 removed. Since the air system BOD_5 removal efficiency began to fall rapidly below sludge retention times of 6 days, whereas the oxygen system efficiency did not, this tended to distort and reduce the apparent sludge production rate difference which was, in fact, due to the higher level of BOD_5 removal for the oxygen system.

Another comparative study was carried out by Jewell and Mackenzie.[31] This study was conducted to compare yields in identical processes with significantly different dissolved oxygen concentrations. The two major biological treatment systems evaluated in this study were the suspended and attached microbial forms. The models employed in this laboratory study consisted of typical bench-scale activated sludge units and specially designed columns with small removable modules for microbial attachment. The synthetic feed containing glucose was varied to give similar hydraulic and mass loading conditions in each unit. The D.O. was maintained at approximately 5 mg/l for the air and 15 mg/l for the oxygen systems. In all comparable cases, the higher D.O. concentration caused a significantly lower sludge yield. The yield difference was generally greater than 20% and was most significant at the higher mass loadings where more active

TABLE 7
Newtown Creek Performance Summary
60% municipal and 40% small industry — weight basis

Performance parameter	Phase I Constant flow		Phase II Diurnal flow		Phase III Diurnal flow		Phase IV Diurnal flow		Phase V Diurnal flow	
	Oxygen	Air	Oxygen	Air	Oxygen	Air	Oxygen	Air	Oxygen	Air
Flow Q, MGD	20.80	10.60	20.30	10.60	25.60	12.00	30.00	11.50	35.40	10.60
Recycle R, MGD	6.10	2.00	9.30	2.20	11.00	2.90	10.20	2.60	9.10	2.50
Retention Time (hr)										
Based on Q only	1.43	2.80	1.46	2.80	1.16	2.47	0.99	2.58	0.84	2.80
Based on Q + R	1.10	2.35	1.00	2.32	0.81	2.00	0.74	2.10	0.67	2.26
Influent Characteristics (mg/l)										
BOD	156.00	156.00	164.00	164.00	238.00	238.00	215.00	215.00	198.00	198.00
COD	356.00	356.00	361.00	361.00	320.00	320.00	290.00	290.00	308.00	308.00
SS	149.00	149.00	158.00	158.00	149.00	149.00	124.00	124.00	131.00	131.00
VSS	119.00	—	120.00	120.00	112.00	112.00	100.00	100.00	108.00	108.00
Temperature (°F)	67.00	67.00	63.00	63.00	69.00	69.00	75.00	75.00	76.00	76.00
pH	7.00	7.00	6.70	6.70	6.60	6.60	6.40	6.40	6.50	6.50
Alkalinity (mg/l as CaCO$_3$)	83.00	83.00	77.00	77.00	78.00	78.00	67.00	67.00	73.00	73.00
Effluent characteristics (mg/l)										
BOD	9.00	24.00	17.00	36.00	23.00	45.00	20.00	36.00	22.00	34.00
Soluble BOD	4.00	11.00	11.00	24.00	13.00	31.00	11.00	18.00	14.00	25.00
COD	61.00	—	76.00	—	71.00	—	64.00	—	62.00	—
Soluble COD	—	—	66.00	—	61.00	—	51.00	—	—	—
SS	12.00	39.00	17.00	45.00	26.00	52.00	17.00	45.00	17.00	47.00
VSS	9.00	—	14.00	—	20.00	—	13.00	—	14.00	—
Efficiency (% removal)										
BOD	94.00	85.00	90.00	78.00	90.00	81.00	91.00	83.00	89.00	83.00
COD	83.00	—	79.00	—	78.00	—	78.00	—	80.00	—
SS	92.00	74.00	89.00	72.00	83.00	65.00	86.00	64.00	87.00	64.00
VSS	—	—	88.00	—	82.00	—	—	—	—	—
Biomass loading (F/M) (lb BOD/day/lb MLVSS)	0.64	0.80	0.86	1.00	1.37	1.47	1.57	1.33	2.26	1.01
Volumetric organic loading, (lb BOD/day/1000 ft^3)	164.00	84.00	168.00	88.00	308.00	145.00	326.00	125.00	354.00	106.00
Waste sludge (lb SS/lb BOD$_R$)	0.78	1.23	1.02	1.32	0.68	0.74	0.65	1.16	0.71	1.23
Sludge production (lb SS/lb BOD$_R$)	0.85	1.53	1.14	1.67	0.80	1.01	0.74	1.43	0.80	1.52
Sludge retention time (days)	2.32	1.16	1.40	0.95	1.30	1.06	1.24	0.94	0.76	0.98

Data from Brenner, R. C., Newtown Creek Project, Environmental Protection Agency report, in press.

organisms should be present and where there is thus more microbial mass that can continue aerobic endogenous respiration due to an increased depth of oxygen penetration into the floc particle.

The data of Hegemann[23] in the Wuppertal, Germany comparative air and oxygen system tests also confirmed lower sludge production for the oxygen system. Tables 4 and 8, for the two different wastewaters tested, show very substantial reductions of sludge production rates by the high D.O. oxygen system. As shown in Table 8, the oxygen system produced approximately 54% less sludge per pound of BOD_5 removed even though the organic loading (F/M) was about twice as high.

All of these tests rather convincingly demonstrate the lower sludge production capability of high D.O. oxygen systems. In many cases it appears that lower sludge production can be obtained even while operating at relatively high organic loadings (F/M).

F. Summary

One of the most important operating parameters influencing the overall performance capability of the activated sludge process, is the mixed-liquor bulk liquid D.O. level. Through the efficient use of high-purity oxygen aeration in the UNOX System, the bulk liquid D.O. level can be reliably and economically maintained at very high values of approximately 6 mg/l at relatively low dissolution power density which is, of course, impossible to attain with air aeration. As seen through consideration of the basic diffusional and biokinetic processes which determine the performance capability of the overall system, operation at high bulk liquid D.O. levels enables a greater fraction of the organisms in the biofloc matrix to remain aerobic and active in the biooxidation process. This greater fraction of active biomass provides several important process benefits that considerably influence the practical design operating ranges of the process and, hence, the overall system costs. The process benefits achievable as a consequence of the ability to operate at high mixed-liquor D.O. levels are:

1. Maintenance of maximum substrate removal rate over a wide range of oxygen uptake rates and biokinetic conditions

2. Ability to operate effectively at considerably higher measured F/M ranges than low D.O. air systems

3. Attainment of significantly increased sludge settling rates and improved sludge dewatering characteristics

4. Substantially decreased sludge production at comparable measured F/M values as compared to low D.O. air systems

5. Improved overall system stability resulting from the ability to handle wide ranges in organic substrate concentration without creating an oxygen deficiency in the system

These basic advantages have been demonstrated in more than a hundred pilot plant and bench-scale studies conducted by Union Carbide with numerous cities and consulting engineers as well as by a large number of independent investigators. These advantages are also now being demonstrated in the dozens of full-scale plants employing oxygen-activated sludge systems around the world.

TABLE 8

Pilot Plant Data Wuppertal, Germany Municipal Sewage

Performance summary	Air	Oxygen
Retention Time, Q, (hr)	6.00	2.00
MLVSS (mg/l)	2540.00	4200.00
D.O. (mg/l)	1—3	14.70
Biomass loading (lb COD/day/lb MLVSS)	0.82	1.47
Biomass loading (lb BOD_5/day/lb MLVSS)	0.33	0.60
Influent Characteristics (mg/l)		
COD	517.00	517.00
BOD_5	211.00	211.00
SS	77.00	89.00
Effluent characteristics (mg/l)		
COD	219.00	219.00
BOD_5	35.00	31.00
SS	17.00	40.00
Removals (%)		
COD	57.60	57.60
BOD_5	83.40	85.30
Clarifier underflow (%)	0.40	1.60
Sludge production (including effluents) (lb ESS/lb BOD_{5R})	0.87	0.40

REFERENCES

1. Pirnie, M., Presentation at Twenty-First Annual Meeting Federation of Sewage Works Associations, Detroit, Michigan, October 18–21, 1948.
2. Okun, D. A., System of bio-precipitation of organic matter from sewage, *Sewage Works J.*, 21, 763, 1949.
3. Budd, W. E. and Lambeth, G. F., High purity oxygen in biological waste treatment, *Sewage Ind. Wastes*, 29, 237, 1957.
4. Okun, D. A., A System of Bio-precipitation of Organic Matter from Sewage, Dissertation, Harvard University, 1948.
5. Okun, D. A., Discussion – high purity oxygen in biological sewage treatment, *Sewage Ind. Wastes*, 29, 253, 1957.
6. Matson, C. V., Characklis, W. G., and Busch, A. W., Oxygen supply limitations in full scale biological treatment systems, paper presented at the 27th Purdue Industrial Waste Conf., Purdue University, Lafayette, Ind., 1972.
7. Chapman, T. D., Matsch, L. C., and Zander, E. H., Effect of High Dissolved Oxygen Concentration in Activated Sludge Systems, *J. Water Pollut. Control Fed.*, 48(11), 2486, 1976.
8. Kalinske, A. A., Turbulence Diffusivity in Activated Sludge Aeration Basins, 5th Int. Water Pollution Research Conf., 1970.
9. Morand, J. M., Physical Factors Affecting the Size of Bacterial Floc, Ph.D. thesis, University of Wisconsin, Madison, 1964.
10. Parker, D. S., Kaufman, W. J., and Jenkins, D., Physical Conditioning of Activated Sludge Floc, *J. Water Pollut. Control Fed.*, 43(9), 1971.
11. Okun, D. A. and Lynn, W. R., Preliminary Investigations into the Effect of Oxygen Tension on Biological Sewage Treatment, *Proc. Conf. Biological Waste Treatment*, McCabe, J. and Eckenfelder, W. W., Jr., Eds., Reinhold, New York, 1956, 192.
12. Gaden, E. L., Aeration and Oxygen Transport in Biological Systems – Basic Considerations, *Proc. Conf. Biological Waste Treatment*, McCabe, J. and Eckenfelder, W. W., Jr., Eds., Reinhold, New York, 1956, 172.
13. Wuhrmann, K., Factors Affecting Efficiency and Solids Production in the Activated Sludge Process, *Proc. Conf. Biological Waste Treatment*, McCabe, J. and Eckenfelder, W. W., Jr., Eds., Reinhold, New York, 1956, 49.
14. Hopwood, A. P. and Downing, A. L., Factors affecting the rate of production and properties of activated sludge in plants treating domestic sewage, *J. Inst. Sewage Purif.*, 5, 3, 1965.
15. Downing, A. L. et al., Aeration and Biological Oxidation in the Activated Sludge Process, *J. Inst. Sewage Purif.*, June 1961.
16. Stamberg, J. B. et al., Activated Sludge Treatment Systems with Oxygen, EPA Report 670/2-73-073, September 1973.
17. Wuhrmann, K., Effect of Oxygen Tension on Biochemical Reactions in Sewage Purification Plants, *Proc. 3rd Conf. Biological Waste Treatment*, Manhattan College, Pergamon Press, New York, 1963.
18. Mueller, J. A., Voelkel, K. G., and Boyle, W. C., Nominal Floc Diameter Related to Oxygen Transfer, *J. Sanit. Eng. Div. Am. Soc. Civ. Eng.*, 92(SA2), 9, 1966.
19. Mueller, J. A. et al., Oxygen Diffusion Through Zoogloeal Flocs, *Biotechnol. and Bioeng.*, 10, 331, 1968.
20. Rickard, M. D. and Gaudy, A. F., Effect of Oxygen Tension on Oxygen Uptake and Sludge Yield in Completely Mixed Heterogeneous Populations, paper presented at Proc. 23rd Industrial Waste Conf., Purdue University, Lafayette, Ind., 1968.
21. Thabaraj, G. J. and Gaudy, A. F., Effect of DO Concentration on Metabolic Response of Completely Mixed Activated Sludge, paper presented at Proc. 24th Industrial Waste Conf., Purdue University, Lafayette, Ind., 1969.
22. Smith, D. B., Measurements of Respiratory Activity of Activated Sludge, *Sewage Ind. Wastes*, 25, 767, 1953.
23. Hegemann, W., Experimental Results on the Application of High Purity Oxygen in Wastewater Treatment, paper presented at Wimpey UNOX® Technical Symposium, London, England, October 31, 1973.
24. Busch, P. L. and Stumm, W., Chemical interactions in the aggregation of bacteria-bioflocculation in waste treatment, *Environ. Sci. Technol.*, 2(1), 1968.
25. McGregor, W. C. and Finn, R. K., *Biotechnol. Bioeng.*, 11, 127, 1969.
26. Jewell, W. J. and Eckenfelder, W. W., The Use of Pure Oxygen for the Biological Treatment of Brewery Wastewaters, paper presented at Purdue Industrial Waste Conf., Purdue University, Lafayette, Ind., 1971.
27. Ball, J. E., Humenick, M. J., and Speece, R. E., *The Kinetics and Settleability of Activated Sludge Developed Under Pure Oxygen Conditions*, Report No. EHE-72-18, Center of Research in Water Resources, University of Texas, Austin, 1972.
28. Water Pollution Research Laboratory, Report on the Settling and Filtration Characteristics of Activated Sludges Taken from the UNOX® and Air Plants at Ray Hall Sewage Treatment Works, Stevenage, England, July and August 1973.
29. Sezgin, M., Jenkins, D., and Parker, D. S., A Unified Theory of Filamentous Activated Sludge Bulking, paper presented at Water Pollution Control Federation 49th Annual Conf., Minneapolis, Minn., 1976.
30. Stamberg, J. B., EPA Research and Development Activities with Oxygen Aeration, paper presented at Technology Transfer Design Seminar for Municipal Wastewater Treatment Facilities, New York, February and March 1973.

31. **Jewell, W. J. and Mackenzie, S. E.,** Applications of Commercial Oxygen to Water and Wastewater Systems, in *Microbial Yield Dependence on Dissolved Oxygen in Suspended and Attached Systems,* Water Resources Symp. No. 6, University of Texas, Austin, 1972.

32. **Brenner, R. C.,** Newtown Creek Project, Environmental Protection Agency report, in press.

Chapter 4

A SUMMARY OF EPA EFFORTS
IN EVALUATING OXYGEN-ACTIVATED SLUDGE

R. C. Brenner

TABLE OF CONTENTS

I. INTRODUCTION

Utilization of oxygen aeration for activated sludge treatment is receiving increasing attention in wastewater treatment plant construction in the U.S. The basic concept, although more than 30 years old, has received commercial consideration only during the last 7 to 8 years with the development of several cost-effective systems for dissolving and utilizing oxygen gas in an aeration tank environment.

The rapid transition from the drawing boards to full-scale implementation has been possible because of intensive government and private research and development programs. The U.S. Environmental Protection Agency (EPA) and its predecessor organizations have contributed significantly to the total research and develop-

ment effort. The purpose of this chapter is to summarize the role of EPA during the period of 1968 to 1977 as the oxygen aeration process progressed to its current level of development.

As outlined in Table 1, EPA has sponsored or cosponsored nine oxygen research and development projects to date, including in-house pilot plant studies to examine process kinetics, extramural feasibility grants and contracts, extramural materials and safety projects, and extramural demonstration grants. The EPA contribution to the projects described in Table 1 exceeded $3.7 million through fiscal year 1977 (ended September 30, 1977). The cost breakdown by project is given in Table 2. Test facilities, experimental programs, and results (where available) for seven of the above projects are summarized in the following sections.

TABLE 1

EPA Research and Development Projects on Oxygen Aeration

Project	Objective
1. Batavia I and II (Union Carbide Corporation)	Establish feasibility of multi-stage, covered-tank oxygenation concept.
2. EPA/District of Columbia (Blue Plains) Pilot Plant	Determine process kinetics over wide range of operating conditions.
3. Newtown Creek (New York City)	Scaled-up demonstration of multi-stage, covered-tank oxygenation system.
4. Las Virgenes Municipal Water District	Demonstration of single-stage, covered-tank oxygenation system.
5. Englewood (FMC Corporation)	Establish feasibility of open-tank oxygenation concept.
6. Metro Denver	Scaled-up demonstration of open-tank oxygenation system.
7. Bureau of Reclamation	Materials of construction corrosion testing.
8. Rocketdyne Division of Rockwell International	Define safety requirements and develop safety manual and checklist.
9. Fairfax County (Air Products and Chemicals, Inc.)	Case history of a municipal oxygen-activated sludge plant.

TABLE 2

EPA Research and Development Expenditures on Oxygen Aeration Through Fiscal Year 1977

Project	Cost to EPA	Type of project
1. Union Carbide Corporation (Batavia I and II)	$ 795,000	Contracts
2. EPA/District of Columbia (Blue Plains) Pilot Plant	500,000	Contracts and in-house
3. New York City (Newtown Creek)	1,574,000	Grant
4. Las Virgenes Municipal Water District	186,000	Grant
5. FMC Corporation (Englewood)	192,000	Grant
6. Metro Denver	235,000	Grant
7. Bureau of Reclamation	165,000	Contract
8. Rocketdyne Division of Rockwell International	92,000	Contract
9. Air Products and Chemicals, Inc. (Fairfax County)	25,000	Contract
Total	$3,764,000	

II. THE BATAVIA PROJECTS

A research and development contract was awarded to the Union Carbide Corporation in October 1968 to evaluate a multi-staged, covered-tank oxygenation system at the Batavia, New York Water Pollution Control Plant. Union Carbide was awarded a follow-up contract in June 1970 to better define soluble organic removals and excess biological sludge production and to undertake initial pilot plant studies on oxygen sludge dewatering and stabilization. The oxygenation system was installed in one of two existing air-activated sludge trains at Batavia. During the first contract, the performance of the oxygen train was evaluated

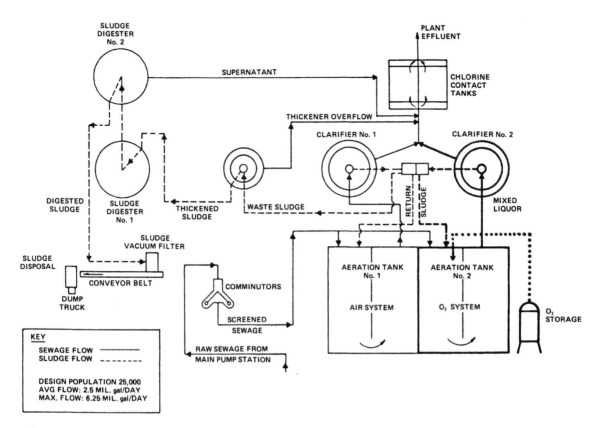

FIGURE 1. Schematic flow diagram for water pollution control plant, City of Batavia, New York. (From Albertsson, J. G., McWhirter, J. R., Robinson, E. K., and Vahldieck, N. P., Investigation of the Use of High Purity Oxygen Aeration in the Conventional Activated Sludge Process, Water Pollution Control Res. Ser. Rep. No. 17050 DNW 05/70, U.S. Department of the Interior, Federal Water Quality Administration, Washington, D.C., May 1970.)

against that of the intact air train. A schematic diagram of the Batavia Plant after installation of the oxygen system is shown in Figure 1.

The oxygen system configuration evaluated at Batavia was the first large-scale embodiment of the now well-known UNOX® process.* A typical three-stage UNOX aerator is shown schematically in Figure 2. The aerator operates as a series of completely mixed stages, thereby approximating plug flow. Oxygen gas is fed under the aeration tank cover at the inlet end of the tank only and flows cocurrently with the liquid stream from stage to stage. Gas is recirculated in each stage by centrifugal compressors which force the gas down hollow shafts out through submerged rotating spargers. Submerged turbines maintain suspension of the mixed-liquor solids and disperse the oxygen gas. A mixture of unused oxygen gas, cell respiration by-product carbon dioxide, and inert

gases is exhausted from the final stage, typically at an oxygen composition of about 50% and a flow rate equal to 10 to 20% of the incoming gas flow rate. Using cocurrent gas and liquid flow to match decreasing dissolution driving force with decreasing oxygen demand has proven to be a very efficient oxygen contacting and utilization technique.

A second-generation, multi-stage process has been developed and utilized both by Union Carbide and Air Products and Chemicals, Inc. This adaptation of the original, covered-tank concept replaces the recirculating compressors and rotating spargers with surface aerators. Oxygen transfer is accomplished by gas entrainment and dissolution. Bottom turbines are also used optionally where tank geometry requires additional mixing capability. As shown in Figure 3, all other aspects of the system are unchanged. The first Air Products and Chemicals version

* Mention of trade names or commercial products does not constitute EPA endorsement or recommendation for use.

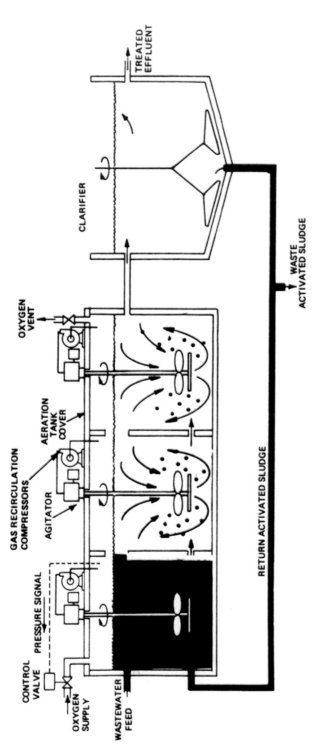

FIGURE 2. Schematic diagram of multi-stage, covered-tank oxygenation system with gas recirculation compressors and submerged turbine/spargers. (Courtesy of Union Carbide Corporation.)

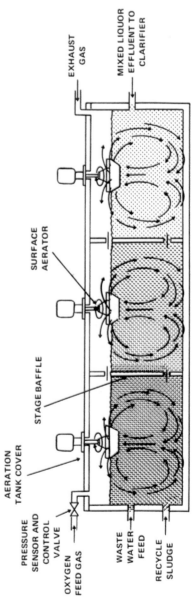

FIGURE 3. Schematic diagram of multi-stage, covered-tank oxygenation system with surface aerators. (Courtesy of Union Carbide Corporation.)

of the covered-tank, surface aerator concept (called the OASES® process) has been operating for approximately 6 years at the Westgate treatment plant in Fairfax County, Virginia (design flow, 14 MGD). Operation commenced in July 1972 at Speedway, Indiana, the first municipal plant to utilize the Union Carbide surface aerator system (design flow 7.5MGD). The surface aerator modification of the basic multistage process has exhibited better cost effectiveness for tanks up to 15 to 20 ft deep and, thus, is much more widely used in full-scale design today than the submerged turbine option. Actual experience to date of firms selling multistage, covered-tank oxygen systems indicates that 80 to 90% of the plants utilizing this oxygen system concept employ surface aerator designs. A report prepared for EPA by Air Products and Chemicals documenting the Fairfax County, Virginia, case history from inception through 2 years of operation with the oxygen system is available.[1]

The results of the Batavia projects have been widely disseminated in two EPA Water Pollution Control Research Series Reports.[2,3] One of the conclusions expressed in these reports is that oxygen aeration can provide equal treatment efficiency to air aeration with only one third as much aeration volume. This conclusion has been subject to widespread criticism. Considering that this generalization was reached by comparing an efficient oxygen contacting system with a relatively inefficient coarse-bubble air aeration system, the criticism appears to be justified. The increasing variety of air aeration equipment being marketed offers a wide range of oxygen transfer efficiencies. Design engineers are urged to investigate and prepare cost estimates for both air and oxygen systems as a basis for process selection. Process selection should be made based on a total integrated system comparison, including aeration, secondary clarification, and excess biological sludge handling and disposal requirements.

Pertinent results of the two Batavia projects relating only to oxygen system performance are summarized below:

1. The feasibility of achieving high oxygen gas utilization (91 to 95%) was established.
2. Efficient biological performance (90 to 95% BOD$_5$ and suspended solids removals,

80 to 85% COD removal) was demonstrated with short aerator detention periods (1.4 to 2.8 hr based on Q) and high organic volumetric loadings (140 to 230 lb BOD$_5$/day/1000 ft^3).
3. High mixed-liquor dissolved oxygen (D.O.) levels (8 to 12 mg/l) were maintained at high mixed-liquor suspended solids (MLSS) concentrations (3500 to 7000 mg/l).
4. Warm weather secondary clarifier performance deteriorated above an average surface loading of 1600 GPD/ft^2.
5. Oxygen sludge exhibited excellent thickening properties during secondary clarification (settled sludge concentration of 1.5 to 3.0% solids).
6. Aerobic digestion of oxygen waste activated sludge with oxygen produced volatile suspended solids (VSS) reduction rates comparable to those given in the literature for air aerobic digestion processes. Reductions in oxygen sludge VSS concentrations of 25 and 40% were achieved with 7 and 15 days of aerobic stabilization, respectively.
7. Direct vacuum filtration of undigested oxygen waste activated sludge was shown to be feasible using 10% ferric chloride for conditioning. Cake yields of 3.5 to 4.5 lb/hr/ft^2 and moisture contents of 83 to 85% were achieved at a cycle time of 2.4 min/rev. Moisture content improved to 75 to 80%, but cake yield dropped to 1.5 to 2.5 lb/hr/ft^2 at a cycle time of 6.3 min/rev.
8. Vacuum filtration of aerobically digested oxygen waste activated sludge proved to be infeasible with the chemical conditioners tested.

III. THE BLUE PLAINS PROJECT

A multi-stage, covered-tank oxygenation pilot system of Union Carbide design (see Figure 4) was operated continuously from June 1970 through September 1972 at the Joint EPA/District of Columbia (Blue Plains) Pilot Plant. Nominal design throughput for the system was 70 GPM (100,000 GPD). The results generated in over 2 years of work (believed to be the single longest continuous oxygenation pilot plant study on record) have been extensively documented and evaluated elsewhere.[4]

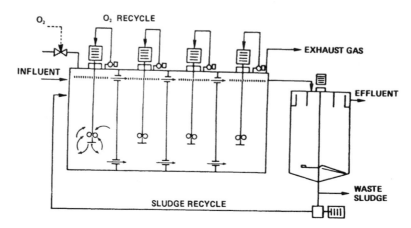

FIGURE 4. Schematic diagram of Blue Plains oxygenation system. (From Stamberg, J., Bishop, D. F., Bennett, S. M., and Hais, A. B., System Alternatives in Oxygen Activated Sludge, Environmental Protection Technology Ser. Rep. No. EPA-670/2-75-008, U.S. Environmental Protection Agency, Cincinnati, April 1975.)

Discussion of the project here is limited to generalized results and observations.

The oxygen system was operated over a wide range of sludge retention times (SRTs)* from 1.3 to 13.0 days. However, on District of Columbia primary effluent, filamentous organisms propagate rapidly with either oxygen or air if the SRT is held below approximately 5 days for any extended period of time, producing a bulking sludge with greatly retarded settling rates. Consequently, the majority of the Blue Plains operation was intentionally restricted to SRTs greater than 5 days. A technique devised by project staff personnel of reducing the incoming flow and twin dosing the sludge recycle stream with 200 mg/l of hydrogen peroxide (based on influent flow) for 24-hr periods at a 1-week interval proved to be an effective method for purging entrenched filamentous bacterial growths from an activated sludge system. The technique provides lasting benefit only if subsequent food to microorganism (F/M) loadings are adjusted to maintain an SRT outside the critical filamentous growth range. The conditions under which filamentous cultures propagate and flourish are unique to each wastewater and location. Most plants can operate in any desired loading range without encountering filamentous problems. Oxygen mixed-liquor at Blue Plains was normally well

bioflocculated and essentially free of fragmented debris between discrete particles.

Above an SRT of 5 days, average system F/M loadings remained in the range of 0.27 to 0.50 lb BOD_5/day/lb MLVSS. On those few occasions when the system was operated at an SRT less than 5 days, F/M loadings rose to levels as high as 1.0 lb BOD_5/day/lb MLVSS. Corresponding average volumetric organic loadings at an SRT above 5 days ranged from 57 to 185 lb BOD_5/day/1000 ft³. Aerator detention times (based on Q) were varied between 1.5 and 2.8 hr throughout the 2-year + period. For all loadings investigated, BOD_5 insolubilization was virtually complete. Effluent soluble BOD_5 residuals were never greater than 5 mg/l and consistently averaged 2 to 3 mg/l. Total BOD_5 and suspended solids removals were a direct function of clarifier performance. Effluent COD and TOC concentrations typically ranged from 35 to 60 and 15 to 20 mg/l, respectively.

During the spring periods of rising wastewater temperature, nitrification was established more slowly in the Blue Plains single sludge oxygen system than in a parallel conventional single sludge air aerated pilot system, probably due to the lower mixed-liquor pH inherent in operation of a covered biological reactor. Once established, however, substantial nitrification was exhibited by the oxygen system during

* Defined as pounds VSS under aeration per pound VSS wasted in the waste sludge and final effluent per day.

warm weather. With decreasing wastewater temperature in the fall, deterioration of nitrification was directly related to SRT. At an SRT of 9.0 days and a wastewater temperature of 63°F, at least partial nitrification was sustained. However, once the wastewater temperature decreased to about 60°F, no nitrification was observed in the Blue Plains oxygen system up to an SRT of 13.0 days.

Phosphorus removal experiments were conducted by adding aluminum sulfate (alum) directly to the oxygen-mixed liquor. At an Al^{+++}/P weight ratio of 1.4/1.0, phosphorus removal averaged 80% with total and soluble phosphorus residuals of 1.8 and 1.6 mg/l (as P), respectively. Increasing the alum dose to an Al^{+++}/P weight ratio of 1.8/1.0 decreased total and soluble residuals to 0.62 and 0.53 mg/l (as P), respectively, but it also lowered mixed-liquor pH from 6.5 to 6.0. At the lowered pH, the oxygen biota eventually dispersed and the experiments were discontinued. For low alkalinity wastewaters such as the District of Columbia's, pH control may be necessary to achieve efficient (90% or greater) phosphorus removal when acidic metallic salts are added directly to oxygen-activated sludge mixed-liquor.

Other important information generated by the Blue Plains project has been the delineation of some of the factors affecting sludge settling at that site and its resultant effect on system design.[4] In addition to the retardant effect on sludge settling previously mentioned due to filamentous infection of mixed-liquor, other factors that affected oxygen sludge settling rates at Blue Plains included solids concentration, bulk sludge density (volatile solids fraction), and wastewater temperature.

In the range of MLSS concentrations at which hindered or zone settling occurs, it has been found that an equation in the form of $v_i = aC_i^{-n}$, where v_i = initial settling velocity, C_i = initial solids concentration, a = intercept constant, and n = slope constant yields a straight line when plotted on log-log paper. Further, it has been shown such a relationship exists for each of the three types of settling: discrete particle, hindered, and consolidation.[5,6] The change in slope between discrete particle settling and hindered settling normally occurs at a C_i between 2000 and 3000 mg/l. The hindered settling zone is characterized by a discrete

subsiding interface and a zone of homogenously mixed settling particles. Clarifiers operating with initial hindered settling are in reality operating as sludge thickeners. It is essential that both hydraulic and mass loadings be considered in the design of secondary clarifiers for high solids systems. In many cases, thickening (mass loading) requirements will control the design. The best available approach for evaluating thickening requirements appears to be the batch flux (mass × settling velocity) method.[5]

Bulk sludge density is a function of volatile solids fraction, i.e., density increases with decreasing volatile fraction. The incorporation of denser inerts into the sludge mass is the primary reason why biomasses developed on raw wastewater will generally settle better than those developed on primary effluent. Sludge density can be temporarily affected by the washing of accumulated inerts into a plant from its sewer system during rain storms. This point was vividly illustrated at Blue Plains during a tropical storm in the summer of 1972, as shown in Figure 5.

The least recognized parameter prior to plant start-up that eventually strongly affected oxygen sludge settling rates at Blue Plains was wastewater temperature. Settling rates decreased significantly from summer to winter operation. For example, during September and October 1971 (a period when the oxygen clarifier was operated with a deep center feed below the sludge blanket to capture unsettleable particles) as wastewater temperature dropped from 81 to 71°F, the initial settling rate (ISR) decreased from 10 ft/hr to 7 ft/hr in a 1-liter graduated cylinder test at an MLSS concentration of 6000 mg/l (see Figure 6). In November of the same year, the center feed was raised above the blanket in an attempt to purge unsettleable particles from the system and increase bulk sludge density. While this technique did increase the sludge density and, temporarily, the ISR, a similar temperature effect was noted over the 2-month period of November and December 1971. As wastewater temperature dropped from 70 to 63°F, the ISR decreased from 14 ft/hr to 9 ft/hr at an MLSS concentration of 4500 mg/l (see Figure 7). Conversely, as Blue Plains wastewater temperature increased in the spring and summer of 1972, substantially increased settling rates were observed, as illus-

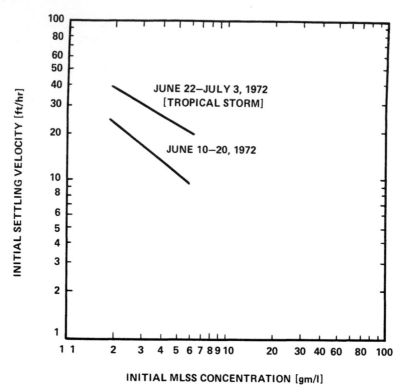

FIGURE 5. Illustration of increased sludge density caused by rain storm and its effect on initial sludge settling velocity. (From Stamberg, J. B., Bishop, D. F., Bennett, S. M., and Hais, A. B., System Alternatives in Oxygen Activated Sludge, Environmental Protection Technology Ser. Rep. No. EPA-670/2-75-008, U.S. Environmental Protection Agency, Cincinnati, April 1975.)

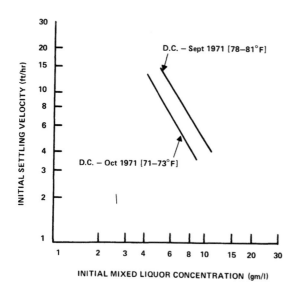

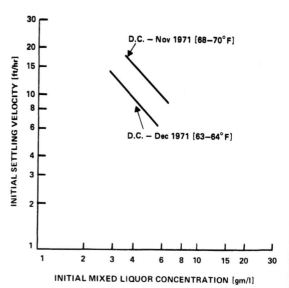

FIGURE 6. Effect of decreasing wastewater temperature on initial sludge settling velocity (September–October 1971). (From Stamberg, J. B., Bishop, D. F., Bennett, S. M., and Hais, A. B., System Alternatives in Oxygen Activated Sludge, Environmental Protection Technology Ser. Rep. No. EPA-670/2-75-008, U.S. Environmental Protection Agency, Cincinnati, April 1975.)

FIGURE 7. Effect of decreasing wastewater temperatures on initial sludge settling velocity (November–December 1971). (From Stamberg, J. B., Bishop, D. F., Bennett, S. M., and Hais, A. B., System Alternatives in Oxygen Activated Sludge, Environmental Protection Technology Ser. Rep. No. EPA-670/2-75-008, U.S. Environmental Protection Agency, Cincinnati, April 1975.)

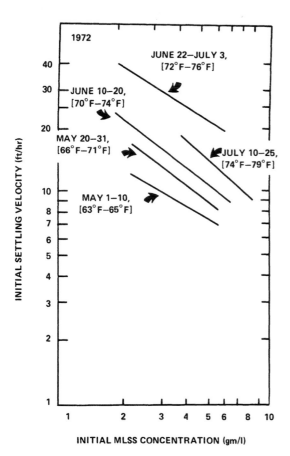

INITIAL SETTLING VELOCITY (ft/hr)

1972

JUNE 22–JULY 3,
[72°F–76°F]

JUNE 10–20,
[70°F–74°F]

MAY 20–31,
[66°F–71°F]

JULY 10–25,
[74°F–79°F]

MAY 1–10,
[63°F–65°F]

INITIAL MLSS CONCENTRATION (gm/l)

FIGURE 8. Effect of rising wastewater temperature on initial sludge settling velocity (spring–summer 1972). (From Stamberg, J. B., Bishop, D. F., Bennett, S. M., and Hais, A. B., System Alternatives in Oxygen Activated Sludge, Environmental Protection Technology Ser. Rep. No. EPA-670/2-75-008, U.S. Environmental Protection Agency, Cincinnati, April 1975.)

perature and sludge settling are not intended to imply that a similar effect will be noted universally.

Results of the long-term Blue Plains work illustrate clearly that oxygen system design, as well as air system design, should be thought of as an integrated package, consisting of a biological reactor, a secondary clarifier, and sludge handling facilities. The system should be designed for the worst anticipated climatic conditions at a given site. Clarifier sizing should be specifically tailored to the design and anticipated operating conditions of the reactor.

IV. THE NEWTOWN CREEK PROJECT

Results of the initial Batavia contract were judged sufficiently encouraging to justify a scaled-up demonstration of the multi-stage oxygen system in a large municipal plant. A research and development grant was awarded to New York City in June 1970 to convert one of 16 parallel bays at its Newtown Creek facility to oxygen using the recirculating compressor/submerged turbine version of the UNOX process. The design flow of the test bay was 20 MGD, roughly ten times higher than the capacity of the Batavia oxygen system. In addition to the $1.574 million EPA grant, New York City provided over $1.2 million in city funds in support of the project.

The Newtown Creek plant was designed on the modified aeration principle for 1.5 hr of aeration time (based on Q) and treatment efficiencies in the range of 65 to 70%. The City is now confronted with an upgrading problem in a landlocked neighborhood (Figure 9), a situation common to many large urban plants in the U.S. Oxygen was believed to be a good candidate for achieving the required 90% BOD_5 removal within the confines of the existing aeration tanks and secondary clarifiers. Future conversion of the entire 310 MGD facility to oxygen was the ultimate objective, provided a removal up to the 90% ± BOD_5 level could be consistently demonstrated in the test bay. Two views of the Newtown Creek test bay are shown in Figure 10.

The test bay went on stream in early June 1972. Extensive mechanical problems and unreliable meters prevented accurate data collec-

trated in Figure 9. The net result of this phenomenon was that a peak oxygen clarifier overflow rate of 1940 GPD/ft² was possible in the summer of 1970 with an MLSS of 8000 mg/l, while the peak overflow rate that could be sustained in either the 1970—1971 or 1971—1972 winters without clarifier failure was 975 GPD/ft² at MLSS levels that varied from 3900 to 5300 mg/l.

Undoubtedly, all of the factors discussed above contributed to the reduced overflow rates necessary at Blue Plains to maintain satisfactory winter clarifier operation. However, it appears that wastewater temperature played a major role at this site. It is strongly emphasized, though, that the conclusions drawn from the Blue Plains project regarding wastewater tem-

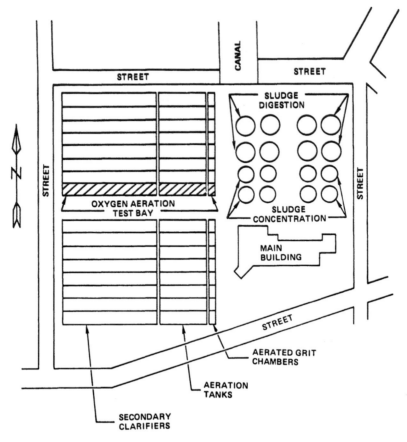

FIGURE 9. Plan layout for Newtown Creek Water Pollution Control Plant, Brooklyn, New York.

tion during the start-up phase (June 4, 1972 to September 16, 1972), during which time the influent flow was increased from 11 MGD to the design level of 20 MGD. Data for this period are limited to effluent quality, as summarized in Table 3.

Metering difficulties were finally resolved by the end of the start-up period permitting commencement of the planned extensive data collection program on September 17, 1972. Although the initial schedule called for terminating field studies 1 year later on September 17, 1973, the ten-phase grant experimental program was extended 7½ months through April 30, 1974. Subsequently, the City conducted two more phases at its own expense between May 1, 1974 and March 11, 1975, the last phase to evaluate a third winter of operation. Fungus organisms in the Newtown Creek influent presented operating problems in the winter that were not observed in warm weather.

The influent flow conditions for these 12 phases are summarized in Table 4. Diurnal peak, average, and minimum flow rates for Phases 4 through 12 are given in Table 5. With the exception of Phase 7, the diurnal fluctuation patterns were selected to simulate the actual influent flow patterns of the Newtown Creek facility.

A performance summary for the oxygenation system for Phases 1 through 12 is presented in Table 6. System sludge characteristics, aerator loadings, and secondary clarifier loadings are summarized in Tables 7, 8, and 9, respectively. Further operating and performance details are available in the literature.[7]

From the beginning of the project, New York City officials considered performance of the oxygen test bay during cold weather the most critical segment of the experimental program. It was during the prolonged severe weather period that the true upgrading potential of oxygen for Newtown Creek would be most evident. A

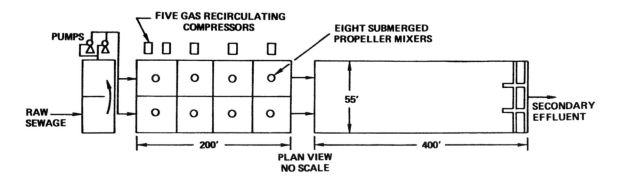

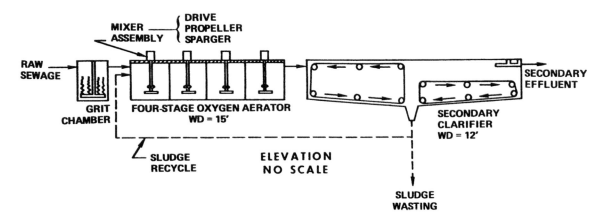

FIGURE 10.   Plan and elevation views of oxygen aeration test bay, Newtown Creek.

TABLE 3

Effluent Quality at Newtown Creek During Start-up (6/4/72—9/16/72)

				Effluent concentration (mg/l)		
Flow (MGD)	Duration (weeks)	Total BOD$_s$	Soluble BOD$_s$	Total COD	Soluble COD	Suspended solids
11±2[a]	4	10	4	68	50	18
14±3[b]	5	8	3	55	41	19
20±4[c]	6	15	4	62	47	18

[a] MLSS = 4865 mg/l, detention time (based on Q) = 2.7 hr±.
[b] MLSS = 5920 mg/l, detention time (based on Q) = 2.1 hr±.
[c] MLSS = 4260 mg/l, detention time (based on Q) = 1.5 hr±.

TABLE 4   TABLE 5

Experimental Schedule for Newtown Creek Project (9/17/72—3/11/75)

Phase	Dates	Influent flow condition (MGD)
1	9/17/72—11/25/72	20.8 (constant)
2	12/10/72—2/1/73	14→20→15
		Average 17.7 (winter upset)
3	2/18/73—4/7/73	8→20
		Average 15.1 (winter restart)
4	4/8/73—6/2/73	20.3 (diurnal)
5	6/3/73—7/7/73	25.6 (diurnal)
6	7/8/73—8/11/73	30.0 (diurnal)
7	8/12/73—9/1/73	35.4 (diurnal)
8	9/16/73—10/8/73	10.1, deliberate underload (diurnal)
9	10/14/73—10/25/73	14.6, transition to winter (diurnal)
10	2/6/74—4/30/74	19.7, winter (diurnal)
11	5/1/74—8/5/74	19.1 (diurnal)
12	12/13/74—3/11/75	20.0, winter (diurnal)

Diurnal Fluctuation Patterns for Newtown Creek (4/8/73—3/11/75)

Phase	Average flow (MGD)	Peak flow (MGD)	Minimum flow (MGD)
4	20.3	24.0	14.0
5	25.6	30.0	17.0
6	30.0	36.0	19.0
7	35.4	37.5[a]	30.0
8	10.1	12.0	7.0
9	14.6	18.0	10.5
10	19.7	24.0	14.0
11	19.1	—	—
12	20.0	—	—

[a] Maximum influent pumping capacity.

TABLE 6

Performance Summary for Newtown Creek (9/17/72—3/11/75)

	Phase											
	1	2	3	4	5	6	7	8	9	10	11	12
Total BOD$_5$ in (mg/l)[a]	156	157	151	168	240	215	198	178	149	136	146	140
Total BOD$_5$ out (mg/l)	9	21	17	17	24	22	21	19	13	26	24	18
% removed	94	87	89	90	90	90	89	89	91	81	84	87
Soluble BOD$_5$ in (mg/l)[a]	84	78	88	96	130	98	82	96	70	77	80	87
Soluble BOD$_5$ out (mg/l)	4	13	12	11	14	12	14	8	6	16	13	10
% removed	95	83	86	89	89	88	83	92	91	79	84	89
Total COD in (mg/l)[a]	356	365	364	365	320	290	308	363	314	307	—	298
Total COD out (mg/l)	61	88	75	77	73	64	62	65	57	87	—	70
% removed	83	76	79	79	77	78	80	82	82	72	—	77
Soluble COD out (mg/l)	50	69	63	69	58	49	46	—	—	—	—	—
Suspended solids in (mg/l)[a]	149	146	145	159	149	119	131	136	135	126	142	101
Suspended solids out (mg/l)	12	22	18	18	26	17	17	18	17	27	26	13
% removed	92	85	88	89	83	86	87	87	87	79	82	87
Sewage temperature range (°F)	77	64	51	53	62	71	72	—	—	42	—	66
	↓	↓	↓	↓	↓	↓	↓			↓		↓
	56	53	61	69	74	77	78	—	—	67	—	47

[a] No primary sedimentation. Concentrations shown are for raw sewage influent to oxygen aerator.

TABLE 7

Average System Sludge Characteristics for Newtown Creek (9/17/72—3/11/75)

Phase	MLSS (mg/l)	MLVSS (mg/l)	Return sludge flow (% of Q)	Return sludge TSS (mg/l)	SVI[a] (ml/g)	SRT (days)
1	4,860	4,130	30	16,200	45	—[b]
2	4,980	4,080	40	13,000	56	—[b]
3	4,010	3,210	50	11,500	83	2.0
4	3,950	3,120	46	13,200	53	1.4
5	4,530	3,580	43	16,300	43	1.3
6	4,150	3,400	34	16,200	43	1.2
7	3,130	2,548	25	13,000	48	0.8
8	4,760	3,760	25	17,300	37	5.5
9	9,040[c]	7,230[c]	26	18,600	50	—[c]
10	2,400	1,870	58	5,200	83	0.9
11	3,870	—	39	12,700	53	—
12	2,630	2,210	29	9,400	67	1.2

[a] Unstirred.

[b] Sludge wasting data not reliable.

[c] Mixed-liquor samples probably contaminated with foam.

TABLE 8

Average Aerator Loadings for Newtown Creek (9/17/72—3/11/75)

Phase	Detention time based on Q (hr)	F/M loading $\left(\dfrac{\text{lb BOD}_s/\text{day}}{\text{lb MLVSS}}\right)$	Volumetric organic loading $\left(\dfrac{\text{lb BOD}_s/\text{day}}{1000 \text{ ft}^3}\right)$
1	1.42	0.63	163
2	1.67	0.55	140
3	1.96	0.57	115
4	1.46	0.88	171
5	1.16	1.38	309
6	0.99	1.53	324
7	0.84	2.22	352
8	2.93	0.39	90
9	2.03	—[a]	109
10	1.50	1.15	135
11	1.55	—	140
12	1.48	1.02	141

[a] Mixed-liquor samples probably contaminated with foam.

TABLE 9

Average Secondary Clarifier Loadings for Newtown Creek (9/17/72—3/11/75)

Phase	Surface overflow rate (GPD/ft²)	Mass loading lb TSS/ft² day	Weir loading (GPD/lineal ft)
1	950	50	130,000
2	810	47	110,600
3	690	35	94,400
4	920	45	126,900
5	1,160	63	160,000
6	1,360	64	187,500
7	1,610	53	221,300
8	460	23	63,100
9	660	—ᵃ	91,400
10	900	28	123,100
11	870	39	119,400
12	910	26	125,000

ᵃ Mixed-liquor samples probably contaminated with foam.

discussion of the data collected at Newtown Creek is first prefaced, therefore, with a summary description of the unforeseen operational difficulties encountered during the 1972—1973 winter season.

During Phase 1 (autumn 1972), operation went smoothly and performance was obviously excellent. On November 25, 1972, the test bay was shut down temporarily to replace a bearing on the sludge recycle pump. What was planned to be a 2-day outage turned into a 2-week shutdown when the stocked spare bearing proved to be the wrong size and a new one had to be located. During the outage, sludge in the reactor was continually oxygenated while the sludge in the secondary clarifier was devoid of oxygen. Due to the imminence of the upcoming cold weather, it was decided to restart the system using the existing sludge rather than empty the tanks and take the time necessary to generate a new biomass. The oxygen test bay was put back into service on December 10, 1972.

For the first several weeks of Phase 2, performance was satisfactory as the influent flow rate was gradually increased to the design level of 20 MGD. Shortly thereafter, sludge settling properties began to deteriorate and effluent BOD$_5$, COD, and suspended solids residuals exhibited a slowly increasing trend. Microscopic examination of the mixed-liquor revealed the appearance of filamentous organisms of both apparent bacterial and fungal origin. Influent flow was then decreased in several increments

during the month of January 1973 in an attempt to starve or "burn out" the filamentous culture or cultures and reestablish a healthy population.

Instead of eradicating the filamentous organisms, reduction of flow and organic loading seemed to have the opposite effect of stimulating proliferation. This proliferation was accompanied by the usual indicators of a bulking sludge, i.e., a substantial increase in sludge volume index (SVI), a rising sludge blanket in the secondary clarifier, increasing suspended solids carry-over in the final effluent, and operational difficulty in managing total system sludge inventory.

In trying to determine the source of the filamentous intrusion, it was postulated that the bacterial species (*Sphaerotilus*) may have developed in the secondary clarifier sludge blanket during the aforementioned shutdown. A local pharmaceutical firm is known to discharge mycelia into the Newtown Creek sewer system, and this was suspected as the source of the fungus organisms. By the end of January 1973, with the influent flow reduced to 15 MGD, the SVI had risen from a summer background level of 45—50 to 85—100, effluent suspended solids were exceeding 30 mg/l, effluent soluble BOD$_5$ had increased to over 20 mg/l, and the clarifier sludge blanket was continuing to rise. At this point, a decision was made to "dump" the entire sludge inventory, hose all settled sludge pockets out of the reactor and clarifier, and start over. This second shutdown began on February 1, 1973.

After taking some additional time to make repairs to the sludge collection mechanism while the clarifier was dewatered, Phase 3 commenced on February 18, 1973. Conservative loading rates were utilized initially based on the premise that the best chance to prevent a recurrence of the filamentous condition was a program of gradual and modest increases in F/M loading until the 20 MGD design flow rate was reached. Relatively high MLSS concentrations of 5000 mg/l ± were maintained as a further measure to minimize F/M loading. However, within several weeks, a repeat of the experience encountered in Phase 2 became evident with the initial appearance of filaments in the oxygen sludge. This time the organisms were definitely identified as fungus. One reason suggested at the time for explaining the higher 1972—1973 winter incidence and enrichment pattern of these fungus organisms in the Newtown Creek oxygen sludge as opposed to that facility's air sludge is the lower mixed-liquor pH inherent to the operation of the covered-tank oxygen system.

With the prospect of impending project failure a real possibility, a joint decision (New York City, Union Carbide, and EPA) was reached to accept the presence and proliferation of the fungus organisms as a cold weather phenomenon and attempt to find an operating mode that would permit satisfactory winter performance at design flow in spite of them. Accordingly, a program of increased sludge wasting was initiated which eventually lowered the MLSS concentration to less than 4000 mg/l and the SRT to slightly more than 1 day. At the same time, influent flow was elevated in several fairly rapid increments to 20 MGD (equivalent to an aerator detention time of 1.5 hr based on Q). These steps yielded an F/M loading at the end of the phase in the range of 0.75 to 0.80 lb BOD_s/day/lb MLVSS, considerably higher than the average of 0.57 for all of Phase 3. The altered operating philosophy proved to be the correct decision, resulting in a controllable clarifier sludge blanket and stable cold weather performance at design flow. Percentage removals for the remainder of Phase 3, although not as high as Phase 1, were within satisfactory limits. As the wastewater temperature increased during early spring, the concentration of filamentous organisms diminished and they eventually disappeared in early May 1973.

During the summer of 1973, the Newtown Creek oxygen system exhibited remarkable capability for absorbing high hydraulic and organic loadings while still producing a high quality secondary effluent. During the four diurnal phases (Phases 4 through 7) conducted from April 8, 1973 through September 1, 1973, the average influent flow was successfully increased from 20.3 to 35.4 MGD. During Phase 7, the average nominal aerator detention time was only 52 min with corresponding average F/M, volumetric, and clarifier surface loadings of 2.22 lb BOD^5/day/lb MLVSS, 352 lb BOD_s/day/1000 ft³, and 1610 GPD/ft², respectively. These results confirmed and exceeded the high-rate loading potential of oxygen-activated sludge first seen at Batavia.

At this point in the project, it was possible to offer the following interim status remarks:

1. The high-rate loading capability (nominal aeration time ⩽ 1 hr) of oxygen aeration operating on Newtown Creek wastewater during warm weather was conclusively demonstrated.

2. Prospects appeared promising that a modified method of operation evaluated late in the 1972—1973 winter could circumvent the negative effects of a potentially indigenous, cold weather filamentous condition with oxygen at Newtown Creek and permit satisfactory performance at a flow rate at least equal to the design level of 20 MGD (1.5 hr of nominal aeration time).

3. The operational measures intentionally employed to effect the improved performance late in the 1972—1973 winter, namely high F/Ms and low SRTs, occurred naturally to an even greater degree during the high loading phases evaluated in the summer of 1973.

4. The above comments provided a tentative basis for speculating that, in some cases, oxygen aeration might most beneficially be employed at ultrahigh loading rates substantially exceeding any which had been approved to date by state agencies.

Because of the importance attached to winter operation and performance, the grant experimental period was extended to the end of April 1974. The two major questions that were to be addressed during the extended 7½-month period were whether filamentous organisms (particularly fungus) would again infest the oxygen sludge as wastewater temperature dropped, and, if so, would the modified method of winter operation previously described permit continuous efficient performance with a diurnal loading pattern centered around an average influent flow rate of 20 MGD.

Data for the period of September 16, 1973 through April 30, 1974 are summarized in Phases 8, 9, and 10. The average influent flow during Phase 8 was maintained at about half the design level to fill a low-loading data gap in warm-weather operation. In October 1973, flow was increased to approximately 15 MGD in a transition step (Phase 9) toward a second winter operation (Phase 10). However, early in Phase 9, a brown froth (later identified as consisting primarily of *Nocardia*) developed in the aeration tank, contaminating the recirculation compressors and some of the oxygen piping and necessitating a system shutdown by October 26 for cleaning. Consequently, Phase 9 lasted only for an abbreviated period of 12 days. The froth disappeared just as suddenly as it appeared, and operation was resumed on November 12. Within 2 weeks, though, the return sludge pump broke down for the second time. Repairs were not completed until early in January 1974. Initial operation following this second 1973—1974 winter outage was satisfactory with rapid establishment of the biomass despite the cold weather start-up. However, by the third week in January, with the flow only at 11 MGD, it became evident that the fungus in the influent wastewater had again concentrated in the mixed-liquor.

Nevertheless, since the intention of the 1973—1974 winter program was to operate at the design rate of about 20 MGD diurnal, influent flow was slowly increased, the fungus notwithstanding. By February 6, 1974, the design flow rate was reached and Phase 10 commenced. Operation during the next 2 months was even more affected by the filamentous organisms than in Phase 3, the previous winter. The sludge density index and the solids concentrations in the mixed-liquor and the return sludge decreased to record lows for the project,

requiring increased wasting rates to prevent sludge blanket failure. Accordingly, removal efficiencies in Phase 10 were somewhat lower than in other phases but still within the bounds of acceptable secondary treatment. In April, as the wastewater temperature increased, the filamentous organism concentration diminished.

The reappearance of large quantities of fungus in the mixed-liquor during the second winter continued to be an unexplained and frustrating experience for the project staff. The encouraging aspect of Phase 10 was that, despite a proliferating fungal culture that would have debilitated most activated sludge systems, the Newtown Creek oxygen test bay was able to operate at design flow and produce satisfactory effluent quality.

Phase 11 was a continuation of Phase 10 into warm weather at full diurnal flow. Although the fungi had completely disappeared from the mixed-liquor, system performance was not quite as good as in the previous summer.

From August 26 until October 28, 1974, the oxygen demonstration module was shut down to maintain and upgrade the capacity of the pressure swing adsorption (PSA) oxygen generation unit. Preventive maintenance was also performed on other equipment components to ready the system for a third and final winter run.

Because of the unanswered questions relating to the fungal proliferation during the first and second winters, a program was begun in November 1974 to: (1) identify the fungus, (2) determine whether it was peculiar to Newtown Creek or was ubiquitous to New York City wastewater, and (3) try to find some means of eliminating it. To accomplish these objectives, a 17.5 GPM oxygen pilot plant was brought in to evaluate potential fungicides. A second pilot plant was stationed at the City's Wards Island plant to compare cold-weather air and oxygen performance on another wastewater.

It was found that the predominant form of the fungus was the arthrospore producer *Geotrichum*. Sampling at three other New York City plants and one drainage area not yet served by a plant revealed the presence in the influents of the same organism, in concentrations generally similar to those at Newtown Creek. At Newtown Creek, the fungus concentration increased moderately in the air plant mixed-liquor but significantly in the mixed-liquors of both the full-scale and pilot oxygen systems. At the three other plants sampled, the fungi did not significantly proliferate in the mixed-liquor. The growth in the Wards Island oxygen pilot plant was about the same as in the full-scale air plant. Apparently, Newtown Creek's waste composition is unique in that it is conducive to the growth of these organisms. None of the fungicides evaluated in the Newtown Creek oxygen pilot plant had any effect on the fungi.

Meanwhile, Phase 12 proceeded (December 13, 1974 to March 11, 1975) at full diurnal flow. During this third winter of operation, which was not interrupted by a breakdown of the return sludge pump, the fungi, although present in the mixed-liquor from beginning to end, did not proliferate to the same degree as in the first two winters. Their effect on the process was, therefore, much less severe than in the two previous winters. Effluent quality was very good throughout the phase, as evident in Table 6.

Another aspect of the project is discussed briefly below. Initially the performance of the four-bed PSA oxygen generator (which was one of the first PSA oxygen plants ever built) was less than satisfactory. During the 1972 summer start-up phase, the unit was out of service, due to mechanical problems, roughly 40% of the time. During 1973 and early 1974, these problems were largely corrected and the generator subsequently functioned with a downtime varying between 5 and 10%. During the 1972 start-up difficulties, however, two of the four beds inadvertently became contaminated with water vapor. Thereafter, the maximum achievable output of the unit was 10 tons oxygen gas/day at 90% oxygen purity vs. a design output of 16.7-ton oxygen gas/day at 90% purity. This necessitated an increase in consumption of and reliance on the back-up liquid oxygen reservoir during peak oxygen demand periods.

To alleviate the need to routinely use back-up liquid oxygen, the PSA unit was completely refurbished in late 1974 at the end of Phase 11, thereby returning output to in excess of the 16.7 ton per day design point. At this time, mechanical modifications were also made to incorporate recently acquired technological advances, including a control system to prevent the recurrence of adsorbent damage due to water vapor, and the generator was operated continuously until termination of the program in 1975 with less than 1% downtime.

In mid-1972, a simplified three-bed (moving parts decreased 50%) PSA unit was installed and started up at Speedway, Indiana. At this writing (October 1977), 20 three-bed PSA units, which are now being used exclusively, have been placed into operation. According to Union Carbide, these units, which represent some 40 plant years of operating experience, have over the past 2 years demonstrated average on-stream times in excess of 98% and have all produced oxygen gas in excess of design capacity and purity.

V. THE LAS VIRGENES PROJECT

A single-stage, covered-tank oxygenation system was designed by the Cosmodyne Division of Cordon International Corporation, Torrance, California, in 1971. The system, given the name SIMPLOX and shown schematically in Figure 11, utilizes an inflated dome-type cover to contain the oxygen-rich atmosphere over the aerator. This concept was developed primarily for upgrading existing air-activated sludge plants with a minimum capital expenditure by utilizing conventional air blowers and coarse-bubble air diffusers to recirculate oxygen gas. Air blowers used in this service must be corrosion proofed and otherwise modified to be compatible with oxygen gas. Virgin oxygen gas is introduced to the aerator through a fine-bubble sparger located on the tank bottom and on the opposite side wall from the conventional air diffusers. Power required for oxygen dissolution is greater for the SIMPLOX process than for the multi-stage systems because: (1) the equipment used for transferring oxygen is modified air aeration equipment and not specifically tailored to oxygen gas kinetics and (2) the gas phase above the mixed-liquor is completely mixed and assumes the same oxygen composition as the exhaust gas stream; thus, the driving force for dissolving oxygen in wastewater is less than in the lead stages of multistage aerators. However, capital costs for converting an existing aerator from air to oxygen service should be significantly less with the SIMPLOX approach because staging baffles and multiple oxygen dissolution equipment assemblies are not required. Since the gas phase

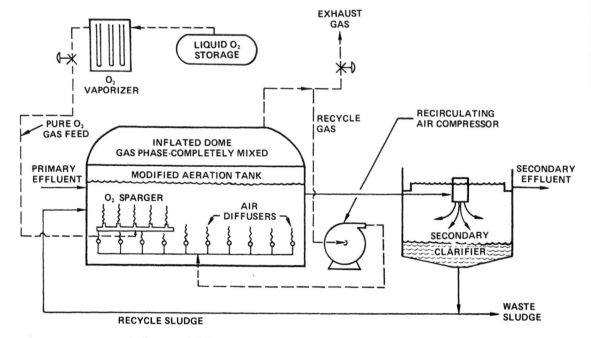

FIGURE 11.   Schematic diagram of diffused air aeration system modified to recirculate oxygen gas, Las Virgenes Project.

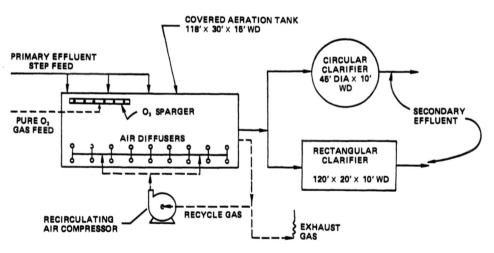

FIGURE 12.   Flow diagram for Las Virgenes Oxygenation system.

is completely mixed, exhaust oxygen, carbon dioxide, and inert gases can be bled from any point of the inflated dome, and any of several activated sludge flow configurations, including plug flow, complete mix, and step aeration, can be used as desired.

A research and development grant was awarded to the Las Virgenes, California (a suburb of Los Angeles) Municipal Water District in June 1971 to evaluate the SIMPLOX system at its Tapia Water Reclamation Facility. The experimental program concluded on September 10, 1973. The District contributed $62,000 in support of the project, supplementing the $186,000 EPA grant. An empty nominal 1-MGD train was available for the oxygen study because of a recent expansion at the Tapia facility. The manner in which the oxygen system was incorporated into this existing train is shown in plan view in Figure 12. The schedule followed during the experimental program for the project is outlined in Table 10. The range of aerator loadings examined was not as broad as at Newtown Creek due to influent flow limi-

TABLE 10

Experimental Schedule for Las Virgenes Project (4/25/72—9/10/73)

Phase	Dates	Influent flow (MGD)	% of aerator in use	No. of clarifiers in use
1	4/25/72—7/31/72	1.0	100	1
2	9/11/72—11/13/72	2.0	100	2
3	1/22/73—3/8/73	1.0	45	1
4	3/9/73—4/3/73	1.13	45	2
5	4/4/73—4/30/73	1.3	45	2
6	5/1/73—5/14/73	1.54	45	2
7	5/15/73—9/10/73	1.85	45	2

TABLE 11

Performance Summary for Las Virgenes (4/25/72—9/10/73)

	Phase						
	1	2	3	4	5	6	7
Total BOD_5 in (mg/l)[a]	82	69	79	107	115	103	95
Total BOD_5 out (mg/l)	2	4	2	5	9	9	10
% removed	97	94	97	95	92	91	89
Total COD in (mg/l)[a]	153	136	170	218	262	242	238
Total COD out (mg/l)	35	35	29	35	37	40	50
% removed	77	74	83	84	86	83	79
Soluble COD in (mg/l)[a]	58	43	76	93	101	101	100
Soluble COD out (mg/l)	16	19	23	26	31	31	32
% removed	72	56	70	72	69	69	68
Suspended solids in (mg/l)[a]	73	67	39	53	63	59	44
Suspended solids out (mg/l)	9	7	4	7	5	4	6
% removed	88	90	90	87	92	93	86
Turbidity out (JTU)[b]	2	3	2	3	2	2	3
NH_4^+-N in (mg/l)[a]	13	6.8	10.7	14.2	15.6	15.8	15.6
NH_4^+-N out (mg/l)	0.4	0.1	0.2	4.1	4.8	2.8	3.1
% removed	97	99	98	71	69	82	80
NO_3^--N out (mg/l)	16.2	15.3	8.8	6.9	5.6	7.5	8

[a] Concentrations shown are for primary effluent feed to oxygen aerator.
[b] Jackson turbidity units.

tations and a weaker aerator feed (primary effluent at Las Virgenes, raw wastewater at Newtown Creek). The experimental program consisted of seven phases characterized by increasing flow and system loadings. To effect a more pronounced increase in aerator loading, only 45% of the available aerator volume was utilized in the last five phases. This was accomplished via the installation of a temporary bulk-

head across the width of the aeration tank after Phase 2.

System performance for the Las Virgenes project is summarized in Table 11. Tables 12, 13, and 14 summarize, respectively, system sludge characteristics, aerator loadings, and secondary clarifier loadings.

A cursory review of Table 11 reveals that effluent quality for the entire Las Virgenes pro-

TABLE 12

Average System Sludge Characteristics for Las Virgenes (4/25/72—9/10/73)

Phase	MLSS (mg/l)	MLVSS (mg/l)	Return sludge flow (% of Q)	Return sludge TSS (mg/l)	SVI (ml/g)	SRT (days)
1	3,700	2,950	30	14,325	99	79
2	3,750	3,050	30	13,295	179	68
3	3,815	2,950	32	12,890	175	46
4	3,570	2,715	32	9,230	200	30
5	3,050	2,485	40	7,105	247	12
6	2,595	2,170	39	6,705	191	9
7	2,535	2,115	40	8,350	117	12

TABLE 13

Average Aerator Loadings for Las Virgenes (4/25/72—9/10/73)

Phase	Detention time based on Q (hr)	F/M loading $\left(\dfrac{\text{lb BOD}_5/\text{day}}{\text{lb MLVSS}}\right)$	Volumetric organic loading $\left(\dfrac{\text{lb BOD}_5/\text{day}}{1000 \text{ ft}^3}\right)$
1	9.56	0.07	13
2	4.78	0.11	22
3	4.30	0.15	27
4	3.81	0.24	42
5	3.31	0.33	52
6	2.79	0.41	56
7	2.32	0.46	62

TABLE 14

Average Secondary Clarifier Loadings for Las Virgenes (4/25/72—9/10/73)

Phase	Surface overflow rate (GPD/ft²)	Mass loading $\left(\dfrac{\text{lb TSS/ft}^2}{\text{day}}\right)$
1	417	17
2	501	20
3	417	18
4	283	11
5	326	12
6	386	12
7	464	14

ject was superb and surpassed that observed at Newtown Creek. This can be attributed to three factors: (1) the lower aerator organic loadings which permitted a high degree of COD insolubilization, (2) the very conservative secondary clarifier hydraulic and mass loadings which promoted highly effective solids capture, and (3) the lack of any significant industrial waste contributions. A major objective of wastewater treatment in the Las Virgenes District is the production of an ultrahigh quality secondary effluent after chlorination suitable for agricultural reuse. The thrust of this project, therefore, was geared not so much to maximizing system loadings (as was the case at Newtown Creek) as to maintaining truly superb quality effluent and determining the effect of a relatively conservative progression in system loadings on single-stage nitrification. As shown in Table 15, virtually complete nitrification was observed with F/M loadings between 0.07 and 0.15 lb BOD_5/day/lb MLVSS. For F/M loadings between 0.24 and 0.46 lb BOD_5/day/lb MLVSS, nitrification was only 69 to 82% complete. Lower wastewater temperatures may also have played a role in the decreased nitrification of the latter four phases.

Another major goal of the Las Virgenes staff was to minimize excess biological sludge production as much as possible. This goal probably led to the most significant problem area encountered on the project, a very evident bulking sludge. No sludge was intentionally wasted from the system during Phases 1 and 2. Wastage of suspended solids in the final effluent and final clarifier skimmings was sufficient to balance net system biomass growth at the low F/M loadings employed. Resulting SRTs were as high as 79 days, and the SVI climbed to a level near 200 ml/g. Despite the instigation of a scheduled wasting program in Phase 3, the sludge continued to bulk and the SVI climbed even higher. It was not until Phase 7 at an SRT of 12 days and an F/M loading of 0.46 lb BOD_5/day/lb MLVSS that a significant drop in SVI occurred. The bulking sludge condition is attributed here to a combination of *Sphaerotilus* filamentous development due to the inordinately high SRTs and the accumulation of other poor settling debris in the floc matrices. It was only because of the low clarifier loadings that efficient overall performance was sustained. Sludge blanket levels frequently rose to within a few feet of the clarifier weirs. The Las Virgenes experience illustrates the potential operating difficulties that can and probably will occur at very low oxygen system loading rates.

One definite conclusion reached during the project was that the inflated tent (dome) concept is not suitable for permanent installation. New leaks developed repeatedly in the polyvinyl material due to separation of the tent/tank in-

terface, abrasion against the tent support structure during high winds, and bullets from pranksters' guns. The gas leak problem made accurate oxygen consumption monitoring impossible, and during the latter higher loading phases, the leaks became sufficiently frequent and large so that it was extremely difficult to maintain a mixed-liquor D.O. above 1 to 2 mg/l. The rationale for using an inflated dome in lieu of a flat cover on this research project was to permit access to the tank interior, a procedure effectively utilized on several occasions. A permanent installation would probably require a flat, more rigid cover for longevity and for minimization of leaks.

The Cosmodyne Division of Cordon International has not attempted to establish a proprietary position with respect to the SIMPLOX system. Despite the attractive capital-cost features of this oxygen dissolution concept for upgrading existing air-activated sludge plants, further development and implementation activities have ceased following the Las Virgenes Project due to the lack of a strong proprietary posture and an associated aggressive marketing effort.

VI. THE ENGLEWOOD PROJECT

In the early 1970s, the Martin Marietta Corporation developed a unique fine-bubble diffuser capable of producing uniform oxygen bubbles of less than 0.2 mm in diameter. The diffuser works on the shear principle of passing a high-velocity liquid stream at right angles to oxygen bubbles discharging into a vertical slot from capillary tubes. Oxygen gas is introduced to the capillary tubes at 30 psig. Diffuser production technology and patent rights were sold to the FMC Corporation in January 1973. FMC subsequently referred to this diffuser as the fixed active diffuser.

A graph provided by FMC showing theoretical water depth required for complete dissolution of varying size oxygen gas bubbles in tap water is reprinted in Figure 13.* The large effect of a relatively small change in bubble size on the water depth required for 100% dissolution is readily evident. For a bubble diameter of 0.10 mm, a 4-ft diffuser submergence would

---

* This graph was prepared using tap water. Dissolution characteristics for various size oxygen gas bubbles may, and probably do, differ for a wastewater undergoing biological treatment.

TABLE 15

Effect of Organic Loading and Wastewater Temperature on Nitrification at Las Virgenes
(4/25/72—9/10/73)

Phase	F/M lb BOD$_s$/day lb MLVSS	SRT (days)	Wastewater temperature range (°F)	% NH$_4^+$-N removed	Final effluent NO$_3^-$-N (mg/l)
1	0.07	79	70—7.7	97	16.2
2	0.11	68	73—79	99	15.3
3	0.15	46	65—67	98	8.8
4	0.24	30	65—67	71	6.9
5	0.33	12	67—70	69	5.6
6	0.41	9	68—71	82	7.5
7	0.46	12	70—75	80	8.0

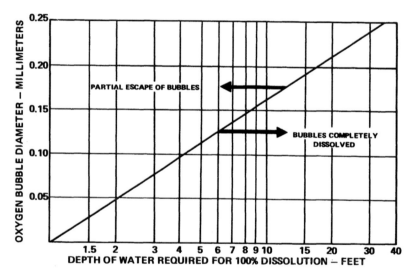

FIGURE 13.   Oxygen gas bubble diameter vs. water depth for complete dissolution. (Courtesy of FMC Corporation.)

be required. The required depth increases to 8.5 ft for a 0.15-mm diameter bubble.

One of the many potential applications projected for the fixed diffuser was its deployment in an aeration system for an open-tank oxygen-activated sludge process. To evaluate the feasibility of the open-tank oxygenation approach, a research and development grant for $192,000 was awarded to Martin Marietta (subsequently FMC) in September 1972 for a nominal 30-GPM pilot plant study. FMC contributed over $85,000 of their funds to the project. The pilot plant was installed on the grounds of the Englewood, Colorado (suburb of Denver) trickling filter plant and received a feed stream of primary effluent from that plant. Pilot plant configuration and dimensions are shown in Figure

14. The aeration tank was provided with two baffles to approximate a plug flow (three-stage) condition. Fixed diffusers were located in each of the stages. Mixed-liquor was recirculated through the diffusers by low-head centrifugal pumps. Pump suction was taken near the liquid surface to promote mixing and tank turnover. Throttling of the oxygen feed was accomplished automatically by D.O. sensing and control.

Major points of research interest in the project were: (1) oxygen utilization efficiency in an open-tank setting, (2) oxygen feed control response based on a D.O. monitoring approach, (3) mixed-liquor recirculation rates and power requirements, (4) diffuser self-cleansing (non-clogging) capabilities, and (5) shearing effect, if any, on mixed-liquor particles caused by con-

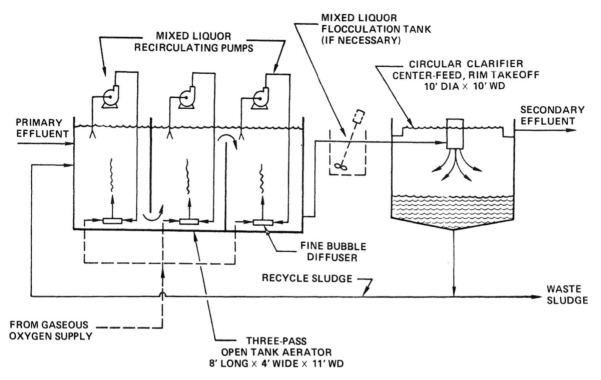

 placement above figure

FIGURE 14. FMC open-tank oxygenation pilot system.

tinuous recirculation through the pumps and diffusers. In the event that floc disruption did occur, a short detention biological reflocculation tank (gentle mixing, no chemicals) was interposed between the aerator and secondary clarifier. Two aspects of system design which could not be adequately defined at the scale of this pilot plant study were diffuser mixing characteristics and additional mixing requirements, if any, for large aeration tanks. This task was addressed separately by FMC in deep tank tests using tap water at both the firm's Englewood and Santa Clara laboratories.

Pilot plant fabrication was completed in late June 1973. System start-up required the first 20 days of July. The experimental program which followed was divided into eight phases and is outlined in Table 16. Performance data for the four highest flow phases (Phases 4 through 7) are summarized in Table 17. Sludge characteristics, aerator loadings, and secondary clarifier loadings for the same four phases are presented in Tables 18, 19, and 20, respectively.

As shown in Table 17, BOD_5 and suspended solids removals during the high loading conditions of Phases 4, 5, 6, and 7 were excellent. One of the significant observations forthcom-

ing from the project was that the feared disruption of sludge settling properties due to floc shearing as the mixed-liquor was continually recirculated through the centrifugal pumps and diffusers did not materialize. SVIs for the above four phases averaged a very acceptable 70 mℓ/g. A highly concentrated float (4 to 6% TSS) approximately 6 in. thick quickly developed on the surface of the aerater after start-up. This combination aeration/flotation effect was anticipated in light of the fine bubbles created by the oxygen diffusers. It was felt at the time that the thickness of the float could be controlled by adjusting the elevation at which the mixed-liquor recirculation suction was taken. FMC suggested that, with proper engineering design, this feature could offer a potentially attractive alternative location for extracting waste sludge from an activated sludge system.

VII. THE METRO DENVER PROJECT

During and following evaluation of its first-generation, open-tank, oxygen-activated sludge system on the Englewood Project, FMC was actively pursuing development of a simplified sec-

TABLE 16

Experimental Program for Englewood Project

Phase	Dates	Influent flow condition (GPM)	No. of clarifiers in use
1	7/21/73—9/5/73	10 (constant)	1
2	9/6/73—9/30/73	10 (diurnal)	1
3	12/6/73—1/28/74	15 (constant)	1
4	4/8/74—4/30/74	25 (constant)	2
5	5/1/74—5/31/74	35 (constant)	2
6	6/1/74—6/30/74	30 (diurnal)	2
7	7/1/74—7/31/74	20 (diurnal)	1
8	10/1/74—10/31/74	15 (constant)	1

TABLE 17

Performance Summary for Englewood Project (4/8/74—7/31/74)

	Phase			
	4	5	6	7
Total BOD_s in (mg/l)[a]	153	159	180	208
Total BOD_s out (mg/l)	13	18	16	16
% removed	92	89	91	92
Total COD in (mg/l)[a]	332	315	259	322
Total COD out (mg/l)	95	74	57	61
% removed	71	77	78	81
Suspended solids in (mg/l)[a]	110	85	85	115
Suspended solids out (mg/l)	13	15	12	15
% removed	88	82	86	87
Turbidity out (JTU)[b]	6	6	4	5
Sewage temperature (°F)	57	62	67	71

[a] Concentrations shown are for primary effluent feed to oxygen aerator.
[b] Jackson turbidity units.

TABLE 18

Average System Sludge Characteristics for Englewood Project (4/8/74—7/31/74)

Phase	MLSS (mg/l)	MLVSS (mg/l)	Return sludge TSS (mg/l)	Return sludge flow (% of Q)	SVI (ml/g)	SRT (days)
4	5,120	4,010	11,585	60	71	2.0
5	4,030	3,435	10,485	52	70	1.5
6	4,745	3,860	12,220	57	67	2.1
7	3,960	3,365	10,850	50	73	2.6

TABLE 19

Average Aerator Loadings for Englewood Project (4/8/74—7/31/74)

Phase	Detention time based on Q (hr)	F/M loading $\left(\dfrac{\text{lb BOD}_s/\text{day}}{\text{lb MLVSS}}\right)$	Volumetric organic loading $\left(\dfrac{\text{lb BOD}_s/\text{day}}{1000\ \text{ft}^3}\right)$
4	1.32	0.69	173
5	0.94	1.17	253
6	1.10	1.01	244
7	1.65	0.91	190

TABLE 20

Average Secondary Clarifier Loadings for Englewood Project (4/8/74—7/31/74)

Phase	Surface overflow rate (GPD/ft²)[a]	Mass loading lb TSS/ft² [a] day
4	514	35
5	720	37
6	617	38
7	823	41

[a] Excludes influent center-well annular area which equals 9.1% of total clarifier surface area.

ond-generation, fine-bubble oxygen diffuser. An elevation view of the key element, the rotating active diffuser, of the second-generation FMC system is shown in Figure 15. As indicated, the basic rotating diffuser consists of a 7-ft diameter submerged rotating disc, mounted to the bottom of a 6 5/8-in. diameter hollow shaft approximately 3 ft above the aeration tank floor. A 7½-in. wide ceramic diffusion medium is inserted into preformed openings, top and bottom, around the periphery of the plate, forming two circular diffusion bands parallel to the outer tapered edge. Approximate 28-in. diameter radial impellers mounted to the top and bottom of the disc provide essential mixing of oxygen, substrate, and biomass.

The relatively low design rotational velocity of 75 to 85 rpm is achieved with a constant speed motor and an appropriate gear reduction unit. The submerged diffuser disc is illustrated in a cutaway perspective view in Figure 16, and the entire diffuser assembly is shown photographically in Figure 17.

A functional flow diagram for a typical FMC system employing rotating diffusers for oxygen transfer is presented in Figure 18. The primary oxygen supply (shown as a cryogenic generator) is supplemented by a liquid oxygen reserve supply and accompanying vaporizer. With a cryogenic generator, unlike a PSA generator, losses occurring from the liquid oxygen backup tank, either through usage or evaporation, can be replenished directly from the primary supply source.

Oxygen gas from the supply system is fed at 20 to 30 psig down through the hollow diffuser shafts and then radially outward through small ducts located inside the diffuser disc bands, to the ceramic medium. As oxygen gas emerges from the upper and lower diffusion bands, the hydraulic shear created by rotation forms micron-size bubbles which do not coalesce as they move outward and pass over the outside tapered edge of the diffuser disc. The primary function of the tapered edge is to prevent turbulence which could induce bubble coalescence. The resulting micron-size bubble dispersion resembles a mist from which oxygen is rapidly and efficiently dissolved in the mixed-liquor. FMC claims the oxygen transfer rate obtained

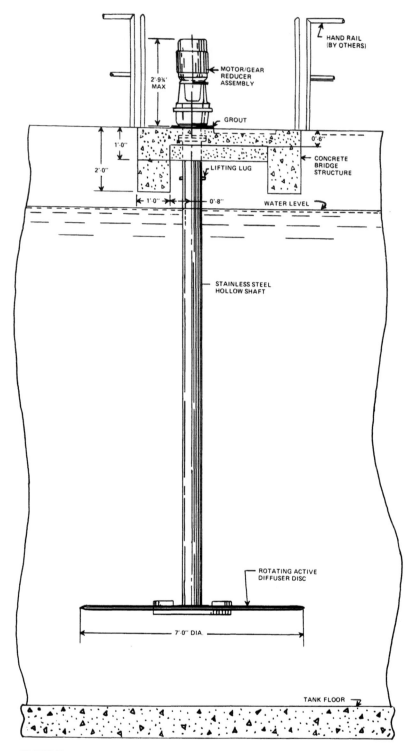

FIGURE 15.   Elevation view of rotating active diffuser and drive assembly. (Courtesy of FMC Corporation.)

with bubbles of this minute size is sufficiently high to sustain an oxygen utilization efficiency greater than 90% in conventional depth uncovered aeration tanks.[8]

A D.O. feedback system is used to control the oxygen feed rate to the dissolution system.

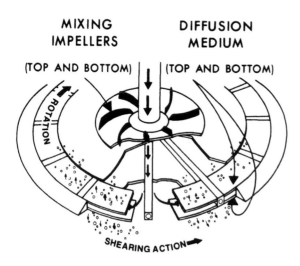

MIXING
IMPELLERS
(TOP AND BOTTOM)

DIFFUSION
MEDIUM
(TOP AND BOTTOM)

ROTATION

SHEARING ACTION

FIGURE 16. Perspective view of submerged rotating active diffuser showing gas flow and bubble formation. (Courtesy of FMC Corporation.)

The control system, consisting of one or more D.O. probes, analyzers, control valves, and electronic controllers, automatically maintains the mixed-liquor D.O. concentration at a predetermined setpoint, within the tolerance range of the equipment. The normal mixed-liquor D.O. operating range is 2 to 4 mg/ℓ; however, higher values can be attained. A one-module control system, i.e., one probe, analyzer, control valve, and controller each, is shown controlling the oxygen feed rate to both diffusers in Figure 18. In a longer tank requiring 10 to 20 diffusers, multiple control modules would be necessary with each module controlling the feed rate to a bank of 3 to 12 diffusers.

The FMC oxygenation system does not utilize tank covers or of necessity for a tank cover enables internal staging baffles. Although most covered-tank systems designed to date have included staging baffles, they are not essential. Both the open- and covered-tank alternatives can be designed compatibly with any of the commonly used activated sludge flow regimes. Covered-tank systems are, however, more naturally adapted to the conventional plug flow regime. Where conventional activated sludge treatment is the flow regime of choice, the staged configuration more nearly approximates ideal plug flow and, other factors being equal, would be expected to deliver an effluent with a slightly lower soluble BOD than an unstaged system.

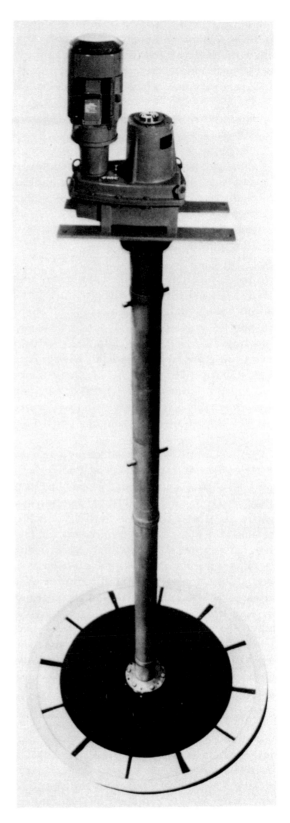

FIGURE 17. Photograph of rotating active diffuser assembly. (Courtesy of FMC Corporation.)

In comparing FMCs two open-tank options, the several inherent advantages of the rotating active diffuser over the fixed diffuser result induce a pronounced preference for the rotating diffuser alternative. These advantages include:

1. No requirement for prescreening of aerator influent
2. No requirement for pumping mixed-liquor through the diffusers to create the necessary shear to produce micron-size bubbles
3. Reduced oxygen dissolution power requirements
4. Simplified installation and
5. Less maintenance

At this writing, the rotating diffuser has completely replaced the fixed diffuser in all applications contemplated or under construction or design.[9]

In June 1975, EPA awarded a $235,000 demonstration grant to Metropolitan Denver (Colorado) Sewage Disposal District No. 1 to evaluate the FMC open-tank oxygenation system. The remainder of the estimated total project cost of $745,000 was shared by the District and FMC.

The evaluation was conducted in a segment of Metro Denver's existing air-activated sludge plant. The plant's secondary system consists of thirty-six 210-ft long, 670,000-gal aeration bays and twelve 130-ft diameter clarifiers. Each of the clarifiers is mated with three aeration bays operated in series to form 12 parallel secondary trains. Several of the bays have on occasion been utilized for aerobic stabilization of waste activated sludge. Sludge is recycled separately for each quadrant of the plant, i.e., settled sludge from the three clarifiers in any given quadrant is transferred to a common collection well from where it is returned for distribution among the three aeration trains in that quadrant.

Approximately two thirds of the average influent flow of 140 MGD receives primary sedimentation before it reaches the plant; the other third is primary settled on site. A new 72-MGD UNOX facility, which became operational in the fall of 1976, has diverted a significant fraction of the primary effluent flow from the existing overloaded air-activated sludge plant.

Prior to grant award, it was mutually decided that the large-scale FMC system to be evaluated

by the District would employ rotating diffusers rather than the older fixed diffusers used in previous pilot-scale studies at Metro Denver and on the Englewood Project. Thirteen rotating diffusers were installed in the first bay of aeration train No. 11 of the existing Metro air plant. The other two bays of this train were taken out of service for the duration of the project. Required hydraulic modifications included the installation of a pipe to transfer mixed-liquor from the end of the first bay to clarifier No. 11 and separate return and waste sludge lines and pumps. The latter step was taken to isolate oxygen sludge from the recycle sludge of the two remaining operating air trains (Nos. 7 and 9) of the plant's northeast quadrant. A liquid oxygen storage tank and vaporizer were installed adjacent to the converted oxygen test bay to supply the system with gaseous oxygen. A process schematic of the Metro Denver FMC oxygen system is given in Figure 19. Dimensioned plan and section views are shown in Figure 20.

As indicated in Figure 20, the rotating active diffusers were located on 21-ft centers. Six of the 13 diffusers were installed in sets of two in the first quarter of the tank, where oxygen demand was greatest. The remaining seven diffusers were located in tandem on the longitudinal center line of the aeration tank. The first 11 diffusers were driven by 10-hp motors and rotated after gear reduction at 85 rpm. The motors for the last two rotating diffusers were 7½-hp units. The rotational speeds of the 12th and 13th diffusers were 80 and 76 rpm, respectively, average oxygen dissolution capability per diffuser was rated by FMC at 90% at an oxygen flow of 1670 lb/day in wastewater for an average system dissolution capacity at Denver of 9.75 ton/day. Previous proprietary tests indicated these diffusers could be operated at oxygen flows up to 33% over their rated capacity with only a 5% decrease in oxygen transfer efficiency. On this basis, assuming an average BOD_5 removal of 140 mg/l and an oxygen requirement of 1.3 lb O_2/lb BOD_5 removed, the maximum theoretical sustained wastewater flow which could be handled by Denver's 13 rotating diffusers was 16.2 MGD. The maximum average daily flow employed on the project was 14 MGD, although peak diurnal flows occasionally reached 19 MGD.

Three D.O. probes and control systems were employed to control oxygen feed to the Metro

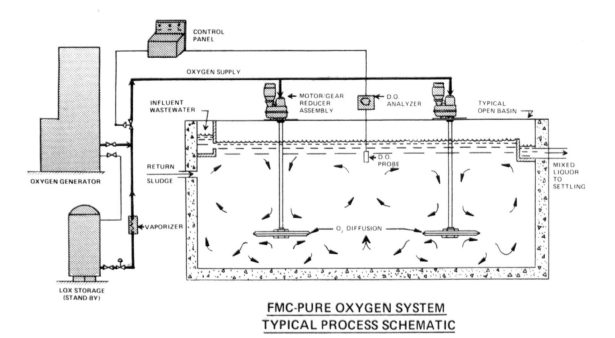

FMC-PURE OXYGEN SYSTEM
TYPICAL PROCESS SCHEMATIC

FIGURE 18. Functional flow diagram of typical FMC system employing rotating active diffusers. (Courtesy of FMC Corporation.)

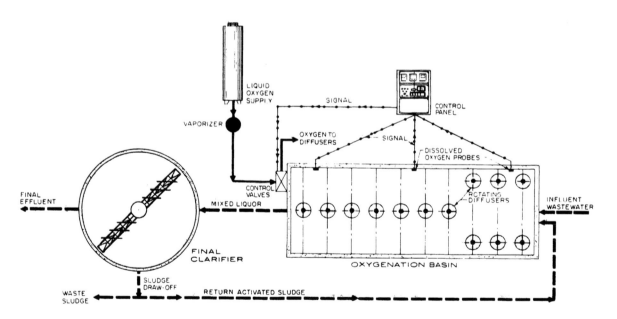

FIGURE 19. Process schematic of Metro Denver FMC test system. (Courtesy of FMC Corporation.)

Denver test bay. One system controlled the feed rate to the first six diffusers, the second to the middle four diffusers, and the third to the last three diffusers. Based on mutual agreement, an initial D.O. setpoint of 3.0 mg/ℓ was selected. During the project, the oxygen control equipment exhibited a variance range of approximately ± 0.7 mg/ℓ from the desired setpoint.

The rotating diffusers and their drives were supported from metal bridges which spanned the aeration test bay, as illustrated in Figure 20. The bridges, in turn, were supported by stanchions (not shown in Figure 20) running to the tank floor. The bridges were tied with minimal

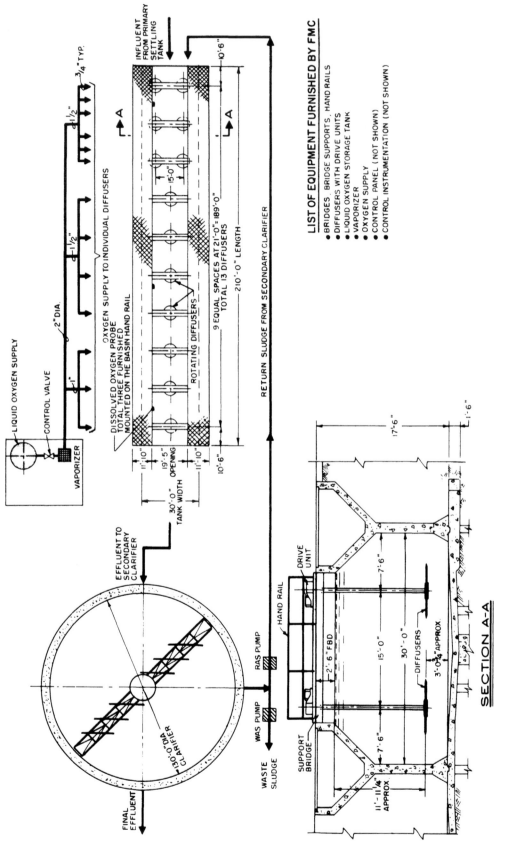

FIGURE 20.   Dimensioned plan and section views of Metro Denver FMC test system.   (Courtesy of FMC Corporation.)

defacing into the side walls of the test bay to prevent lateral movement. Following delivery of the key components of the oxygen supply and dissolution systems to the project site, the entire installation, including piping modifications, was completed in 6 weeks. Due in part to the short period in which its system components can be installed and the minimum structural modifications required, the upgrading of existing air-activated sludge plants as exemplified by the Metro Denver demonstration project is expected to become an important FMC application.

The major objective of the project from the District's standpoint was to determine the technical feasibility and attendant costs of converting its existing air-activated sludge plant to a higher capacity (i.e., two to three times higher), open-tank, oxygen-activated sludge system. If successful, the District believed it could potentially avert another major secondary plant expansion for the foreseeable future, with the exception of the additional clarifiers which would be needed to handle increases in influent flow. EPAs primary project objectives were: (1) to demonstrate at a representative field scale an alternative oxygenation concept which had been extensively and successfully evaluated at pilot scale and (2) to define reliable design criteria, operating conditions and costs, and performance expectations for a system embodying that concept for use by the engineering community.

Equipment installation and piping modifications were completed in early May 1976. The remainder of the month was devoted to facility shakedown and adjustments. June was utilized as a process start-up period for training operators and refining a data logging and retrieval system. The five planned phases and corresponding dates of the original evaluation program were as follows:

Phase I, July 1976
    Constant flow at 9 MGD; warm wastewater temperatures; one clarifier only in use.
Phase II, August—September 1976
    Diurnally varied flow at 7—14 MGD; warm wastewater temperatures; second clarifier available, if necesssary.
Phase III, October 1976
    Constant flow at 9 MGD; cool wastewater temperatures; one clarifier only in use.

Phase IV, November—December 1976
    Diurnally varied flow at 7—14 MGD; cool wastewater temperatures; second clarifier available, if necessary.
Phase V, January—April 1977
    Constant flow increased in increments to failure; cool wastewater temperatures; two clarifiers in use.

For reference purposes, baseline operating conditions are summarized below for the 9-MGD constant flow phases, assuming an average primary effluent BOD_5 concentration of 140 mg/l, a return sludge rate equal to 40% of the influent flow rate, and average MLSS and MLVSS concentrations of 4000 and 3200 mg/l, respectively:

Nominal aeration time (based on Q) = 1.79 hr
Actual aeration time (based on Q + R) = 1.28 hr
F/M loading
    = 0.59 lb BOD_5 applied/day/lb MLVSS under aeration
Volumetric organic loading
    = 117 lb BOD_5 applied/day/1000 ft^3 aerator volume
Secondary clarifier overflow rate (based on, surface total area) = 678 GPD/ft^2
Secondary clarifier overflow rate (based on useful surface area; excludes effluent launder area)
    = 746 GPD/ft^2
Secondary clarifier mass loading (based on total surface area and Q + R)
    = 31.7 lb MLSS/day/ft^2

Due to an unexpected problem which emerged in the third full month of operation (August 1976) (described in the following two paragraphs), the planned 10-month evaluation program was extended 4½ months to mid-September 1977. Because of this extension, project performance data had not been fully reduced and analyzed and were, therefore, not available for publication at the time of this writing (October 1977).

The existing Metro Denver aeration tanks are equipped with submerged orifice outlet pipes for transferring mixed-liquor to the final clarifiers. Without an avenue for exiting the aeration test bay, a 4 to 5-in thick, highly concentrated float (3 to 4% TSS), similar to that encountered at Englewood, accumulated on the mixed-liquor surface within 1 month following start-up. No effort was expended initially to remove or disperse the float since it did not appear to negatively impact system operation or performance. By the third month of operation, however, filamentous organism concentrations in both the bulk mixed-liquor and the trapped float began to increase substantially, resulting in high SVIs and a poor-settling, bulking sludge. It was conjectured that the float which acted as a concentrator of grease and floc debris as well as undesirable organisms might be implicated in the development of the bulking sludge condition. Several unsuccessful temporary measures were tried in the next 4 to 5 months to disperse the float uniformly throughout the bulk mixed-liquor and promote its regular transfer to the final clarifier and incorporation in the waste activated sludge mass.

The float problem was finally resolved in March 1977 by constructing an overflow weir to replace the submerged orifice outlet in the effluent and of the aeration test bay. With the continuous renewal of the mixed-liquor surface thereby provided, no further float accumulation occurred in either the aeration test bay or the test system final clarifiers for the duration of the project. Bulking sludge conditions disappeared during the 5½ months (April to mid-September 1977) of operation following installation of the overflow weir. At this point, however, the project staff has reached no definite conclusion as to whether the bulking sludge conditions that prevailed throughout the fall of 1976 and the 1976—1977 winter were due to the presence of the float, changing wastewater characteristics (which happens frequently with the Metro Denver influent), decreasing wastewater temperatures, or a combination thereof.

VIII. THE BUREAU OF RECLAMATION PROJECT

The Bureau of Reclamation's Engineering and Research Center in Denver, under an interagency agreement with EPA, has completed a 3-year project to test many different materials of construction to evaluate their suitability for use with oxygen aeration wastewater treatment systems. The materials tested included several different types of concrete; many different metals; and numerous protective coatings, linings, joint sealers, and gaskets.

The materials were exposed for varying lengths of time to oxygen-rich mixed-liquor, oxygen-rich vapor above the mixed-liquor, and to the interface between the two phases and then withdrawn for examination. Oxygen reactors utilized for these tests were located in Las Virgenes, California; Speedway, Indiana; and Fairfax County, Virginia.

The final report for this project was completed and approved in late summer 1977. At the time of this writing (October 1977), the report is in the process of being published by EPA.

REFERENCES

1. **McDowell, C. S. and Gianelli, J.**, Oxygen-activated Sludge Plant Completes Two Years of Successful Operation, Environmental Protection Technology Ser. Rep. No. EPA-600/2-77-040, U.S. Environmental Protection Agency, Cincinnati, July 1977.
2. **Albertsson, J. G., McWhirter, J. R., Robinson, E. K., and Vahldieck, N. P.**, Investigation of the Use of High Purity Oxygen Aeration in the Conventional Activated Sludge Process, Water Pollution Control Res. Ser. Rep. No. 17050 DNW 05/70, U.S. Department of the Interior, Federal Water Quality Administration, Washington, D.C., May 1970.
3. Union Carbide Corporation, Continued Evaluation of Oxygen Use in Conventional Activated Sludge Processing, Water Pollution Control Res. Ser. Rep. No. 17050 DNW 02/72, U.S. Environmental Protection Agency, Cincinnati, February 1972.
4. **Stamberg, J. B., Bishop, D. F., Bennett, S. M., and Hais, A. B.**, System Alternatives in Oxygen Activated Sludge, Environmental Protection Technology Ser. Rep. No. EPA-670/2-75-008, U.S. Environmental Protection Agency, Cincinnati, April 1975.
5. **Dick, R. I.**, Thickening, in *Water Quality Improvement by Physical and Chemical Processes*, Vol. 2, University of Texas Press, Austin, 1970.
6. **Duncan, J. and Hawata, K.**, Discussion of evaluation of activated sludge thickening theories, *J. Sanit. Eng. Div., Am. Soc. Civ. Eng.*, 94(SA2), 431, 1968.
7. **Nash, N., Krasnoff, P. J., Pressman, W. B., and Brenner, R. C.** Oxygen Aeration at Newtown Creek, *J. Water Pollut. Control Fed.*, 49(3), 388, 1977.
8. FMC Corporation, FMC Pure Oxygen Wastewater Treatment in Open Tanks, FMC Bulletin 8000-A, Itasca, Illinois, 1976.
9. **Stetzer, R.**, personal communication.

Basic Process Design Considerations

Chapter 5

OXYGEN-ACTIVATED SLUDGE SYSTEM PROCESS DESIGN

R. J. Grader, M. A. Miller, and E. H. Zander

TABLE OF CONTENTS

I. PROCESS DESIGN OBJECTIVES

The primary objective of the activated sludge process is the removal of biologically oxidizable materials. The organic pollutants present in municipal, industrial, or combined municipal-industrial wastewater streams are synthesized through microbiological action into insoluble cell mass or are oxidized into carbon dioxide and water. The biochemical reactions, however, only partially represent the success of meeting secondary effluent standards, as the separation of the resulting cell mass and other entrapped insoluble material (generally through sedimentation techniques) is of prime importance. Suspended solids removal, as well as residual organic pollutants, significantly affects the efficiency of activated sludge treatment because the carry-over of biological solids from the secondary clarifier increases effluent BOD. It is generally accepted that, even provided that hydraulic overloading does not occur, effluent suspended solids from a final clarification unit (sedimentation tank) will rarely be below 10 to 20 mg/l. It is therefore obvious that an integrated design philosophy must be adhered to in order to achieve desired effluent quality.

The degree of treatment required varies, as do the contaminants present in wastewater streams, their specified removal fractions, the assimilative capacity of the receiving waters, and the potential uses of these streams. This is to say that, for example, discharge requirements from a heavy industrial complex may need treatment only to the extent that the treated wastewater stream is suitable for discharge into a municipal sewerage system. In another instance, the treated wastewater may be required to have undergone sufficient removal of pollutants such that it may be employed for irrigation water. In general, the relationship of pollution abatement and the economic and social well-being of the population must be considered as well as the environmental and ecological effects. Effluent standards are becoming increasingly rigorous, and it is the responsibility of the design engineer to provide the most effective and advantageous design in order that specified discharge quality is met at all times.

Selection of the most advantageous process scheme to achieve biological oxidation to the degree required is necessarily influenced by consideration of initial investment, operating expenses (including utilities, labor, chemicals, and maintenance and repair), process stability (including instrumentation and process control), and the ability to achieve the required treatment efficiency. Several factors must be considered before the designer can judge with confidence the best alternative approach and the appropriate design point recognizing that economy, reliability, and efficiency should ultimately govern. Full economic optimization must also be consistent with plans for future expansion, as well as integrated with design plans providing for existing primary and/or future tertiary treatment. As an example, designs may have to be structured such that second step (separate sludge) biological nitrification may be added at some future date or that an initial carbonaceous removal system be utilized as the first step of a two-step system coincident with an increasing load pattern.

II. SELECTION OF DESIGN POINT

Planning, design, and construction of wastewater treatment systems usually involves years of effort from design and consulting engineers. The performance and capacity requirements are fundamental to the process design. The size of any wastewater treatment facility must be large enough to meet the growing needs of the municipality or industry involved within a reasonable time period and also remain economically justified. This, of course, necessitates the coordination of all unit operations. Past, current, and future growth patterns should be considered.

Two major problems face engineers in the selection of the design point for municipal wastewater treatment facilities. These are (1) the ability to forecast changes in population and wastewater flow and (2) estimation of changes in wastewater character due to rapid industrialization of an area. The uncertainty involved with respect to these factors is quite burdensome, especially in relatively undeveloped areas. Consideration of planning studies, zoning ordinances, industrial wastewater discharge ordinances, topographical constraints, and political boundaries can be extremely useful in judging the future expected hydraulic and organic loads.

Before the design point can actually be determined, consideration must be given to the overall contribution of all wastewater streams, including in-plant recycle streams from other unit operations. Each stream must be weighted and variations in characteristics noted. Factors to be considered for each component stream are temperature, pH or acidity, chloride content, toxic metallic or organic constituents, grease and/or oils, sulfides, surfactants, nutrients such as nitrogen and phosphorus, the concentrations of biochemical oxygen demand (BOD), both soluble and total, chemical oxygen demand (COD), suspended solids, and flow.

For industrial wastes which contain an unusually high percentage of organic or inorganic material compared with typical domestic wastewater, or for those wastewaters containing toxic components, pretreatment standards may be imposed. At times equalization of industrial wastewater discharges may be employed either at the industrial site or in-plant in order that flow variations and shock loads can be minimized. The population equivalents method is a reasonable method for equating industrial contributions to the ordinary per capita contributions present in domestic wastewater streams. These equivalents should be based on COD analysis rather than BOD_5 due to the fact that strong or toxic streams may exhibit artificially low BOD_5 values.

Existing sewage treatment plant records provide

a historical record of influent flow variations and loadings as well as information related to in-plant wastewater streams. Significant in-plant concentrated flows stem from various sludge-handling facilities, including aerobic digestion, anaerobic digestion, thickener overflow, centrifuge concentrate, thermal conditioning sludge supernatant vacuum filter filtrate, scrubber waters, etc. The return or recycle of some of these streams to the feed end of a treatment plant can impose high oxygen demands on the system and add significant amounts of solids. Periodic discharge of these flows can create shock load conditions and, therefore, best efforts should be employed to equalize the additional loading.

The assemblage of all required hydraulic and organic loadings is utilized for determination of the design point. The coincident relationships that exist between flow and strength can be determined through considering such factors as seasonal variations (e.g., canning vs. noncanning periods), daily diurnal patterns (e.g., small facilities with high day shift industrial fraction), and midweek peaks (e.g., considerable 5 day/week industrial contribution). In addition, consideration is given to minimum flow days, maximum rain days for combined systems, annual maximum 4-hr sustained peaks and averages, in-plant return streams, infiltration, agricultural runoff, dampening effects of the sewerage system, the use of equalization basins, the plant size itself, etc., which all enter into final selection of the design point. Statistical analysis of available data may show pronounced differences among the various flow and strength averages. Unknown factors are reduced or eliminated as more data are assembled and design criteria are established and accepted.

It becomes readily apparent that the annual average flow and strength usually does not constitute the design point. Completeness and consistency are of utmost importance in the selection of the design point. Once this milestone has been reached, the detailed process design procedure can be initiated with confidence.

III. INFORMATION REQUIRED FOR PROCESS DESIGN

In general, the information required to perform a process design can be classified as (1) wastewater quantity, (2) wastewater quality, (3) effluent quality required, and (4) definition of the other

unit operations or existing facilities in the treatment plant and their relationship to the secondary system. Table 1 generally depicts the required data, while the following paragraphs elaborate on each of the aforementioned categories. From the discussion, it will be apparent that it is not always possible to obtain all the required information. Therefore, it is usually necessary to make some

TABLE 1

Definitional Parameters

1. *Wastewater Quantity:*
   Average flow
   Maximum 4-hr sustained peak flow
   Design flow

2. *Wastewater Quality:*
   Average BOD_5, mg/l
   Maximum sustained (coincides with max. flow), mg/l
   Design BOD_5, mg/l
   COD/BOD_5 all cases
   Source — industrial contribution — nature of quantity
   Preceding treatment
   Average suspended solids, mg/l
   Average volatile suspended solids, mg/l
   Temperature, °C
   pH
   Alkalinity, mg/l as $CaCO_3$
   Alpha, beta
   Nutrient content, mg/l N, P
   Heavy metals
   Other toxic components

3. *Required Effluent Quality:*
   BOD_5, mg/l
   SS, mg/l
   N, mg/l
   P, mg/l
   COD, mg/l
   Other

4. *Site Limitations:*
   New or existing tankage
   Land area available
   Piling required
   Maximum tank depth, ft

5. *Solids Handling Equipment:*
   Dewatering
   Sludge disposal

6. *Secondary Clarifier Specifications:*
   Overflow rate at design flow
   Depth
   Feed
   Take off
   Sludge return
   Circular, rectangular, square

judgments and consequently build some conservatism into the design to account for both inadequate and potentially inaccurate data. However, the degree of conservatism required can be better assessed by a recognition of those parameters which affect process performance and the resultant appreciation for the quantity and type of information that is lacking.

Wastewater quantity is, of course, characterized by the rate of flow and is obviously a major parameter in the design of the activated sludge system. The phrase "wastewater quality" is used herein to describe characteristics of the wastewater which result from the components contained in it and includes such familiar parameters as 5-day Biochemical Oxygen Demand (BOD_5) and Chemical Oxygen Demand (COD). These parameters measure the oxygen requirements of the wastewater. The diurnal variations and annual average and peak values should therefore be known to accurately quantify the instantaneous and average oxygen demands in the secondary system. Another parameter which helps to characterize the wastewater is whether the source is municipal or partially or totally industrial and what type of industry is involved. Also important, especially as it affects the expected settling characteristics of the mixed-liquor and the quantity and quality of waste-activated sludge, is the treatment preceding the secondary system. If the waste is only degritted or comminuted, the mixed-liquor suspended solids will exhibit better settling and compacting properties and will contain a lower volatile fraction than if the waste had received primary settling. Also, greater quantities of waste-activated sludge will leave the secondary system if primaries are not employed. Other important "quality" parameters are temperature (average and range), pH, alkalinity, total and volatile suspended solids level, alpha and beta values for the sewage, nutrient content, and the expected concentration of heavy metals or other components which are potentially toxic to the secondary system biological mass.

Another "quality" parameter which is best determined through pilot plant operation is the oxygen consumption requirements in terms of the amount required for biomass synthesis and endogenous metabolism, both of which occur during the removal of the pollutant material. However, in many cases, the waste stream does not presently exist or it is known that additional streams will alter its character in the near future, so pilot plant results would not necessarily be representative of those to be expected in full-scale operation. It is situations such as these where estimations based on pilot plant results on similar wastes become particularly useful as the only reliable source of data.

The same discussion used in regard to the oxygen requirement can also be used to describe the mixed liquor settleability and the biokinetic characteristics of the wastewater, which must also be known and can best be determined from pilot plant data. They too can be estimated based on other pilot plant experience and the known source and type of the wastewater.

The effluent quality required pertains to specific site parameters and prior and subsequent unit operations which are to be a part of the overall wastewater treatment facility. Among the site information required is whether the job is a conversion of existing tankage to upgrade the plant capacity or installation of new tankage. Also, it must be known whether land area is limiting and what limitations must be placed on tank depth and area due to water table level and piling requirements.

As previously mentioned, the prior and subsequent unit operations are important in that they influence the character of the waste fed to the secondary system. Most sludge-handling systems have supernate streams which are returned to either the primary or secondary system, and, depending on the system in question, these may have a noticeable effect on the character (nitrogen, phosphorus, and carbonaceous material) of the combined wastewater entering the secondary system. Particularly important in this regard are the wet oxidation techniques often practiced to reduce sludge quantity and improve dewaterability without the need for chemical addition. The supernate streams from these operations can increase the organic load on the activated sludge system by 10 to 30%, depending on the strength of the influent waste and the degree of sludge quantity reduction attained. Anaerobic digestion supernates and gravity thickener supernates also add appreciably to the load (~10 to 15%), while returns from well-operating aerobic digesters, centrifuges, and vacuum filters generally do not add significantly to the secondary system organic load. The main point of this portion of the discussion is that the effect of recycle streams from other unit operations within the facility must be recognized and quantified.

The importance of the secondary clarifier in the design of an activated sludge system (air or oxygen) cannot be overemphasized. Both the theory and practice of clarifier design and the interrelationship of the clarifier and aeration tank will be discussed in subsequent chapters of this book. Suffice it to say here that a knowledge of the clarifier overflow rate, side water depth, weir loading, geometrical shape, and mixed-liquor feed and sludge takeoff mechanisms is required to properly integrate the clarifier and aeration tank in an activated sludge system.

IV. FACTORS AFFECTING DESIGN APPROACH

A. Limitations on Activated Sludge System Performance

Once the designer of an oxygen-activated sludge system has established the design point and gathered all the necessary information, he must decide what will be the best approach to designing the system. There are many potential starting points for design and although there is a "best approach" for most typical situations, there can be some variation. The purpose of the following discussion is to develop and present the "best approach" to process design of oxygen-activated sludge systems and to discuss circumstances under which variations in this approach should be considered. The proper design approach should be one which, considering all design limitations, results in the most direct engineering path to a design which most economically meets the wastewater treatment requirements.

There are internal and external process factors which include various process limitations. Internal factors include limitations on mass transfer and mixing capabilities, settling and thickening of the biomass generated in the system, and the rate of substrate metabolism in the system. External factors may include restrictions on tankage size or configuration for both the reactor tank and the clarifiers and extremely unusual waste stream characteristics. The former category determines the "best approach" for most typical classes of wastewaters, while the latter group, when present, results in exceptions to that approach.

Of primary interest are the three important internal process limitations mentioned previously.

These limiting factors are biological kinetics, solids settling and thickening, and mass transfer.

The biological kinetics of the system determine the rate at which substrate can be metabolized in the system. It will be shown in a later section of this chapter that the substrate removal rate is dependent on the nature of the waste, the hydraulic retention time in the aeration tankage, and the amount of active biomass in the system. The reactor retention time is, of course, determined by the reactor size. The active biomass concentration is influenced by the settling and thickening characteristics of the biomass produced in the system. Because the change in biomass concentration due to biomass growth in the aeration tankage is small compared with biomass concentration in the recycle stream, the reactor solids concentration can be related, through a simple mass balance, to the underflow concentration from the secondary clarifiers and the rate of recycle of solids from the clarifiers to the reactor.

The product of the reactor retention time and the active biomass concentration in the reactor is an important parameter in biological reactor kinetics and will hereafter be referred to as the time concentration product. This time concentration product is inversely related to the more familiar parameter food to biomass ratio.

The food to biomass ratio (F/M) is defined as mass of substrate applied per day per mass of volatile suspended solids in the system. The value of the F/M loading ratio is calculated as follows:

$$\frac{\text{pound substrate applied}}{\text{day}} = 8.34 \, Q \, S_A$$

pounds MLVSS aerated = 8.34 (MLVSS)V

$$F/M = \frac{8.34 \, Q \, S_A}{8.34 \, (\text{MLVSS})V}$$

$$F/M = \frac{24 \, S_A}{t_Q \, (\text{MLVSS})} \tag{1}$$

If it is assumed that the MLVSS is an indicator of the active biomass concentration, then the time concentration product is defined by:

$$t_Q(\text{MLVSS}) = \frac{24 \, S_A}{F/M}$$

It can be shown that for a particular waste, the above assumption is good since the active biomass concentration is constantly proportional to the MLVSS over a fairly wide range of F/M values. Further discussion of the active biomass fraction of the MLVSS and methods for its determination are given in later sections of this chapter.

The time concentration product necessary to produce a particular substrate removal rate is determined by the growth rate of the biomass. Growth rates vary widely for different waste-waters, and for carbonaceous systems are a function of the type of substrate, the level of toxic materials present, the temperature, the pH, and maintenance of an aerobic environment. Thus, the biomass time concentration product requirements for a particular substrate removal rate can be expected to vary considerably, particularly for industrial wastes. This fact brings to light the necessity of accumulation of large quantities of data on wastewaters of various types. With such data available, process designs can generally be successfully completed without detailed test work for each design application. An activated sludge system can be defined as kinetics limited when variation of the time concentration product results in corresponding variation in the effluent soluble substrate level. When kinetics is not limiting, this product can vary without noticeably changing the effluent quality. The level of time-concentration product required to achieve a desired level of treatment for a particular waste must be determined experimentally or estimated based on experience.

As mentioned, a system is limited by kinetics from a process design standpoint when the size of the reactor tanks is determined by the required time-concentration product rather than other limiting factors. Generally, the growth rate for biological masses growing on municipal waste-water is so high that kinetics is not limiting. That is, from a kinetics standpoint, the system could produce the desired effluent quality with a lower time-concentration product (or higher F/M) than can in fact be used due to other limitations.

The settling and thickening character of the biomass is dependent on the nature of the waste-water, the dissolved oxygen concentration in the reactor, the substrate loading level (weight of substrate/weight of biomass/day), and the level of turbulence and shear in the reactor. It is well known that the settling and thickening characteristics of the biomass depend directly on the size

and character of the bacterial floc particles formed in the reactor. All of the four factors mentioned above directly affect the nature of the floc particles formed.

Floc formation is postulated to occur as the result of secretion of a polymeric substance by endogenously respiring microorganisms.[22-24] When a sufficient amount of this polymeric substance is being generated in the microbial population to result in the proper ratio of polymer to microorganism, the polymer serves as a "glue" for the formation of a good flocculent sludge mass. If the substrate loading is very high, the degree of synthesis (rate of production of new organisms) will be much higher than the relatively constant level of endogenous activity, and the quantity of polymer being generated may become insufficient to result in good floc formation. If, on the other hand, the biomass loading is too low, the substrate level will be so low that the biological population will essentially be starving and the polymeric substance may be consumed as substrate. Floc formation will be poor under these circumstances as well.

The dissolved oxygen level in the reactor affects the degree of penetration of oxygen into the floc particles and thus the fraction of the particles which can be kept aerobic. Even if the substrate loading is in a range which should result in large, dense floc particles, if the D.O. level is too low, the size, shape, and density will be adversely affected. An explanation for this is that when a significant portion of the floc particle is not aerobic, endogenous respiration does not occur in that portion. The quantity of polymer generated would be insufficient for the formation and maintenance of the floc particle.

If the turbulence and shear levels are too high in the activated sludge reactor, the large, dense, aerobic floc particles which might otherwise exist can be physically torn apart.[25,26] This also can result in small, poorly settling floc.

The effect of the nature of the wastewater on floc formation is less specific than the previous factors. It can be said that the wastewater character establishes a best attainable floc size and density, and only when none of the previous three factors influences the floc formation will this "best attainable" level be reached.

The maximum level of biologically active solids attainable in the reactor will determine the minimum tank volume which can be employed. If this resultant minimum tank volume is satisfactory and not limited by other factors (to be discussed), the design can be said to be thickening limited. The tank volume is determined only by the degree to which the biological mass can be thickened and thus the concentration of biological mass which can be maintained in the reactor. A design can be thickening limited regardless of whether it is kinetics limited.

The second concern regarding settling does not relate directly to maintenance of a reactor solids concentration or kinetics but is, in fact, even more directly influential on the activated sludge system performance since it determines the system final effluent suspended solids level. This is referred to as the settling limitation factor and is placed on an equal importance level with kinetics limitation so far as determination of the design approach is concerned.

The clarifier effluent suspended solids level is directly dependent on the settling characteristics of the biomass and thus on the quality of the floc particles generated in the reactor. The settling characteristics as related to the clarifier effluent suspended solids level will be affected by the same parameters as for thickening and reactor solids concentration. It must be noted, however, that settling as it determines the effluent suspended solids concentration and thickening, and as it determines the clarifier underflow and reactor solids concentrations, are two relatively distinct phenomena. Although both are affected by the same factors, the ranges and degree of effect of each of the factors are different.

The system is said to be settling limited when the desired design loading (F/M) falls outside the range which results in good flocculation for the particular wastewater in question. The result of using that loading would be poor effluent quality due to high effluent suspended solids. Under these circumstances, the loading must be adjusted to fall within the range necessary for good settling. For municipal systems this means reducing the loading and increasing the time-concentration product above the value indicated by kinetics considerations. The design then is no longer kinetics limited but solids settling limited. This is typically the case for municipal designs.

The third and final major limitation factor to be considered is mass transfer of oxygen. An aerobic biomass requires oxygen to maintain the metabolic and synthesis activities which result in substrate removal. The microorganisms contained in the floc particles obtain this oxygen via mass transfer from the bulk liquid into and through the floc matrix and finally into the microbial cells themselves. It has been shown that the controlling step in this transfer sequence is through the floc matrix.[27-29] The driving force for this transfer step is the difference between the bulk liquid dissolved oxygen concentration at the center of the floc particle. The larger the floc particles are, the higher the bulk liquid dissolved oxygen (D.O.) level must be to assure that the particle is entirely aerobic. The maintenance of a high D.O. level is critical to the formation of large, dense floc particles and is necessary to maintain a high level of aerobic activity in a system operating with such floc particles.

The maintenance of a bulk liquid dissolved oxygen concentration is dependent on mass transfer from the gas phase to the bulk liquid. The driving force for this liquid phase mass transfer controlled step is the difference between the potential liquid phase oxygen concentration in equilibrium with the bulk gas phase oxygen concentration and the actual bulk liquid dissolved oxygen content. Thus, the higher the gas phase oxygen concentration, the higher the mass transfer driving force and the higher the bulk liquid D.O. level which can be maintained for a given biological oxygen demand rate and system energy input.

The rate of oxygen transfer from the gas phase to the bulk liquid also depends, in addition to the transfer driving force, on the surface area generated in the system for mass transfer as well as the "completeness" of mixing of the volume into which oxygen is being transferred. These factors depend on the type of mass transfer and mixing device used and on the power input to the device. In general, higher volumetric power inputs in an activated sludge reactor mean higher turbulence and, in most cases, increased shear levels.

Other factors which can affect a system's ability to transfer oxygen, besides the driving force, the transfer device, and the power input, include the reactor tank size and configuration, the solids concentration in the reactor, and the reactor temperature. The tank size and configuration affect the mixing efficiency of the mass transfer device. Higher temperatures decrease the mass transfer capacity of a particular system because the oxygen solubility in the mixed-liquor is decreased, thus decreasing the mass transfer driving force more than the transfer coefficient is increased.

Mass transfer limitation can be described as occurring when the oxygen mass transfer system cannot meet the oxygen demand of the biooxidation process while maintaining the desired D.O. level. The oxygen mass transfer system is defined here as the aeration devices operating in the tank volume and configuration and under all other process conditions established by the process design.

Consider a system which is neither kinetics nor settling limited in which the loading is moderately high and the solids concentration is quite good, resulting in a low retention time. First, it is possible that the volumetric oxygen demand for this system will be so high that it cannot be met by any practically available aeration device. Secondly, if an aerator can be found which has the capability of transferring the required oxygen, it is possible that when placed in the small tankage and this particular configuration, its ability to transfer oxygen will be reduced because of poor mixing of the tank to the point that it will not satisfy the system demand. Thirdly, if a transfer device can be found which will meet the oxygen demand in the designed tankage, it is still possible that it will not be capable of maintaining the consistently high D.O. level which is necessary to assure good settling and thickening. All three of these cases represent mass transfer limited situations; the first was because the volumetric oxygen demand was just too high, the second was because the tankage configuration was not favorable for good mixing necessary to efficient mass transfer, and the third because the required D.O. level could not be maintained in the system.

The interrelationship of all the above limiting factors under the particular conditions prevalent in oxygen-activated sludge systems combine to define the best approach to design of such systems.

The use of high-purity oxygen as the aeration gas in an activated sludge system has several distinct advantages. The key advantage is that the mass transfer driving force from the gas phase to the liquid phase is much greater than if the aeration gas were air. The effect of this is that substantially higher D.O. levels can be maintained at equivalent volumetric oxygen demand rate and power input. In fact, considerably higher volumetric oxygen demand levels than tolerable for air systems can be readily handled by oxygenation

systems even while high bulk liquid D.O. levels are maintained. In general, the system can tolerate much higher volumetric oxygen demand without becoming mass transfer limited, as previously discussed.

The maintenance of high bulk liquid D.O. levels has at least three benefits. First, high D.O. concentrations result in formation and maintenance of larger, denser floc particles than are developed in low D.O. systems. This results in better settling and thickening characteristics and thus less probability of the system becoming settling or thickening limited. Secondly, the high D.O. levels assure that the biomass is more nearly completely aerobic due to greater penetration of oxygen into the floc particles. This fact has a significant effect on the biological sludge production in the system as well as settling and thickening. Thirdly, the high D.O. level allows greater tolerance for short-term fluctuations in the substrate loadings on the system. During periods of peak oxygen demand, the D.O. level in the system decreases, increasing the mass transfer driving force until the peak demand is satisfied.

Better settling and thickening results in two distinct advantages in the design of an oxygenation system. First, the range of food to biomass loadings within which the system can operate and not be limited in performance by settling is greater for high D.O. (better settling) systems. Secondly, higher clarifier underflow concentrations and thus higher reactor solids levels can be obtained for a particular clarifier design. The latter point means that a time-concentration product required for the desired substrate removal can be achieved using a smaller aeration retention time. The former point makes further decreases in retention time possible since higher food to biomass loadings can now be tolerated without increasing the effluent suspended solids.

It is important to fully understand why an oxygenation system can operate at a higher food to biomass ratio than a conventional air activated sludge system before becoming either settling or biokinetically controlled. As previously mentioned, one of the critical requirements for good performance of an activated sludge process is the maintenance of a flocculant biomass that settles well and does not allow significant quantities of solid material to escape in the effluent stream. It is this requirement for a flocculant biomass that results in the need to consider diffusional resistance in the design of the system. Because the floc particle contains numerous organisms and because each organism requires nutrients, carbonaceous substrate, and dissolved oxygen for it to be aerobically viable, diffusion of these components into the floc particle is of paramount importance. The degree of penetration (fraction of the floc particle that is viable) of these components will depend on the size of the floc particle, the concentration levels of the components in the bulk mixed liquor, the rate at which the organisms consume the components, and the nature of the component itself. The hypothesis that diffusion of components into the floc particle is quite significant in an activated sludge system is supported by the fact that several investigators have found that increased turbulence in a system sometimes improves the rate of removal of the carbonaceous substrate. The increased turbulence decreases the size of the floc particle which in turn reduces the diffusional resistance to nutrients, substrate, and oxygen, and thereby increases the number of organisms taking part in the reaction. Further evidence of the importance of diffusion in an activated sludge process is the observation made by Eckenfelder[12] that temperature dependence on the apparent biokinetic rate constant in a flocculated growth system is substantially less than that in a dispersed growth system when both are operating at conventional air system D.O. levels of 0.5 to 2 mg/l. As explained by Eckenfelder, this reduced dependence on temperature in a flocculated system is apparent because in a warm environment the metabolic activity is faster, and because of diffusional resistances, the required components do not penetrate as far into the floc particle as they would if the metabolic uptake rates were slower. Under colder conditions, the metabolic uptake rate of the organism is reduced and penetration into the floc particle increases. The net result is that under warm conditions there are fewer organisms acting faster and under cold conditions more organisms acting slower. This results in a smaller temperature dependence on the apparent biokinetic reaction rate than occurs in a dispersed growth system where diffusional resistances are not a factor.

It is, therefore, quite well established that diffusional resistances play an important role in the performance of an activated sludge system, and some discussion is required to determine which of the three components, nutrients, sub-

strate, or dissolved oxygen, is most likely to be the controlling one, in the diffusional sense. As mentioned, the degree of penetration into the floc particle depends on floc size, bulk liquor concentration of the component, the rate of uptake of the component by the organisms, and the nature (diffusion coefficient) of the component itself. Since the molecular weight of most nutrients is substantially less than that of most organic substrates found in wastewater, the diffusion coefficients can be expected to be greater for the nutrients than for the substrates. Therefore, if the concentration of nutrients in the system is as great as it would need to be when diffusion was not a factor (dispersed growth system), the nutrients would be expected to penetrate the floc at least as far as the substrate, since the product of the use rate and concentration in the bulk liquor would be the same as for the substrate material, while the diffusion coefficient should be greater.

The diffusion coefficient of oxygen is also substantially greater than that for most organic substrate materials; however, since the uptake rate is very nearly the same for oxygen as for the substrate, the ratio of the bulk liquor concentrations for substrate and oxygen should be about the same as their diffusion coefficients so that neither one is controlling over the other. Matson et al.[7] have shown that for many substrates characteristically contained in domestic wastewaters, this ratio is about 5 to 10. That is, if the substrate material is present in concentrations between five and ten times that of the dissolved oxygen, the oxygen will be the diffusionally controlling component. Of course, if the bulk concentration of both substrates and oxygen are sufficient, the floc particle will be completely penetrated by both, and diffusional limitations will not be a factor. Matson et al.[7] show that for floc sizes typically encountered in activated sludge systems an oxygen concentration of from 3 to 5 mg/l is sufficient for complete floc penetration. Thus, diffusional resistance should not be a factor in a system with this D.O. level until the substrate concentration reaches from 15 to 50 mg/l. However, in a system where the D.O. level is less than 3 mg/l, it can be concluded that the system is not as "active" as it could otherwise be due to the fact that the entire floc particle and, therefore, the whole body of organisms present do not have sufficient oxygen available to remain aerobically active.

In a more recent and comprehensive investigation, Scaccia and Lee[8] have studied this phenomenon much more vigorously. Their work consisted basically of two parts. The first dealt with the determination of the average floc particle size existing in aeration basins. They came to the conclusion that in addition to the basic characteristic of the floc particle, the size of the flocs depends on the turbulence intensity in the mixed-liquor and the absolute basin size. In fact, the effect of the basin size is more pronounced than that of the power density, since the average floc diameter increases with the one third power of the basin size and decreases only with the one sixth power of the power density. This means, e.g., that in full scale aeration basins, the average floc diameter can be expected to be about 2.4 times larger than in pilot plants.

The second part of their study dealt with the penetration of the substrate and the nutrients into the floc particles. By using a double-Monod kinetic model, which assumes that the substrate, as well as the oxygen nutrient, obey Monod kinetics and also takes their interaction into account, a relationship is developed between the depth of penetration of the oxygen, the bulk D.O. level, the oxygen uptake rate, the floc diameter, and other parameters. From this relationship, it was shown that, in full-scale operation, oxygen is always the limiting nutrient — even at very low rates of food supply — and that, at average oxygen uptake rates, a bulk D.O. level of 2 mg/l results only in a modest penetration of oxygen into the floc particles. On the other hand, when high D.O. levels exist in the bulk liquid (greater than 5 mg/l D.O.), the major portion of the floc is viable and will participate in the degradation process.

Another important finding of this study[8] is the fact that one has to be extremely careful in interpreting the results of pilot plant operations for designing full-scale plants. The major reason for this is that even if one carefully attempts to simulate full-scale plant behavior on pilot scale by maintaining the same power density and the same modest D.O. level in the bulk mixed-liquor of, say, 2 mg/l, one can be misled. The reason for this is that the average floc particle diameter in the pilot plant will be only about half of the particle diameter of the full-scale plant and, therefore, a much larger portion of the flocs in the pilot operation will be exposed to oxygen and be able to remain "active." At high D.O. levels in the mixed-liquor, this difference due to scale will disappear because the oxygen will penetrate most of the flocs, both in the pilot operation as well as in full-scale aeration.

As mentioned in the definition of the food to biomass ratio, the mixed liquor volatile suspended solids are assumed to be a measure of the active biomass in the system. Certainly it is this active biomass that is important in the operation of the activated sludge system, and the only reason that it is not used in the definition of the F/M is that it is too difficult a parameter to measure and thus would not be a very useful basis for design or control of the activated sludge process. The assumption that the mixed liquor volatile suspended solids is a good approximation of the active biomass is a reasonably good one for wastes of a similar nature as long as the dissolved oxygen levels are nearly the same in the various designs or if the dissolved oxygen levels are always high enough to insure complete penetration of the floc so that oxygen diffusion is not a factor. However, when comparing the design F/M for systems that will operate at low D.O. levels vs. those operating at high D.O. levels, the fact that the high D.O. system has a larger fraction of aerobically active biomass in the mixed liquor suspended solids should be accounted for. If experience teaches that a certain range of F/M is required for a system with low, or conventional, D.O. levels to function properly and have a good flocculating biomass, this experience should be used in design of systems with similar D.O. levels. However, with systems carrying D.O. levels high enough to completely penetrate the floc particles, it should be recognized that the true mass of applied substrate per day per mass of active organisms is lower in such a "high D.O." system than in a "low D.O." system at the same F/M. Thus, the system at "high D.O." will hold a good settling biomass at higher system F/M's than a "low D.O." system because the mass of substrate per day per mass of active biomass does not become equal until higher F/M's are reached for the "high D.O." system.

B. Design Approach

The first step to oxygenation system design should be the sizing of the clarifier to produce the highest reasonable and practical reactor solids concentration while optimizing the underflow concentration to attain a high waste sludge concentration. The clarifier is chosen as the starting point because typical designs are not kinetics limited, and it can be assumed that the substrate loading level can be chosen within the range resulting in good settling and thickening. It is desirable to use high reactor solids concentrations because this reduces the required tank volume and increases the economic attractiveness of the system without decreasing performance. However, there are limits to this. First, the reactor solids concentration is increased either at the expense of larger clarifiers than would normally be used or by using higher recycle rates, which have the effect of decreasing the waste solids concentrations. Secondly, the tank volume can be reduced to the point that mass transfer and mixing become problems, even in an oxygenation system. This means either costly increases in horsepower or increasing the retention time in spite of the high solids concentrations just to reduce the volumetric oxygen demands and configurational constraints which will have to be met by the mass transfer system.

The second major step after sizing the clarifiers and determining the reactor solids concentration is to determine the tank size. For typical nonkinetics-limited cases this is done by first judiciously selecting, based on data and experience, a biomass loading which is within the range assuring good settling and is not so high that it results in mass transfer limitation. For nonkinetics-limited systems the loading should be chosen near to the upper end of the allowable ranges but including due consideration of peak loadings. The system must be capable of maintaining good effluent suspended solids levels and reasonably high D.O. levels through any long duration peak flows or loads.

For cases where the kinetics is limiting and the design loading will be established by this requirement, the loading should be established before the clarifier is sized. This is because if the necessary loading is near the lower limit normally used for assurance of good settling, due conservatism will need to be incorporated into the expected settling and thickening characteristics.

The choice of whether to size the clarifier or reactor first will also be influenced by the use of existing tankage for one or the other of these units. This is one of the special cases influenced by external factors. Which unit is sized first will depend on the particular situation.

The next major step in oxygen system design should normally be the determination of oxygen demand. This depends on the quantity and nature of the substrate and the design biomass loading rate. It should be noted that higher biomass loadings will reduce the overall oxygen demand but the concurrent penalty of increased sludge production must be paid.

Next, the dissolution and oxygen supply equipment is sized to meet the oxygen demand. Two key points must be remembered here. First, it was assumed that the solids settling would not be limiting in the chosen biomass loading range. This assumption was based on the presumption of high D.O. levels in the reactors. The mass transfer system must be sized to meet the oxygen demand while maintaining the high D.O. Secondly, it was assumed that the volumetric oxygen demand resulting from the chosen biomass loading would not be limiting. Once the mixer power requirements are determined, care must be taken to check all configurational restrictions and requirements to assure that the selected units will operate as predicted in the tanks as they have been sized.

The final step in the process design is to determine the sludge production rate for the process design point already established. The solids concentration to be wasted is equal to the clarifier underflow concentration previously determined, and this together with the required poundage to waste determines the system volumetric sludge wasting.

V. PROCESS DESIGN

Previous sections of this chapter have discussed the groundwork for process design: the gathering of design information and the formulation of an approach to process design. This section presents the detailed process considerations which must go into design of a secondary oxygen-activated sludge wastewater treatment system. The order of presentation of the design steps is consistent with the overall design approach just discussed.

A. Clarifier Design

Design of a secondary clarifier system requires consideration of many factors and parameters. These will be discussed individually to provide a basis for the recommended design procedure.

The type of clarifier to be utilized is of particular significance because the process performance (thickening and clarification) of all types and configurations is not equivalent.

Although there are many brands of clarifiers, there are only a few general categories as far as process performance is concerned. The major difference is between rectangular and circular clarifier configurations. The rectangular units have a longitudinal flow pattern with the influent at one end and the effluent at the other. The sludge collection and pickup mechanism generally consists of a chain flight scraper system which moves the sludge along the bottom to a collection trough. Most circular units have a radial flow pattern with the influent in the center and the effluent at the periphery. Another circular configuration increasing in popularity utilizes peripheral feed and effluent withdrawal. Sludge collection can be either by a scraper mechanism which moves the sludge to a collection and withdrawal well or by a suction system which withdraws the sludge directly from the clarifier bottom sludge blanket. In the former case, the bottom of the unit is generally sloped toward the center, while the latter type will have a flat bottom.

The primary performance differences between rectangular and circular units are in clarification and thickening characteristics. The circular unit typically provides better thickening than the rectangular unit, while the rectangular clarifier has an improved capacity for clarification or reduction of effluent suspended solids. The thickening advantage of the circular unit is quite significant, while, for most recommended design ranges, the disadvantage in clarification is practically negligible. The circular clarifier is generally recommended for activated sludge systems because of its better thickening characteristics. For sludge collection, a scraper system is recommended for units up to 50 to 60 ft in diameter. For larger circular clarifiers, the suction collection system or rapid sludge return is recommended. The suction system will work satisfactorily in the smaller clarifiers but requires more maintenance than the scraper system. The scraper pickup mechanisms are difficult to use in clarifiers larger than 60 ft because the rate of transport of solids to the collection well becomes limiting for such large diameters and results in reduction of the clarifier underflow solids concentration. The models discussed in the remaining portion of this section have been shown to apply well to circular clarifiers which have a uniform solids flux over the bottom area, but they do not apply for the prediction of the performance of rectangular units which do not have a uniform solids flux.

Four key parameters in determination of the process performance of a clarifier are the sludge settling characteristics, the surface overflow rate, the recycle rate, and the clarifier underflow concentration. It is shown in Chapter 7, Volume I that these four parameters are related by the following equation:[31]

where

$$RSS = \left[\frac{K}{(OR)(R/Q)} \right]^{1/n} \tag{2}$$

$$K = \frac{n}{(n-1)} [180a \, (n-1)]^{1/n} 10^6 \tag{3}$$

where OR is expressed in gallons per day per square foot.

The two constants a and n are determined from sludge settling characteristics according to the settling velocity equation

$$v_i = ac_i^{-n} \tag{4}$$

as discussed in Chapter 7, Volume I. The values of these constants vary widely with the character of the wastewater. Some typical values have been established for municipal wastes, but the settling characteristics of each of the wide variety of industrial wastes must be evaluated individually. The a and n constants must be experimentally determined for those cases where no previous experience is available. The methods used for this determination are discussed in Chapter 7, Volume I, In general, municipal wastes can be subdivided into those which have primary clarification and those which do not (i.e., raw degritted or comminuted wastewaters). The primary clarified municipal wastewater tends to settle and compact somewhat more poorly than raw waste because the primary clarifier selectively removes dense inert particulate material which enhances biological sludge settling. Some typical values for the a and n constants for municipal wastes are as follows:

	a (ft/hr)	n (dimensionless)
primary effluent	$3.5 * 10^{-5}$	2.26
raw waste	$4.0 * 10^{-4}$	1.87

Although the n values have been found to be fairly constant at the indicated values, the given values for a are averages of values which vary over a considerable range.

The clarifier surface overflow rate, designated (OR) in the above equation, is inversely dependent on the surface area of the clarifier.

$$OR = \frac{Q\ 10^6}{A_x} \qquad (5)$$

The overflow rate is also referred to as the rise rate and can be expressed as a rise velocity in ft/hr. If this value is too high relative to the settling characteristics of the sludge, large quantities of solids can be lost over the effluent weir. To prevent this possibility, it is recommended that the sustained peak overflow rate not be allowed to exceed 2000 GPD/ft² for typical oxygenation system sludges.

The design average overflow rate should normally be between 500 to 1000 GPD/ft² for typical oxygenation system sludges. This corresponds to meeting the maximum (OR) requirement specified above for a sustained peak to average flow rate ratios of 2.5 to 1.0. If this ratio is higher, the sustained peak (OR) limitation should be used as the design criteria; if lower, the average (OR) range should be used for design. In general, for wastes producing poorly settling sludges, the lower end of the design ranges should be applied.

Equation 2 indicates that the product of (OR) and R/Q determines the clarifier underflow (RSS) concentration for a particular wastewater. This product, however, is nothing more than the recycle rate per unit clarifier area, which means that for a particular waste, the RSS is dependent on only A_x, the clarifier area, and R, the recycle flow. Thus, for a particular clarifier, the underflow concentration will not change if the recycle flow does not change. However, if the flow changes in a particular system and the recycle fraction (R/Q) rather than R is kept constant, the underflow concentration (and therefore the reactor MLSS concentration) will vary. The details of this interrelationship and the effect of R/Q on mixed liquor concentrations are discussed in Chapter 8, Volume I.

Designing a system with a higher R/Q when all other parameters have been defined has several effects. The clarifier underflow concentration is decreased, while the reactor solids concentration increases, although the rate of increase in reactor solids concentration decreases as the (R/Q) is increased. Above an R/Q of 50 to 60%, very little increase in MLSS can be gained by increasing R/Q. Other effects of increasing R/Q include lower actual reactor retention time for a given influent flow and decrease in the sludge blanket solids retention time for a constant overflow rate. It should also be realized that decreasing the underflow concentration at the design point decreases the concentration of the sludge wasting stream.

Determination of the actual design point recycle flow is a balance between desired reactor solids concentration and retention time, waste solids concentration, clarifier size, and recycle pumping costs. The recommended design point range for the recycle ratio in an oxygenation system is between 25 and 75% of the design flow. It is further recommended that pumping flexibility be installed to supply recycle capacities up to 50% of the sustained peak flow. This provides the ability to adjust the recycle rate and to vary the reactor solids concentration at design flow if this becomes desirable.

It is necessary to apply certain limitations to the underflow concentrations predicted by Equation 2. This is because the settling velocity correlation used to develop the model does not apply over the full range of concentrations.[31] At very low concentrations, the settling velocity tends to approach a constant value independent of the concentration. This is probably due to transition from the hindered zone settling regime to the discrete particle settling regime at low concentrations. At very high concentrations, on the other hand, the particles become so close together that a maximum concentration value is reached and the settling velocity drops to essentially zero. At very low OR values and/or R/Q values, the model will sometimes predict extremely high RSS values. This is not realistic and is probably due to the limitations of the model just discussed. Available data substantiate the applicability of the model to predict underflow concentrations of from 3000 to 40,000 mg/l. Use of predicted values outside this range is not recommended.

The parameters discussed are key in sizing of the clarifier to produce a desired underflow concentration consistent with the design of the rest of the reactor clarifier system. There are some additional clarifier operational parameters which should be considered to assure proper performance of the unit. These include sludge blanket level variations, clarifier water depth, sludge blanket solids retention time, and weir loadings.

For a particular system design, the sludge blanket level in the clarifier will vary as a function of variations in flow (overflow rate), variation in recycle rate, and variation in the sludge wasting rate when the system is not at steady state. Steady state requires that the sludge wasting rate from the system be whatever is required to result in no net accumulation or depletion of solids in the reactor clarifier system. The wasting rate which will be required to maintain this state condition depends on the nature and quantity of the influent waste as well as the process operating conditions. The sum of the net sludge produced and the influent inert and nondegraded solids must be equalled by the rate of wasting in order for "steady state" to be maintained. If the wasting rate is too low for a particular set of operating conditions, the sludge blanket level will rise due to accumulation of solids in the system. If the sludge wasting rate is too high, the sludge blanket will drop due to depletion of solids. Further implications of this operational problem are discussed in Chapter 8, Volume I. For process design, it must be assumed that the sludge wasting rate for the system is so adjusted, on a time average basis, to result in "steady state" operation.

Even at steady state, variations in flow or recycle ratio can also cause changes in the sludge blanket level by causing a redistribution of the steady state quantity of solids in the system between the clarifier and the reactor. The details of this phenomenon are also discussed in Chapter 8, Volume I. In this chapter, a method of estimating the fluctuation of the blanket level, at steady state, as a function of flow and recycle is developed.

For the design average condition, the normal operating clarifier blanket depth should be between 2 and 3 ft. From that average level, the blanket level should not be allowed to drop below 0.75 ft during minimum flows or channeling may occur in the sludge blanket, resulting in low recycle concentrations. During peak flow periods, the blanket should not be permitted to rise higher than 2 ft below the liquid surface or large quantities of solids may be lost over the effluent weir. The minimum and maximum blanket levels are determined by applying the previously mentioned relationship at the appropriate flow and recycle conditions. Adjustments in the recommended clarifier liquid depth or, less desirably, in the average blanket level can be made if necessary to accommodate an expected blanket level variation range.

Another important parameter relating to the sludge blanket is the average solids retention time in the clarifier. This parameter is determined by dividing the mass of solids in the clarifier blanket at average conditions by the average rate of withdrawal of solids from the clarifier. In general, this solids retention time should not be allowed to exceed 3 to 4 hr, or anoxic activity may develop in the sludge blanket, resulting in poor clarifier performance. For most typical oxygenation system designs, the previously recommended blanket depth of 2 to 3 ft falls within the recommended solids retention time limits.

A final point is that the design parameters and limits given here apply to circular clarifier designs and assume that the hydraulics and sludge handling mechanisms of the clarifier have been properly designed.

One key factor in influencing the hydraulics in the clarifier is the weir loading and weir placement. Recommended weir loadings vary widely for different clarifier designs and applications, but, in general, they must be low enough to prevent short-circuiting of the flows through the unit.

Based on this discussion of the process and operating characteristics of secondary clarifiers for oxygenation systems, a procedure for their design is presented. For clarity, the procedure has been divided into the following series of steps:

1. From data and/or experience, establish the settling constants a and n and calculate the constant K of Equation 3.
2. Establish the maximum overflow rate corresponding to the maximum sustained peak flow within the recommended range.
3. Based on the ratio between the maximum sustained peak flow and the design average flow along with recommendations previously discussed, establish the design average overflow rate.
4. Select a recycle flow for the design average case and operational limits for design of the recycle system considering the previous discussion.
5. From the equations presented here and developed in Chapter 7, Volume I, calculate the clarifier underflow concentration at both the design average and peak conditions. Check to assure that this calculated value is consistent within the

recommended ranges. This calculation and procedure assumes the clarifier is a circular unit as described.

6. Calculate the variation in clarifier sludge blanket level from the average level (selected) to maximum and minimum and make proper adjustments in the clarifier design or operating point to assure that the blanket level variation will be within prescribed limits.

7. Check the average clarifier sludge blanket solids retention time to assure that it falls below the allowable maximum.

8. Be sure that the expected clarifier weir loading will be within the range consistent with good hydraulic design.

9. Assuming the clarifier is properly designed mechanically and hydraulically, determine the effluent suspended solids level based on experience and data on the wastewater in question. A typical level for well-designed clarifiers operating in oxygenation systems and on municipal wastewater is 10 to 25 mg/l.

B. Oxygenation System Reactors

As was discussed in the section on design approach, when the clarifiers have been sized and their performance evaluated at both the average and design maximum conditions, the next design step should typically be the sizing of the reactor basins. The method of sizing the reactor basins will depend on whether the system is expected to be kinetics limited, oxygen mass transfer limited, or settling limited with respect to effluent quality. That is, the required tank volume could be controlled by the time-concentration product necessary to effect removal, the tank size necessary to accommodate the required size of mass transfer devices, or the range of operational biomass loadings necessary to assure good settling characteristics of the biomass. The question of which of the above will finally determine the design must be answered through experimental data or previous experience on the wastewater to be treated.

The typical design situation for oxygenation systems operating on municipal wastes is that either mass transfer or settling rather than kinetics will be limiting and will determine the tank size. As was previously stated, a system can be said to be nonkinetics limited if the effluent substrate concentration does not vary over a wide range of operating time concentration products. It has also

been explained why this is typically the case for municipal wastewaters in Section IV of this chapter. Equation 1 relates the time-concentration product to the biomass loading of (F/M). From this equation, it can be seen that changing the time-concentration product at a constant applied substrate concentration is equivalent to changing the F/M ratio. Figure 1 is a plot of the substrate removal efficiency vs. the operating F/M for data taken from several municipal oxygenation system pilot plant and full-scale programs. Each data point represents an extended period of operation at a constant loading and process condition. The range of F/M values shown is the result of broad variations in time-concentration product and applied BOD_5 levels ranging from 56 to 1010 mg/l, with an average of 265 mg/l. The effluent substrate level remains fairly constant over this broad range of F/M or time-concentration products, indicating that these municipal systems are not reaction kinetics limited over the range of F/M loadings observed.

If it is determined that the system will not be kinetics limited in normal design loading ranges, the design point F/M will be established by mass transfer and settling considerations. These considerations do not usually determine a particular value for design but rather a range of allowable design values. The final design loading within this range can be selected based on economic or other considerations such as anticipated load fluctuations.

The design F/M or time-concentration product for the nonkinetics-limited case cannot be so high that the volumetric oxygen demands exceed the volumetric mass transfer capabilities of obtainable equipment, nor can they exceed the range where good settling characteristics can be expected. The loading at which settling begins to be affected must also be established from either data or experience. For municipal wastewater, undesirable effects on flocculation and possibly high effluent suspended solids have been noted in systems operating continuously at extremely high biomass loadings in excess of 2.75 lb substrate applied/lb biomass aerated/day. Mass transfer limitations are generally encountered before these biomass loading levels are reached. As was previously discussed, this limitation may manifest itself as the inability of the mass transfer system to maintain the dissolved oxygen levels necessary for good floc formation and settling as the required mass trans-

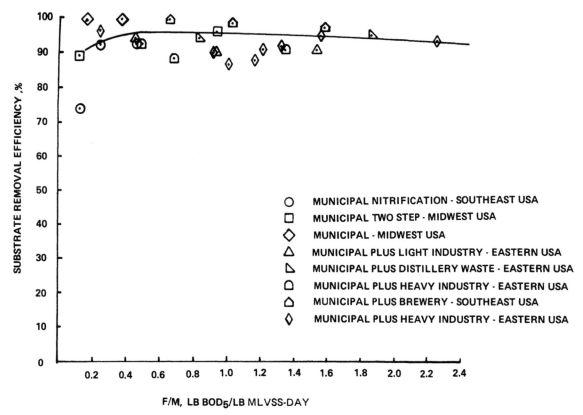

FIGURE 1. Biomass loading (F/M) vs. percent substrate removal. (Courtesy of Union Carbide Corporation.)

fer devices may become too large to be properly accommodated in the resulting aeration tank volumes.

Extensive design experience has shown that for municipal wastewaters, a good design point loading range is 0.5 to 0.9 lb BOD_5/lb MLVSS-day. For most design situations, this range results in good settling and volumetric oxygen demands which are readily attainable. The given range is for the design point range and includes adequate allowance for normal fluctuation in biomass loading on the system. This range, however, may not apply for many industrial wastes. The design range for such cases must be individually established from data on the particular wastewater.

It should be stressed at this point that there is a great deal of data available from oxygenation system pilot plant and full-scale plant experience, demonstrating that diurnal and even longer-term peak loadings far exceeding the above design point range can be readily handled in oxygenation systems. This is primarily due to the maintenance of D.O. levels. Figures 2 and 3 represent phase (2 weeks to 6 months) average operating data from both pilot plant and full-scale oxygenation activated sludge facilities.

At peak loadings, when the volumetric oxygen demand is also at its peak, the D.O. level in the system simply drops from 6 to 8 mg/l to perhaps 1 to 2 mg/l, depending on the magnitude of the peak. This allows the installed mass transfer system to meet the oxygen demands, while still maintaining the biomass in an aerobic condition. Such data also serve to verify the contention that most municipal systems are far from kinetics limited since even at the high loadings the time-concentration product normally does not need to be increased in the oxygenation system to maintain the effluent quality.

To size the reactor basin, for the nonkinetics-limited case, the design point F/M must first be selected. The range previously discussed can be used for municipal wastes. For industrial wastes, the allowable design range may be different and must be separately determined. If large fluctuations in biomass loading (more than 2.5 to 3.0 times the average) are expected, the design loading F/M should be selected toward the lower end of the allowable range.

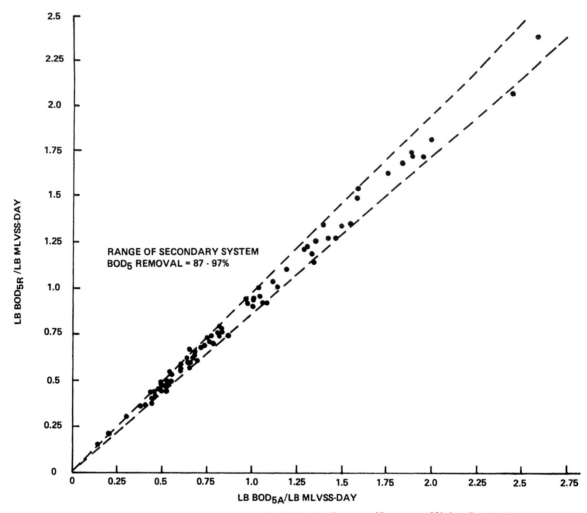

FIGURE 2. Specific BOD_5 removal rate vs. specific BOD_5 loading rate. (Courtesy of Union Carbide Corporation.)

Once the design F/M is selected, and thus the time-concentration product is established, the reactor mixed-liquor volatile suspended solids must be evaluated before the required tank volume can be established. The total suspended solids in the reactor can be related to the previously determined recycle solids flow (R) and concentration (RSS) through a mass balance around the reactor. Since production and destruction of solids within the reactor are negligible compared to those in the influent and recycle stream, this balance can be written as:

$$Q(SS_A) + R(RSS) = (Q + R)\ MLVSS$$

Since $Q * SS_A$ will generally be very small compared with $R * RSS$, it can usually be neglected. The balance can then be rearranged to give

$$MLSS = \left(\frac{R/Q}{1 + R/Q} \right) RSS \qquad (6)$$

from which the mixed-liquor suspended solids in the reactor can be determined. To obtain the mixed-liquor volatile suspended solids concentration, the MLSS must be multiplied by the estimated VSS/TSS ratio. Typical values of the VSS/TSS ratio in the reactor for municipal wastes are 0.65 to 0.8 for raw degritted wastewater and 0.70 to 0.85 for primary effluent wastewater. The reactor VSS/TSS for industrial applications can vary quite widely, and its estimated value must be based on experience with the particular waste in question. The actual VSS/TSS ratio at which the reactor will operate depends on the quantity of inert and nondegradable material entering the system and the biomass loading at which the system operates. The actual VSS/TSS ratio can be

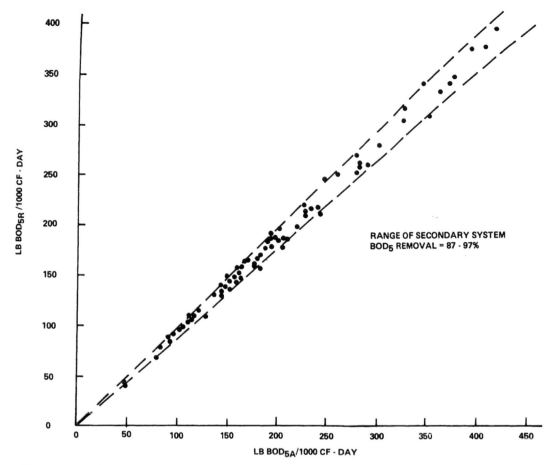

FIGURE 3. Volumetric BOD_5 removal rate vs. volumetric BOD_5 loading rate. (Courtesy of Union Carbide Corporation.)

calculated at a later stage of process design using volatile and total solids mass balances. This calculation is discussed in Section V.D. The calculated value should be compared against the estimated value used for design. Design experience will minimize the discrepancy between the chosen and calculated values and thus minimize the number of iterations necessary to make the two values equivalent.

When the MLVSS has been determined using the estimated VSS/TSS value, the reactor retention time can be determined using the following equation:

$$t_Q = \frac{24 \ BOD_5}{(MLVSS)F/M} \qquad (7)$$

and the basin volume is then

$$V = t_Q \left(\frac{Q}{24} \right) \qquad (8)$$

where V is in millions of gallons.

Having determined the tank volume, it is recommended that the expected biomass loading under maximum sustained loading conditions be checked. From the overflow rate and R/Q at maximum conditions, the RSS and MLSS can be determined assuming the same settling characteristics. Assume that the VSS/TSS will be the same as at the design point (this assumption should be checked later, as previously mentioned). Using the BOD_5 and retention time at the maximum sustained peak, the F/M can be calculated. Note that the maximum loading (lb BOD_5/day) point may not be the same as the maximum hydraulic peak (MGD) which was used in sizing the clarifier and evaluating its performance. For typical municipal designs, a general rule using reasonable conservatism is that the maximum sustained BOD$_5$ loading should not be allowed to exceed 2.8 lb BOD$_5$/lb MLVSS-day. The ability of the system to meet the mass transfer requirements at this peak condition should also be checked, as should the system kinetics. Municipal wastes in the loading ranges discussed will usually remain in a nonkinetics-limited condition at the peak loading.

In addition to the maximum biomass loading, the minimum retention time should be checked at the maximum sustained hydraulic peak flow. For normal municipal wastes, experience has shown that this value should not be below about 20 min. At this level, some deterioration in effluent quality may occur during the peak flow.

It is of interest at this point to discuss the method used for aeration tank sizing in a kinetics-limited situation. Recall that a system is determined to be kinetics limited by experimentally determining that the effluent soluble substrate concentration varies with changes in the time-concentration product or the biomass loading. The relationship between the time-concentration product and the effluent soluble substrate level must be determined experimentally. As shown in Section V.E, this relationship for a single-stage, completely mixed system can be modeled as follows:

$$S = \frac{S_A}{[1 + k(\psi\delta)(MLVSS)t_Q]} \qquad (9)$$

The value of the reaction rate coefficient (k) must be determined for a particular case from the data. The active fraction of the volatile solids in the reactor must be estimated and is dependent on the influent wastewater character and the biomass loading at which the system is operating. For typical municipal wastewaters and the design loadings recommended here, the value of $\psi\delta$ will generally be between 0.3 and 0.8 lb active organisms/lb volatile solids. The actual value of $(\psi\delta)$ at which the system will operate can be calculated for a particular system using the mass balance presented in Section V.D. It is recommended that $\psi\delta$ be calculated upon completion of the design work and back-checked against the value estimated during the design work.

Once k and $\psi\delta$ have been established for the particular wastewater stream, the value of the time-concentration product required to produce a particular effluent soluble substrate can be determined from Equation 9 for a single-stage system. Equation 57 (Section V.E) should be used for a staged system. F/M is then determined from Equation 1, and from this point, the same techniques as previously discussed for determining the aeration basin size in the nonkinetics-limited case can be applied. Since the design is kinetics limited, however, the effluent soluble substrate level will be higher at the peak sustained loading. Its value can be determined from k and $\psi\delta$ and the peak time-concentration product and applied substrate values.

The basin configuration used to provide the reactor volume just calculated depends on the oxygen requirements in the reactor, the type, size, and number of mass transfer devices employed, and the mixing criteria utilized. The basin configuration should not be determined until after the above aspects of the design have been established. Occasionally, it will be determined that no configuration of the established basin volume will accommodate the mass transfer devices required by the system. This is an example of a mass transfer limited design case. Under these circumstances, the design must be changed to increase the tank volume and decrease the volumetric oxygen demand to an acceptable level. This must, of course, be done without violating the numerous design limitations discussed previously.

There is another somewhat different approach to process design which has application in determination of the required aeration tank volume. This design technique uses the SRT or sludge retention time rather than the F/M as the key design parameter. The definition of the SRT is as follows:

$$SRT = \frac{\text{lb volatile solids under aeration}}{\text{lb volatile solids leaving the system/day}}$$

$$= \frac{V(MLVSS)}{W(RVSS) + (Q - W)VSS_E} \qquad (10)$$

The SRT can be related to oxygen consumption, sludge production, and substrate removal kinetics, as can the F/M. This parameter is particularly useful in growth or kinetics-limited applications as it is equal to the inverse of the rate of growth of the average microorganism in the reactor when the system is operating under steady state conditions. Its inverse can also be interpreted as the fraction of the organisms in the system which leave the system per unit of time.

It is shown in Section V.C of this chapter that the steady state wasting rate (W) depends not only on the quantity and nature of solids in the influent wastewater but also on the quantity of substrate (BOD_5) and on the biomass loading lb BOD_5/lb MLVSS/day in the reactor. The effluent suspended solids level is, as has been discussed, also dependent to some degree on the biomass loading as well as the design and performance characteristics of the clarifier. These two terms are both critical to the value of SRT determined by the above equation. However, of even more importance is the basic assumption used in the definition and derivation of SRT that is the requirement for steady state operation.

As has been discussed, the influent solids and the sludge generation rate determine the total quantity of solids which must be removed from the system in order to maintain steady state conditions. However, the system will operate, at least for a while, when the wasting plus effluent solids rates are not equal to the influent plus produced solids rates. The result will be an increase or depletion of the quantity of solids in the reactor clarifier system. In most cases, a change in the clarifier blanket level will occur. Thus, the SRT, if evaluated under nonsteady state conditions, using the above equation, will have no relation to the performance or operation of the process. At nonsteady state conditions, the SRT is meaningless as a design and operation parameter.

In order to meaningfully use the SRT in design of a secondary activated sludge system, additional design steps must be used to assure that the SRT used for design will be the steady state value at the design point. To clearly illustrate this point, the expression for reactor retention time (t_Q), which also determines reactor tank volume, will be derived below as a function of the SRT. In Section V.C of this chapter, an equation relating SRT to various solids balance parameters is discussed and derived very simply from a detailed steady state volatile solids material balance. That equation (Equation 18) is as follows:

$$\frac{1}{SRT} = \left[(1 - f_I + f_I f) \frac{VSS_A}{(S_A - S)} + Y\delta (1 - P) \right] q$$
$$- K_D \psi (\delta - \gamma) \qquad (18)$$

Recognizing that q, the specific substrate removal rate, is defined as the F/M times the substrate removal efficiency ($[S_A - S]/S_A$) and recalling the definition of the F/M from Equation 1, the above relationship can be rearranged to give the following equation:

$$t_Q = \frac{[(1 - f_I + f_I f)(VSS_A)24]SRT}{[1 + K_D \psi (\delta - \gamma)SRT](MLVSS)} +$$

$$\frac{[(1 - P)Y\delta(S_A - S)24]SRT}{[1 + K_D \psi (\delta - \gamma)SRT](MLVSS)} \qquad (11)$$

While the terms in this equation are discussed in detail in Section V.C, it is pertinent to this discussion to point out that the term $(1 - f_I + f_I f)$ VSS_A [24]) defines that fraction of the applied VSS which is not degraded in the system and thus contributes to the sludge production of the system. The parameters Y (sludge yield coefficient) and Kd (endogenous decay coefficient) are necessary to define the net biological sludge production of the system. The incorporation of these terms in this equation, which was derived from a steady state mass balance, serves to assure that the resultant reactor retention time will produce a solids wasting rate consistent with the design point SRT at steady state. The design point SRT, just as was the case for the F/M, must be determined based on experimental data and other design limitations (settling and oxygen transfer) regardless of whether the system is kinetics or nonkinetics limited.

It is interesting to note that using the F/M method of design previously described does not require the prior establishment of steady state operating conditions. Substrate removal, oxygen consumption, and sludge production are all related directly to the biomass loading regardless of whether the system is in a steady state. Clearly, in operation of the plant, "steady state" is very important, however, and can be attained by adjusting the wasting rate to maintain a constant bed level in the clarifier. This, then, is the only basic difference between the two design parameters. In order to be valid, SRT must be applied under steady state conditions, which requires the assumption of sludge production parameters before it can be applied for design. F/M does not have this restriction as, unlike SRT, it applies directly to the reactor performance regardless of whether the system is at steady state. The SRT of a system is dependent on the steady state sludge production rate, which is determined by the F/M of the system as well as influent characteristics and degree of substrate removal. That is, the SRT is a dependent variable in an activated sludge system, while the F/M is an independent variable dependent only on parameters external to the reactor system. Thus, F/M is a more direct and simple parameter for use in design of a secondary oxygen-activated sludge system.

C. Solids Balance

An estimate of the solids which will have to be wasted from an activated sludge system is an essential part of the process design. Not only is it essential for the sizing of subsequent dewatering and disposal facilities, but the technique utilized to estimate sludge production, a solids balance, can also be used to calculate the expected volatile fraction of the mixed-liquor solids given the system loading. This VSS/TSS ratio result is then checked against the value which had previously been assumed. In addition, the solids balance can also be used to estimate the activated sludge system viability, which is a necessary parameter in the estimation of the system's biokinetic behavior. Finally, a solids balance is utilized to determine the sludge retention time in the system. This parameter is directly related to the food-to-biomass ratio and is preferred by some for use in design of carbonaceous removal systems. It is also utilized in design of biological nitrification systems, where it is the controlling design parameter. The sludge retention time is also an important parameter to be aware of in system design because, as can easily be shown, the ratio of it to the hydraulic retention time is the factor by which the inert and settleable material concentration will increase in the aeration tanks over that in the influent wastewater. As an example, if a fibrous or filamentous material is present in the influent at a concentration of 25 mg/l and the system characteristics are such that with a hydraulic retention time of 2 hr the sludge retention time (SRT) is 6 days, the concentrating

factor is $\frac{(6)(24)}{2} = 72$, and the concentrations of the fibrous or filamentous material can build up to $(72)(25) = 1800$ mg/l in the aeration tankage. At this level it is quite likely that a significant deterioration in settling properties of the activated sludge would result. It should be pointed out that this 1800 mg/l is a steady state value which assumes that the sludge retention time stays at 6 days; that is, the effluent suspended solids do not appreciably increase. The fact that it is a steady state value is also important since it will take a period of time to reach it. In the example cited, more than 6 days would be required to reach the steady state condition.

1. Sludge Retention Time

The expression for the sludge residence time in terms of the system characteristics can be determined from a volatile suspended solids balance around the aeration tank and clarifier and from the definition of the SRT. The SRT is defined as the average time a volatile solid particle spends in the system under aeration and can be expressed by the mass of volatile solids under aeration in the mixed liquor divided by the mass of volatile solids leaving the system (including the waste and effluent solids) per day. It is, in the steady state, the inverse of the average net growth rate of organisms in the system. The solids balance must include the solids entering the system, those destroyed in the system, and those accumulating in the system. For the volatile solids balance, we can express these terms with Equation 12.

$$\frac{VSS_{acc}}{day} = \frac{VSS_{in}}{day} + \frac{VSS_{prod}}{day} - \frac{VSS_{des}}{day} - \frac{VSS_{out}}{day} \quad (12)$$

The entering volatile solids can readily be expressed by Equation 13, where the units must, of course, be consistent.

$$\frac{VSS_{in}}{day} = Q(VSS_A) \quad (13)$$

Both the wasting and effluent volatile solids contribute to the VSS leaving, which is given by Equation 14.

$$\frac{VSS_{out}}{day} = (Q - W) VSS_E + W(RVSS) \quad (14)$$

The mass of volatile solids destroyed within the system includes that portion of the entering volatile solids which are solubilized and subsequently degraded as well as those viable solids destroyed in providing energy for viable cell mobility, maintenance of osmotic balance, and regeneration of protein materials (referred to as endogenous respiration). If we let f_I represent that fraction of the influent volatile solids which are inactive, that is, are not viable biomass, and let f represent that fraction of the inactive influent volatile solids which are not degraded in the system (either because they are nondegradable or because they do not remain in the system long enough to be degraded), the expression for the portion of the destroyed VSS which results from the degradation of influent volatile solids is given by $Qf_I (1 - f) VSS_A$. If we now define a "pure" endogenous decay coefficient (K_D) as the mass of viable organisms lysed per day per mass of viable organisms in the system, a viability (ψ) as the mass of viable organisms per mass of mixed-liquor volatile solids, and recognize that organisms are not completely volatile such that we have a term (δ) representing the mass of VSS per mass of organism, the expression for the VSS destroyed by endogenous respiration per mass of mixed-liquor volatile solids is given by $\psi\delta K_D$. There is material remaining following cell destruction which may not be degraded and therefore results in a somewhat smaller net VSS reduction than would otherwise be the case.[33] This can be accounted for by defining a parameter, γ, as the mass of nondegraded VSS which remains per unit mass of organisms destroyed. Thus, the net mass of VSS destroyed by endogenous respiration per unit mass of mixed-liquor volatile suspended solids is $\psi K_D (\delta - \gamma)$. The resultant expression for the volatile solids destroyed per day is then given by Equation 15.

$$\frac{VSS_{des}}{day} = Qf_I(1 - f) VSS_A + V(MLVSS)\psi\delta K_D -$$

$$V(MLVSS)\psi\gamma K_D \tag{15}$$

In most carbonaceous removal systems, the fraction of the incoming volatile solids that is active biomass is extremely small, and f_I will therefore usually be essentially equal to unity. However, when we are considering the second step of a two-step activated sludge system or an activated

sludge system treating a trickle filter effluent, a large fraction (perhaps essentially all) of the influent VSS will be biomass, and f_I can approach zero for these cases.

The volatile solids produced within the system are those synthesized due to the metabolic activity of the organisms. Assuming a constant yield coefficient (Y) and defining it as the mass of organisms produced per mass of substrate material metabolized, Equation 16 represents the VSS produced per day.

$$\frac{VSS_{prod}}{day} = Q (Y\delta) (S_A - S) (1 - P) \tag{16}$$

The term δ is required to convert the units from mass of organisms produced to mass of VSS. The P term represents the fraction of substrate material that is removed physically. That is, some of the solid substrate entering the system will settle out in the clarifier and be removed with the waste stream before it can be solubilized and metabolized by the organisms. The term $(1 - P)$ is therefore the fraction of the substrate that is metabolized by the organisms in the production of new organisms.

We can now combine Equations 13, 14, 15, and 16 in the manner indicated by Equation 12, and if we assume steady state (the accumulation term is zero), Equation 17 describes the system volatile solids steady state material balance.

$$Q(VSS_A) + Q(Y\delta)(S_A - S)(1 - P) -$$

$$Qf_I(1 - f)VSS_A - V(MLVSS)K_D\psi(\delta - \gamma)$$

$$= (Q - W)VSS_E + W(RVSS) \tag{17}$$

If we divide both sides of Equation 17 by V * MLVSS (the mass of VSS under aeration), the right side of Equation 17 becomes by definition the reciprocal of the sludge retention time and Equation 17 can be written as below.

$$\frac{1}{SRT} = \left[(1 - f_I + f_I f) \frac{VSS_A}{(S_A - S)} + Y\delta (1 - P) \right] q -$$

$$K_D \psi (\delta - \gamma) \tag{18}$$

The above equation recognizes that the hydraulic detention time based on influent flow (t_Q) is given by the ratio V/Q and that the system's substrate removal rate (food-to-biomass ratio times removal fraction — given by q) is $(S_A - S)$ divided by t_Q * MLVSS. Equation 18 simplifies somewhat when f_I = 1.0 (as is the case for most raw degritted or primary effluent wastewaters) in that $(1 - f_I + f_I f)$ becomes f.

It is interesting to compare Equation 18 with the classical equation which assumed constant yield and endogenous decay coefficients but did not tie them to the organism, nor did it consider nondegraded material in the system. This classical equation is given as Equation 19 below.

$$\frac{1}{SRT} = a'q - b \tag{19}$$

From Equation 19, it is seen that a plot of 1/SRT vs. q should give a straight line with a slope equal to a', the "yield" coefficient, and an intercept of b, the "endogenous decay" coefficient. This plot has been used by many investigators to determine yield and decay coefficients for a variety of wastewaters. Reference to Equation 18, however, reveals that a plot of 1/SRT vs. q will only yield a straight line if the fraction of nondegraded influent solids remains the same, the ratio of the influent volatile solids to substrate removed remains unchanged, the amount of physical removal is constant, and the system viability also does not change. This would be expected to occur only over a rather narrow range of system loadings.

2. Unit Sludge Wasting

If we add an inert solids balance to the volatile solids balance given by Equation 17, it is possible to derive an expression for the solids which will have to be wasted from the system each day per mass of substrate removal, assuming that steady state (no accumulation, positive or negative) persists. In the general case there are no net nonvolatile solids produced or destroyed, and the only terms requiring consideration are the non-

volatile solids entering and leaving the system. These terms are given simply by $Q (SS - VSS)_A$ and $(Q - W)(SS - VSS)_E + W (RSS - RVSS)$, respectively. Adding these to the volatile solids balance, dividing by the mass of substrate removed daily, and recognizing that RSS is much greater than SS_E yields Equation 20.

$$\frac{SS_{wasted}}{Substrate\ removed} = \frac{(SS - VSS)_A}{\Delta S} + \frac{(1 - f_I + f_I f)VSS_A}{\Delta S} +$$

$$Y\delta \quad (1-P) - \frac{K_D \psi (\delta - \gamma)}{q} - \frac{SS_E}{\Delta S} \tag{20}$$

In some cases, usually with industrial wastes, the influent solids level will be very low. When this is the case, it may be necessary to account for the "production" and "destruction" of nonvolatile solids in the system. If we assume that the nonvolatile fraction of the influent SS is in equilibrium with some inorganic dissolved solids, the nonvolatile portion of the organism can be envisioned as being formed by drawing on the dissolved solids. This results in dissolution of some of the nonvolatile suspended solids to maintain the saturation equilibrium conditions. In essence, then, nonvolatile suspended solids are produced (formation of inorganic portion of the cell) and they are destroyed (by dissolution) in equal quantities, with the net effect being zero. This zero net effect is inherent in the derivation of Equation 20. If the nonvolatile portion of the influent suspended solids is low enough such that the production of organisms uses up more nonvolatile solids than are in the influent, the solution will not remain saturated and there will be a depletion of dissolved solids. When this happens, there is a net production of nonvolatile suspended solids in the system. The total production of the nonvolatile portion of the organisms can be expressed as $Y(1 - \delta)$ and the net production as $Y(1 - \delta) - (SS - VSS)_A/\Delta S$. Thus, in its more general form, Equation 21 is written below.

$$\frac{SS_{wasted}}{Substrate\ removed} = \frac{(SS - VSS)_A}{\Delta S} \text{ or } Y(1 - \delta)(1 - P) + \frac{(1 - f_I + f_I f)VSS_A}{\Delta S} + Y\delta (1-P) - \frac{K_D \psi (\delta - \gamma)}{q} - \frac{SS_E}{\Delta S} \tag{21}$$

whichever is larger

3. Mixed-liquor Volatile Suspended Solids Fraction

From the previous discussion and derivation of Equations 17 and 21, we can write the volatile and total solids balances around the system as follows (as long as $Q \gg W$, as is nearly always the case):

$$Q(VSS_E) + W(RVSS) = Q(VSS_A)(1 - f_I + f_I f) + Q(Y\delta)(1 - P)(S_A - S) - K_D\psi(\delta - \gamma)V(MLVSS) \tag{22a}$$

$$Q(SS_E) + W(RSS) = Q(VSS_A)(1 - f_I + f_I f) + Q(SS - VSS)_A + Q(Y\delta)(1 - P)(S_A - S) - K_D\psi(\delta - \gamma)V(MLVSS) \tag{22b}$$

Equation 22b applies in the usual case where the inert solids in the influent exceed those resulting from the formation of organisms. If this is not the case and $Y(1 - \delta)$ is greater than $\frac{(SS - VSS)}{S}_A$, then Equation 22c applies.

$$Q(SS_E) + W(RSS) = Q(VSS_A)(1 - f_I + f_I f) + Q(S_A - S)Y(1 - P) + Q(Y\delta)(S_A - S)(1 - P) - K_D\psi(\delta - \gamma)V(MLVSS) \tag{22c}$$

Also, the ratio of these masses must equal the volatile solids fraction in the mixed-liquor. This can be expressed by Equation 23.

$$\frac{VSS}{TSS} = \frac{Q(VSS_E) + W(RVSS)}{Q(SS_E) + W(RSS)} \tag{23}$$

Substituting Equations 22a and 22b into Equation 23 and reinstalling the definition of SRT, an expression for the VSS/TSS ratio is obtained.

$$\frac{VSS}{TSS} = \frac{1}{1 + \dfrac{(SS - VSS)_A \text{ or } (S_A - S)Y(1 - \delta)(1 - P)}{(MLVSS) t_Q/SRT}} \tag{24}$$

4. System Viability

The viable fraction of the mixed liquor can be represented by the term $\psi\delta$, which is the fraction of the volatile suspended solids in the mixed liquor that is viable organisms. A material balance on these viable organisms can be performed to give an expression for system viability. Viable organisms entering the system are given by $Q * (1 - f_I) * VSS_A$, those generated by $Q * Y\delta * (S_A - S) * (1 - P)$, and those destroyed by $V * K_D * \psi\delta * MLVSS$. Since, as previously discussed, the reciprocal of the SRT can be interpreted as the fraction of the solids present under aeration which leave the system daily, the viable organisms leaving the system is given by $V * \psi\delta * MLVSS/SRT$. Combining these terms gives Equation 25.

$$Q(1 - f_I)VSS_A + Q(Y\delta)(1 - P)(S_A - S) = \frac{V(\psi\delta)(MLVSS)}{SRT} + V(MLVSS) K_D\psi\delta \tag{25}$$

Recognizing that $t_Q = V/Q$ and solving for $\psi\delta$ gives

$$\psi\delta = \frac{Y\delta(S_A - S)(1 - P) + (1 - f_I)VSS_A}{t_Q(MLVSS)\left(\dfrac{1}{SRT} + K_D\right)} \tag{26}$$

D. Oxygen Demand Determinations

The ability to accurately determine the total oxygen requirements for the activated sludge system is of paramount importance. Sufficient oxygen must be supplied such that the system does not become oxygen starved and suffer a resultant loss in treatment efficiency. In addition, mass transfer equipment must be selected such that it possesses the capability to dissolve the required oxygen and maintain adequate dissolved oxygen levels at design and peak conditions. Finally, the knowledge of total oxygen demand requirement allows selection and design of the oxygen supply system.

Activated sludge system oxygen demand may be separated into four basic areas as follows:

1. Instantaneous oxygen demand
2. Elevation of dissolved oxygen level
3. Biomass synthesis
4. Biomass endogenous respiration

1. Direct Demands

Instantaneous oxygen demand (IOD) is a result of the direct oxidation of chemical constituents in the wastewater when exposed to an oxygen-rich environment. The oxygen demand is determined by the stoichiometric requirements for oxidation of the substance involved. Hydrogen sulfide represents one of the more common IOD substances, and from Equation 27 it is calculated that 2 lb mol of oxygen are required per lb mol of hydrogen sulfide oxidized. The impact of IOD on total system oxygen demand is usually small.

$$H_2S + 2O_2 \rightleftharpoons SO_4^= + 2H^+ \qquad (27)$$

The dissolved oxygen level of influent wastewaters is generally low, if not zero, unless some form of preaeration takes place in preceding unit operations such as an aerated grit chamber. The oxygen required to elevate the wastewater dissolved oxygen level to the activated sludge system design D.O. level must be included in the determination of the total oxygen demand. Incremental oxygen for D.O. elevation is low, amounting to only 8.34 lb of oxygen per million gallons per day per part per million D.O. elevation. Although small, this can, for example, amount to one ton of O_2 per day for a D.O. elevation of 6 mg/l for a 40-MGD treatment plant.

2. Biological Demand

The major fraction of the system oxygen demand is due to biological synthesis and endogenous respiration.

Experimental data are usually plotted as lb O_2 consumed/(lb MLVSS) (day) vs. lb BOD_R/(lb MLVSS) (day). The straight-line relationship is described by the following equation:

$$\frac{lb\ O_2}{(lb\ MLVSS)\ (day)} = A \frac{lb\ BOD_R}{(lb\ MLVSS)\ (day)} + B \qquad (28)$$

This slope, A, lb O_2/lb BOD_R, is related to the rate of oxygen utilization for synthesis or the amount of oxygen required for a microorganism to convert organic material into new cells. New cell mass is created through enzymatic breakdown (hydrolysis) of substrate to the point where the products can be assimilated by the microorganisms and synthesized. This hydrolysis and synthesis process requires energy. The source of this energy is the oxidation of a fraction of the organically degradable substrate to carbon dioxide and water. This oxidized fraction is no longer available for cell material. The more difficult a substance is to degrade, that is, to hydrolyze and assimilate, the more energy is required per pound of new cell mass formed and the larger the fraction of the substrate used for energy generation. This results in high oxygen consumption and low sludge yields for difficult-to-degrade organics such as petrochemical wastes and low oxygen consumption and high sludge yields for readily degradable organics such as food processing waste. The slope, A, will therefore increase with decreasing substrate degradability and decrease with increasing substrate degradability.

In actual practice, some COD and BOD_5 removal occurs physically, that is, without biological activity and the concomitant consumption of oxygen. The effects which account for this physical removal are the removal of insoluble and nondegraded substrate (colloidal materials) with the waste sludge and the adsorption or absorption of soluble substrate with subsequent wastage before oxidation. This phenomenon is most prevalent in high rate systems since the biological activity is much more time dependent than is the physical removal, which occurs by adsorption on and enmeshment in the floc particles.

The rate of oxygen or energy consumption per unit of MLVSS related to the requirements other than synthesis is determined from the intercept B of the experimental data plot. These requirements (for mobility, maintenance of osmotic balance, and regeneration of protein materials) are collectively termed endogenous respiration.[32] If the active biomass in a system is proportional to the MLVSS, as is typically assumed, B will be essentially constant for a particular system loading range and be dependent mainly on temperature. As with the A coefficient, the value of B obtained from experimental data is usually only good for that particular wastewater stream from which the data were collected. For the same reason as mentioned for the slope and intercept of the 1/SRT vs. q plot in the previous section, the values of A and B would only be expected to be valid over a narrow range of system loadings.

Figure 4 depicts the typical variation of oxygen consumption per unit of BOD_5 removal as a function of biomass loading for different types of wastewaters. It indicates that oxygen consumption characteristics can differ significantly from one wastewater to another. At high food-to-microorganism ratios (F/M), the oxygen requirements per unit of BOD_5 removal are governed

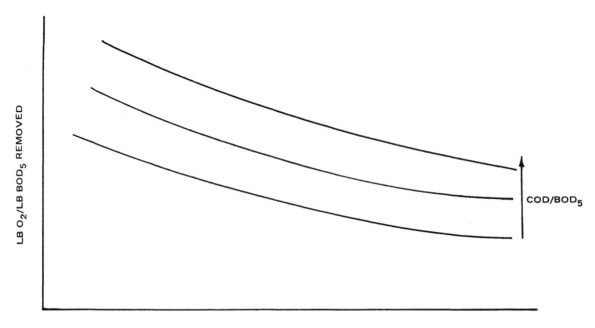

FIGURE 4. Variation of oxygen consumption per unit substrate removal as a function of applied biomass loading and COD/BOD$_s$ ratio. (Courtesy of Union Carbide Corporation.)

mainly by the oxygen required for cell synthesis since the degree of endogenous respiration is relatively low. At low food-to-microorganism ratios, the degree of autooxidation increases relative to synthesis activity and the total oxygen requirements per unit of BOD$_5$ removal increase. The variation of oxygen requirement at any F/M in Figure 4 is due to the variation of the coefficient A with substrate degradability. More difficult-to-degrade wastewaters have high ultimate biochemical oxygen demand to 5-day biochemical oxygen demand (BOD$_5$) ratios. Since COD is generally proportional to the ultimate demand, an increase in the COD/BOD$_5$ ratio results in an increase in the oxygen consumption per unit of BOD$_5$ removal.

In other words, the COD test measures the total oxygen demand of the wastewater substrate. This test includes biodegradable oxygen demand as well as nonbiodegradable or refractory substances. The BOD$_5$ test measures some fraction of the biodegradability relative to the COD. In an activated sludge process, the amount of substrate removed biologically generally will exceed the indicated BOD$_5$. This incremental removal in excess of the measured influent BOD$_5$ represents biodegradable COD not entirely detected in the standard BOD$_5$ test simply because it does not show up in 5 days

in the BOD$_5$ bottle. If the ultimate BOD were measured, it would be identical to the biodegradable COD.

Thus, the COD/BOD$_5$ ratio is indicative of the relative (biological) toughness and total organic strength of the waste as well as being a measure of the system's oxygen requirements.

3. Direct Data Correlation

There are various methods available to experimentally determine oxygen consumption and consequently evaluate the constants A and B. Experimental data may be collected from treatability studies, pilot plant programs, or full-scale installations. Oxygen consumption may then be determined from oxygen uptake rate measurements, gas phase oxygen material balances for closed systems, and through COD balances.

One of the most widely employed techniques is the oxygen uptake rate determination, which measures the decrease in dissolved oxygen level as a function of time in the mixed-liquor sample. Reliable equipment and consistent experimental techniques are extremely important if worthwhile results are desired. This test does not usually represent wholly accurate information for design purposes due to measurement device lag times, variances between aeration tankage and sample-contained environmental conditions (e.g., energy), problems associated with collecting representative and sufficient numbers of samples, the inherent instantaneous nature of the measurement, etc.

Gas space oxygen material balances are easily utilized on closed aeration tank systems. A complete oxygen material balance is made in order to determine the mass of oxygen consumed per unit time. This material balance relies on the measurement of influent and effluent gas phase flow rates and their corresponding oxygen purities. This method is usually quite accurate provided the system is relatively free of gas leaks and especially when employed at full-scale installations. Pilot units which contain inordinate amounts of gas phase experimental plumbing at times experience difficulty since very small leaks create rather large measurement errors when dealing with small gas feed quantities.

A method gaining wide acceptance is the chemical oxygen demand balance, which is a stoichiometric balance in which all components are expressed in terms of oxygen. By definition, COD is a measure of chemical oxygen equivalents. Therefore,

$$\text{lb } O_2 \text{ consumed} = \text{lb COD oxidized} = \text{lb COD}_{inf}$$
$$- \text{lb COD}_{eff} - \text{lb COD}_{wasted} \pm \text{COD}_{accumulated}$$
$$(29)$$

or

$$\text{lb } O_2 \text{ consumed} = \Delta \text{ COD} - \Delta \text{ COD}_{biological solids}$$
$$(30)$$

where Δ COD is defined as the total COD of the influent minus the total COD of the effluent and Δ COD (biological solids) is the net change in biological solids expressed in terms of oxygen equivalents. The chemical oxygen demand of the solids may be either directly measured through COD analysis or estimated utilizing an empirical formula for the oxidation of cells to CO_2 and H_2O.

$$C_5H_7NO_2 + 5O_2 \rightarrow 5CO_2 + 2H_2O + NH_3 \quad (31)$$

$$(113) \frac{mg}{mg \, mol} \quad (160) \frac{mg}{mg \, mol}$$

The oxygen equivalency of the cells is then calculated through ratioing the weights for cells and oxygen. Therefore, each mg/l of biological solids is equivalent to 160/113 = 1.42 mg/l of oxygen. This estimate ignores the COD material other than organisms in the solids and thus must be considered as only an approximation when applied to actual systems. The greatest difficulty involved in this test is accurately determining the quantities of waste-activated sludge and the changes in solids inventory due to changes in reactor MLVSS levels and changes in clarifier sludge inventory. The COD balance is not useful for instantaneous demand determinations but is extremely reliable for long-term, steady-state evaluations.

Once oxygen consumption has been determined along with BOD and COD removals and MLVSS levels, the coefficients A and B are readily determined as previously discussed.

4. Empirical Correlations

It is not always convenient to perform experimental work in order to determine oxygen requirements, nor are the waste streams always available if new systems or combined systems are under design. Therefore, it is necessary to possess the ability to estimate the A and B coefficients when only wastewater characteristics are known.

The amount of oxygen utilized for synthesis is that which would have been used if all the substrate was oxidized to carbon dioxide and water minus that which was not used because some of the substrate was converted to cell material and some was removed physically. The relationship between the yield coefficient, Y, and the synthesis oxygen consumption may be expressed by Equation 32a on a BOD_5 basis:

$$A = (1 - P)(\phi - Y\delta\epsilon) \quad (32a)$$

Since by definition chemical oxygen demand is a measure of oxygen equivalents, the value of ϕ depends on how many pounds of COD would be consumed when one pound of BOD_5 is oxidized to CO_2 and H_2O. Therefore, if the chemical oxygen demand and the ultimate biochemical oxygen demand are equal, then ϕ is equivalent to the COD/BOD_5 ratio. Typically, this is not the case; the value of ϕ is less than the COD/BOD_5

ratio and is related to the fraction of COD and BOD_5 removal

$$\phi = \left(\frac{COD}{BOD_s} \right) M \qquad (33)$$

where

$$M = \frac{\% \, COD_{removed}}{\% \, BOD_{s \, removed}}$$

When COD is the substrate, the lb of O_2/lb COD removal for synthesis, A_c is expressed as:

$$A_c = (1 - \delta Y_c \epsilon)(1 - P) \qquad (32b)$$

since, by definition, $\phi = 1$ when COD is the substrate.

The yield coefficient, Y, may be expressed as:

$$Y = Y_c \left(\frac{COD}{BOD_s} \right) M \qquad (34)$$

Through proper substitutions, it may be shown that the oxygen required for synthesis on a BOD_5 basis may be expressed in terms of the oxygen required for synthesis on a COD basis and the COD/BOD_5 ratio.

$$A = \left(\frac{COD}{BOD_s} \right) M(A_c) \qquad (35)$$

Extensive data analysis from numerous pilot and full-scale plants has confirmed the relationship between A, M, and the COD/BOD_5 ratio. Figure 5 depicts a qualitative variation of A with type of wastewater. The parameter M can be as low as 0.5 (e.g., pulp and paper mill wastes) and as high as 0.97 (e.g., low molecular weight acid wastes and brewery wastes). M is usually between 0.8 and 0.9 for municipal wastewaters.

The endogenous respiration coefficient, B, expressed as pounds of oxygen consumed for autooxidation per pounds of MLVSS under aeration per day, may be represented by

$$B = K_D \psi (\delta - \gamma) \epsilon' \qquad (36)$$

Since the fraction of "active mass," ψ, for typical oxygen system design F/M's is nearly a constant and the value of ϵ' is fixed, it is readily apparent that the value of B varies proportionate to changes in K_D, which itself varies with temperature.

Present correlations for oxygen consumption are based on the dependence of the consumption ratio on the substrate to microorganism ratio and the COD to BOD_5 ratio. Consumption coefficients may be determined either experimentally or through empirical correlations. Adequate oxygen supply is one of the primary keys to the successful operation of an activated sludge treatment facility, and adequate accounting of all oxygen demanding sources, including IOD and D.O. elevation, must be checked and accounted for by the designer.

E. Activated Sludge Kinetics
1. General Discussion

The activated sludge process is a microbiological enrichment culture[5] in which a rather extensive ecological system develops. The system is enormously complex due partially to the myriad of substrate materials in the influent sewage and the nature of these materials. Both the physical (soluble, suspended, colloidal) and chemical (fats, carbohydrates, proteins, inorganic salts, etc.) properties of the sewage are important. In addition, the variations in these properties are also important as the microbial populations existing in the treatment plant (and depending on the sewage constituents for their sustenance) require a period of acclimation or adaptation for optimum efficiency. This is accomplished either by producing the necessary enzymes (through activation of latent characteristics) or through a genetic change that may arise spontaneously (by stimulation from environmental factors) or by the selection of the species which can best utilize the substrate under the given environmental conditions. The other species are then outnumbered and ultimately outlived. In an activated sludge, the species capable of utilizing the substrate most efficiently and having the shortest generation time will tend to predominate. The selection of these species depends on the waste characteristics and the environment (pH, presence of nutrients or toxic materials, temperature, dissolved oxygen level, etc.).

In addition to the variation and inherent complexity of the nature of the wastewater and microbial populations, the physical nature of the microbes themselves must also be considered in any attempt to model the kinetics of an activated sludge system. The bacterial surface itself is

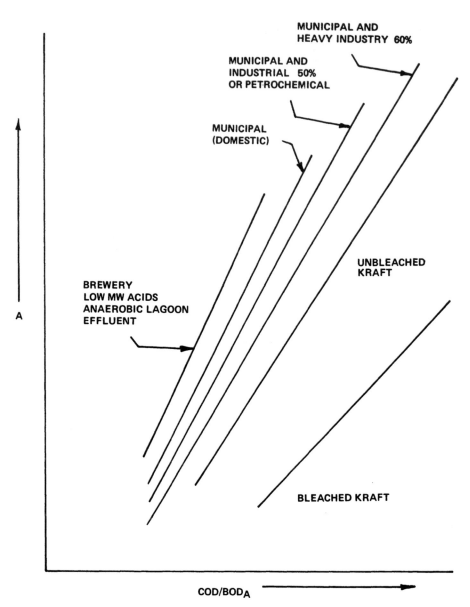

FIGURE 5. Variation of oxygen demand for synthesis with COD/BOD$_5$ ratio and waste type. (Courtesy of Union Carbide Corporation.)

composed of as many as four layers.[1] The outside layer is composed of an ion atmosphere which helps determine which particles or substances penetrate to inner layers. It is this layer that is thought to be important in determining the flocculation properties of the organisms. The settling properties of activated sludge are indeed improved by adding large concentrations of inorganic cations, an effect which may be ascribed to the reduction of the Zeta potential which appears to exist at the surface of most microorganisms. Inside the ion layer there is frequently a capsule layer secreted as a waste product of the cell under certain environmental circumstances. This layer is very viscous and thought to be ion permeable. It dissolves into the surrounding medium and may be quite important in the dynamics of activated sludge due to diffusional resistance which it presents. Within the capsule layer is a lipid material (essential to the cell) which controls the flow of materials through the cell. Only materials soluble in this layer gain admittance. Within the lipid layer lies a rigid membrane which gives structure and form to the cell. This layer is held responsible for controlling the flow of inorganic ions across the membrane.

The previous discussion of the complexities inherent in the description of an activated sludge system makes it apparent that for practical reasons, simplifications and approximations must be made in order to reduce the problem to one of a somewhat more treatable nature. In the activated sludge process there are several restrictive conditions[5] which tend to favor the establishment of a dynamically balanced flora, thus simplifying the problem to a slight extent. The first of these conditions is that the culture is usually being continously inoculated with microorganisms by the incoming waste. Secondly, even though short-term variations in the composition of waste streams are common, long-term variations are not frequently encountered. Finally, the successful operation of the activated sludge process depends on the retention of microorganisms within the system for periods longer than the hydraulic residence time. This is usually accomplished by the use of gravity separation in settling basins and recycle of the settled organisms to the aeration tank. Hence, those organisms having characteristics which make them amenable to sedimentation tend to predominate. The culture is therefore enriched because relatively rigid environmental conditions are maintained, and the desired organisms are selectively encouraged by the chosen technique of cell recycle.

Because recycle is such an important part of the process, it is necessary to have flocculated organisms which settle reasonably well for the efficient operation of the system. Indeed, the range of operation of an activated sludge system is generally determined by the settling properties of the biomass rather than by the most efficient and effective removal of BOD. Given the fact that a flocculated system is desirable rather than a completely dispersed system of cells, it is reasonable to suspect that transport phenomena may be as important as the actual biokinetics. Therefore, assuming an excess of nutrients and no inhibitory substances present, the following rate processes should be considered:

1. Diffusion of substrate (soluble or colloidal) and oxygen through the liquid boundary layer surrounding the floc particle
2. Diffusion through the floc particle to the organism surface
3. Adsorption on the surface of the organism
4. Secretion of appropriate enzymes by the organisms

5. Enzymatic hydrolysis of the substrate to a size molecule which can be absorbed
6. Absorption into the interior of the cell
7. Conversion to new cell mass and energy
8. Secretion from the cell of waste products
9. Desorption of waste products from the cell surfaces
10. Diffusion of the waste products through the floc particle to the surface
11. Diffusion of the waste products through the liquid boundary layer and into the bulk liquid

2. Rate Processes

The listing above reemphasizes the need to make several assumptions in order to develop a tractable model for the activated sludge process. An initial assumption is that steps 8 through 11 can be neglected. This assumption should be valid since the metabolic waste products must be reasonably small molecules to get through the cell membrane and will therefore have reasonably rapid diffusion rates. For the same reason, step 6 can likely be ignored. The remaining rate processes can be grouped into diffusional (1 and 2), adsorption (3), and cellular biokinetic reactions (4, 5, and 7).

Matson et al.[7] discuss the diffusional resistances of the liquid boundary layer and the floc particle itself to both oxygen and substrate material, either of which can cause a limitation in the overall rate of substrate removal. From this work, they conclude that the diffusion through the floc particle itself is more significant than that through the boundary layer. They also show that although the diffusional limitation of the system depends on the degradability of the substrate material, the concentration of viable organisms in the aeration tankage, and the energy input into the system, for overall practical ranges of these parameters in an activated sludge system, oxygen diffusion will not limit the rate of the biological activity if the dissolved oxygen level is maintained (at all times) above about 3 to 5 mg/l in the bulk liquor. Since it is common practice to have dissolved oxygen levels in excess of this limitation in an oxygen-activated sludge system, it can be concluded that oxygen diffusion will not limit the rate of substrate removal in oxygen-activated sludge systems. Matson et al. have also shown in this work that even though most substrate materials have lower diffusion coefficients than oxygen, more moles of oxygen are required at the reaction site so that the oxygen limitation controls over that of the substrate at lower concentrations of substrate than might intuitively be expected. For the many common compounds found in wastewater that Matson et al. examined, the substrate concentration became limiting only below about 5 to 15 mg/l COD. This concentration is lower than that usually found in activated sludge effluents and is at such a level that it can be assumed that the bulk of the substrate removal occurs under conditions where the diffusion of substrate also does not limit the rate.

The phenomenon of adsorbing substrate materials on the surface of activated sludge flocs is well known.[1-3] It has been used commercially in the contact stabilization process where the activated sludge adsorbs substrate in a small first stage and completes the degradation in another stage following a larger settling tank. The very fact that this process works is strong evidence that the rate of adsorption should not be the rate-controlling step, thereby leading to the conclusion that the rate is most likely controlled by substrate or oxygen diffusion, or by cellular biokinetics.

A great deal of the early pioneering work on the use of oxygen in activated sludge systems focused on the effect of high dissolved oxygen levels on the rate of substrate removal. For the most part, this work was carried out in bench scale or small pilot plant complete mix reactors treating domestic sewage. The studies were conducted at the normal food to biomass ratios usually encountered in conventional air activated sludge practice and operating at mixed-liquor suspended solids levels and residence times adequate to give reasonably high quality effluents. Under these conditions, the organic substrate is at relatively low bulk liquid concentration levels and the overall oxygen uptake rate is limited by organic substrate diffusion into the biomass particles. This is to say that although the oxygen may be completely penetrating the floc particles, the substrate is not. Although this does not alter the fact that the

higher D.O. systems could have coped with a higher F/M than a low D.O. system before becoming biokinetically controlled, it did lead the investigators to conclude that there was no effect of high D.O. on the biokinetic rate constant of the system. In fact, most recent pilot plant and field studies show that effluent substrate concentration is essentially the same for both oxygen- and air-activated sludge systems when treating domestic sewages at comparable system loadings; this is because of the fact that for most domestic wastewaters, the basic biokinetic oxidation rate is sufficiently high and the effluent organic substrate levels are sufficiently low that these systems operate in the organic substrate diffusion limiting regime. Here, these systems are not biokinetically controlled or oxygen diffusion controlled and effluent quality will be nearly the same for air and oxygen systems operated over a wide range of loadings as long as the system flocculates well and dispersed growth does not occur.

However, there are many types of industrial or combination municipal industrial wastewaters where materials that are difficult to degrade cause the system to be biokinetically controlled at loadings when the system flocculates well. The effluent quality under these circumstances will depend directly on the quantity of aerobically viable organisms in the system and, thus, directly in the degree of oxygen penetration into the biofloc particle. An oxygen system operates at higher dissolved oxygen levels, enabling more of the floc particle to be aerobically active; therefore, the oxygen activated sludge system would be expected to produce a better quality effluent at the same value of the time-concentration product.

3. Biokinetics

The reaction pathways occurring in the cell can be illustrated diagrammatically by a greatly simplified model as in the sketch below.[14]

$$
\begin{array}{c}
\text{Oxidant} \\
\downarrow \\
\text{Reactant} \dashrightarrow \begin{array}{c}\text{Complex} \\ \text{reaction} \\ \text{pathways}\end{array} \dashrightarrow \text{Intermediate product} \rightarrow CO_2 + H_2O \\
\nwarrow \\
\text{Microorganisms}
\end{array}
$$

Individual sequences in the reaction path are controlled by different enzymes, and if it is assumed that one stage is controlling, it should be possible to apply the Michaelis-Menton equation[16] to that stage.

$$\text{Rate of reaction} = \frac{US}{(K_m + S)} \qquad (37)$$

Monod[17] found in his study of pure culture kinetics that the specific growth rate of the microorganisms could be described by an equation of the same form as the Michaelis-Menton equation. The Monod equation, as it has come to be called, may be written:

$$\frac{1}{X}\frac{dX}{dt} = \frac{u_m S}{K_s + S} \qquad (38)$$

In the past it has found wide acceptance due to its similarity to the Michaelis-Menton equation, and thus its intuitive theoretical basis, and to its success in describing the growth of pure bacterial cultures at the expense of certain organic substrates when all other essential growth factors were in surplus. Recently, it has also been found[13] that the relationship between specific growth rates in heterogeneous populations and the substrate concentration is better represented by the Monod equation than by any of the others tested. It is thus considered reasonable to assume that the Monod equation can be used to describe the specific growth rate of cells as determined by some growth-limiting substrate which is immediately available to the organism (not limited by diffusional resistances) and when no inhibitory substrates are present.

In order to express the effectiveness of the organisms in utilizing the substrate (the major concern in the activated sludge process), it is necessary to relate the rate of cell growth to the rate of substrate consumption. This is usually accomplished through the use of the "cell yield" coefficient, Y. Monod[17] found mass yields to be constant for each organism-substrate combination examined but to vary among different organisms and to vary for the same organism grown on different substrates. Monod thus postulated that for a given organism-substrate combination, the yield is a constant. More recent studies[15] have shown that cell composition, and thus Y, is

variable and depends on the conditions of growth, but when the environmental conditions are controlled within reasonably narrow ranges (as in the activated sludge process), the yield of microorganisms does appear constant, and Monod's postulate seems justified. Expressed mathematically this postulate is

$$Y = \frac{dX}{dS} \qquad (39)$$

Therefore,

$$-\frac{dS}{dt} = \frac{1}{Y}\left(\frac{dX}{dt}\right)$$

and

$$\frac{dS}{dt} = \frac{u_M}{Y}\left(\frac{SX}{K_s + S}\right) \qquad (40)$$

It is apparent from Equation 38 that the Monod equation does not include a term for the endogenous decay of the organism and that an improved version[11] of the equation would be

$$\frac{1}{X}\frac{dX}{dt} = \frac{u_M S}{(K_s + S)} - K_D \qquad (41)$$

However, if it can be assumed that the endogenous respiration of the organisms neither creates nor destroys substrate material, the only term affecting the substrate removal is the first one on the right and Equation 40 remains valid as it stands. This assumption is, of course, not true since lysing of cells does produce substrate which can be utilized for further metabolism. However, the error introduced by making the assumption is quite small.

Separating variables in Equation 40 and integrating yields

$$K_s \ln \frac{S}{S_A} + S - S_A = -\left(\frac{u_M X}{Y}\right) t \qquad (42)$$

$$S_A - S - K_s \left(\frac{S/S_A - 1}{S/S_A}\right) = \left(\frac{u_M X}{Y}\right) t \qquad (43)$$

Further mathematical manipulation yields

$$(S)^2 + S\left[\frac{u_M X t}{Y} + K_s - S_A\right] - K_s S_A = 0 \qquad (44)$$

The classical solution to this quadratic is

$$S = - \frac{\left(\frac{u_m \, xt}{Y} + K_s - S_A\right) \pm \left[\left(\frac{u_m \, xt}{Y} + K_s - S_A\right)^2 + 4K_sS_A\right]^{1/2}}{2} \qquad (45)$$

4. Mixing Model for Continuous-flow System

The model used for mixing within the continuous-flow reactor for the activated sludge system will determine the resulting form of the equations to be used to describe the system. Historically, mixing models have been combinations of or perturbations around the two idealized systems of "plug flow" and "Continuous Stirred Tank Reactor" (CSTR). It has recently[9,10] been recognized that these models, which are a measure of the residence time or exit-age distribution, are descriptive of the macromixing within the system. Macromixing involves the mixing of fluid elements and cells on a macroscopic scale such as that involved when clumps, groups, or aggregates of molecules or cells are mixed. In addition to macromixing, what has come to be called micromixing is also important. Micromixing involves the mixing of individual fluid molecules and individual cells. When individual molecules and cells are free to move about the medium to collide and intermix with all other cells and molecules, micromixing can occur and fairly uniform concentrations of cells and substrate molecules can be obtained. However, when the cells tend to aggregate and the culture contains clumps of biomass rather than individual cells or solid substrate rather than soluble, it is much more difficult for micromixing to occur. It would therefore appear that the micromixing phenomenon would be important in the activated sludge process. Mathematically, two extremes of micromixing have been proposed — complete segregation (no micromixing) and maximum mixedness (complete micromixing). These limits are chosen as all systems should fall somewhere in between.

For the macromixing model, the CSTR or series of CSTRs is chosen since the plug flow model can be approximated by setting the number of CSTRs (each of infinitesimal size) to infinity, and any number of CSTRs between 1 and infinity can be used to describe a system in between a CSTR and plug flow system.

In a CSTR where maximum complete mixing is assumed, the equations reduce to those for the conventional CSTR. When complete segregation is the case, it has been shown[9] that

$$S_{seg} = \int_0^\infty (E(t)) (S_B(t)) \, dt \qquad (46)$$

where S_{seg} is the average concentration in the effluent of the reactor, $S_B(t)$ is the concentration predicted by the kinetics of the reaction under consideration in a batch reactor, and $E(t)$ is the exit-age distribution function describing the distribution of the lengths of stay of the fluid elements in the reactor. For a CSTR

$$E(t) = \frac{1}{t_Q} e^{(-t/t_Q)} \qquad (47)$$

where t_Q is the residence time in the reactor based on flow and volume. Therefore, the effluent concentration is expressed by

$$S_{seg} = \int_0^\infty \frac{1}{t_Q} e^{(-t/t_Q)} S_B(t) \, dt \qquad (48)$$

where S_{seg} is the effluent concentration from the complete mix reactor and S_B is the concentration expressed by Equation 45.

5. Simplification for Application

Both Eckenfelder[12] and Eckhoff and Jenkins[4,5] have shown that a simplified form of Equation 40 provides satisfactory representation of activated sludge biokinetics. This form is the same as would be obtained if it were assumed that the value of K_s in Equation 40 is much greater than S. Equation 40 then reduces to

$$\frac{ds}{dt} = - \frac{u_m}{Y K_s} (SX) = - k \, SX \qquad (49)$$

In a single-stage complete mix system, the value of S that is used in Equation 40 is the concentration of substrate in the reactor. This concentrais the same as that in the effluent and should be quite low if the activated sludge system is performing adequately. It has been experimentally determined by several investigators referenced by Eckhoff and Jenkins[5] that the magnitude of the saturation constant, K_s, in Equation 40 exceeds 200 mg/l for a large number of substances often found in common wastewaters. Therefore, it is not surprising that the simplified version given by Equation 49 seems to represent the data well except for wastes with effluent substrate concentrations in excess of 100 mg/l.

As has been previously mentioned, the activated sludge process requires a recycle stream of organisms to the aeration tank. It can easily be shown that this recycle stream far exceeds the influent stream and the aeration tank growth of cells in terms of adding solids to the aeration tank. It can thus be assumed that if the waste nature does not fluctuate widely, the concentration of organisms, X, in the aeration tank will not change once a constant recycle fraction and clarifier underflow concentration are maintained. When this is the case, Equation 49 becomes a pseudo-first order reaction equation with the rate-determining parameter being the substrate concentration alone. Integrating Equation 49 and substituting it into Equation 48, we obtain

$$S_{seg} = \int_0^\infty \frac{1}{t_Q} e^{(-t/t_Q)} S_A e^{-kXt} \, dt \qquad (50)$$

Integrating Equation 50 between the designated limits, we obtain

$$S_{seg} = \frac{S_A}{1 + kXt_Q} \qquad (51)$$

This is the same expression for the average concentration in the effluent as would be derived assuming maximum mixedness and utilizing Equation 49 in a complete mix continuous flow system. Therefore, it can be concluded that in the *special case* of a first order reaction, the effects of micromixing limitations can be ignored and we can assume that the conventional CSTR equations will apply.

Up to this point, we have been speaking in terms of substrate concentration and biomass concentration. In order to use Equation 51, we must identify these terms by association with some measurable commodity. Because wastewaters have such a large number of different organic materials in them, a "catchall" term known as the five day biochemical oxygen demand (BOD_5) has been used to describe the organic content of wastewater. To blatantly say that we will use BOD_5 as the substrate concentration in Equation 51 would really be in violation of the assumption of the growth-limiting substrate. However, if we assume that each individual substrate will cause an organism to dominate which utilizes it as its growth-limiting substrate, we can write a Monod equation (after Eckhoff and Jenkins) for each substrate material as below.

$$\frac{dS_i}{dt} = -k_{ij} X_j S_i \qquad (52)$$

In Equation 52, i refers to the ith substrate and j refers to the jth microbial species which is metabolizing that substrate. If we now assume that the feed composition is constant (in constituents, not necessarily strength), it should also be safe to assume that the biological population will not change appreciably and

$$k_{ij} X_j \simeq k_{ij} C_j X = k_i X. \qquad (53)$$

If we assume that k_1 is essentially the same for all substrates (consistent with the concept of "survival of the fittest"), we can sum up all the i equations as represented by Equation 52 and Equation 54 results, where S now can be the BOD_5 and $(\psi\delta)$ MLVSS is the concentration of biomass in the system ($\psi\delta$ is the viable fraction of the mixed-liquor volatile suspended solids).

$$\frac{dS}{dt} = -k\psi\delta(MLVSS)S \qquad (54)$$

A substrate material balance around the complete mix reactor must include the substrate entering both in the influent and recycle streams, the substrate leaving in the total effluent flow from the reactor, the substrate being destroyed in the reactor, the substrate being created in the reactor, and the substrate accumulating in the reactor. In the steady state the substrate accumulation term is definitionally zero, and if we also assume the generation of substrate by endogenous decay of the organisms is small, the material balance becomes

$$Q(S_A) + R(S) - (Q + R)S - V(MLVSS)k\psi\delta(S) = 0$$

$$(55)$$

Solving Equation 55 for the effluent substrate or BOD_5 concentration we obtain Equation 56, which is used to calculate effluent quality for a complete mix activated sludge system.

$$S = \frac{S_A}{(1 + k\psi\delta(MLVSS)t_Q)} \qquad (56)$$

In this equation, S is considered to be the soluble BOD_5 in the effluent rather than the total. The reason for this is, of course, that some solid material escapes the clarifier in the effluent stream and there is BOD_5 associated with those solids. Since the solids are mostly biomass, the BOD_5 associated with them should not be considered to be determining the rate of substrate removal.

6. Kinetics of Staged Systems

Having established the kinetic expression for the single-stage completely mixed reactor, extending it to a series of complete mix reactors requires merely performing a material balance around each reactor. If each reactor is the same size, this material balance results in Equation 57 below.

$$\frac{S}{S_A} =$$

$$\frac{1}{(1 + R/Q)} \left[\frac{1}{(1 + k\psi\delta(MLVSS)\dfrac{t(Q + R)}{m})^m - R/Q} \right]$$

$$(57)$$

In this expression, m is the number of equally sized stages.

By comparing Equations 56 and 57 it can readily be shown that a staged system can attain the same removal of substrate with less total tankage volume than a system with only a single stage, provided that all other parameters are the same. The reason for this is, of course, that the rate of removal of substrate is proportional to its concentration in the system. In a single-stage system, this concentration is that of the effluent, which is by definition quite low. In a staged system only the rate in the last stage is that low, with the other stages having a more rapid removal rate due to the higher substrate concentration in them.

Further experimentation with Equation 57 will reveal that as we increase the number of stages, the effect of adding additional stages begins to diminish, and once five or six stages are present, very little additional benefit is possible. This is to say that once six stages have been reached, the mathematics are such that, given the assumption of no back mixing, the six-stage system very closely approximates a plug flow system (infinite stages). In a biological system the number of stages which produces the approximation to plug flow kinetics is less because as the stage numbers increase for the same total volume, the stage volumes decrease and the substrate concentrations in the mixed-liquor of the initial stages increase. Once the stage concentrations reach a high enough level such that the assumption of the saturation constant being substantially greater than the substrate concentration is no longer valid, the removal rate is no longer first order. It approaches zero order and the beneficial effects of staging are thus reduced. In addition, since interstage openings must be large enough so as not to introduce significant head loss and so as to permit passage of suspended solid materials without plugging, some degree of back mixing will occur. This further reduces the biological kinetic benefits of staging. It has been shown by both bench-scale and pilot plant operations conducted by Union Carbide Corporation over the past 4 years on a variety of industrial and municipal wastes that the point at which additional staging no longer appears to help from a biological kinetics point of view is between a three- and four-stage system. It should be noted, however, that additional stages may be beneficial from an operating power viewpoint since gas phase purities may be more closely matched with the required stagewise uptakes and available dissolution equipment when additional gas stages are utilized.

7. Experimental Determination of "k"

Either Equation 56 or 57 can be rearranged to solve for the rate constant depending on the type of system being utilized. However, care must be taken to be sure that the system is in a biokinetically limiting situation before the parameters are measured and the rate constant calculated. As has been previously mentioned, most activated sludge systems are controlled by flocculation and sedimentation or mass transfer and not by biokinetics. This is to say that the required system loading to ensure a consistently well-flocculated sludge is lower than that required biokinetically to produce the desired effluent. If this is the case in the experimentation, an unrealistically low value of k will be determined. This is because once low effluent concentrations are obtained, further increases in detention time no longer result in reduced effluent concentrations because diffusional resistances of substrate material begin to dominate and the kinetic expressions no longer apply, or because of the fact that the BOD_5 produced by endogenous respiration of organisms begins to become appreciable relative to the low effluent concentrations. In either case (or both), unless one has established the system as being kinetically controlled (a significant change in effluent quality occurs with a change in detention time), a low value for the rate constant will result.

F. System Power

Total system installed power, power usage, and operating expenses are key factors involved in selection of the treatment scheme.

The total power required for an oxygen-activated sludge system consists of dissolution equipment and oxygen generation equipment. Oxygen dissolution may be accomplished through the use of surface aeration devices, sparged-turbine aeration devices, or diffused aeration equipment. Oxygen generation may be accomplished either through cryogenic distillation of air or through the use of pressure swing adsorption (PSA) processes.

The interrelationship between oxygen dissolution equipment and oxygen supply equipment represents a complex nonlinear system which must be augmented by variable constraints. The final selection of the dissolution equipment and the oxygen supply system are directly affected by wastewater strength, temperature, oxygen consumption characteristics, dissolved oxygen levels, site elevation or atmospheric pressure, oxygen feed mole fraction, percent oxygen utilization, mixing criteria for adequate solids suspension, land limitations, the cost of energy, investment, amortization rate, and system load variations, including diurnal and peak periods.

Once the design engineer has selected the design point, reviewed site constraints, set system geometry, and established system oxygen demand, the selection of the dissolution equipment and corresponding operating power becomes dependent on:

1. Feed gas oxygen mole fraction
2. Oxygen utilization or efficiency
3. Dissolved oxygen level
4. Mixing constraints
5. Energy costs and investment

It is readily apparent that the investment and energy required for oxygen production and dissolution are directly related to efficient oxygen usage. The degree of usage hinges heavily on the relative costs of power and the dollar outlay involved.

For any oxygen feed gas mole fraction and constant system oxygen demand, optimum oxygen utilization may be determined as a function of annual operating expense, where annual expense includes power for generation and dissolution, amortization of investment, and maintenance repair costs. Figure 6 presents such a relationship. There exists an optimum — a low point — where power and investment have been matched such that lowest total annual expense occurs. If attempts are made to achieve very high oxygen utilization, dissolution power requirements and investment increase. If low utilizations are employed, oxygen is being unduly wasted and generation power and investment are increased. It is evident that as the price of required equipment and/or power costs increase (oxygen becomes more expensive, dollars/ton), the optimum percentage oxygen utilization curve will move to the right, and, conversely, if oxygen becomes less expensive, it will move to the left for lower efficiencies.

Figure 7 presents a qualitative relationship between optimum percent oxygen utilization and oxygen feed gas mole fraction, with the cost of oxygen as a variable. As the feed gas mole fraction increases, its value also increases (90% oxygen costs more than air), and it is, therefore, desirable to waste as nominal a quantity as is economically attractive. Also, as the relative cost of any set purity increases, it becomes increasingly attractive to design for higher utilizations.

The effective transfer efficiency of the dissolution system may be described by the following equation:

$$ETE = \frac{\alpha(STE)}{DO^*_{20}} \left[\frac{Yo_2}{0.21}(DO^*_{T,\rho}) \beta - DO \right] \Theta^{(T-20)}$$

(58)

If it is assumed that the α, β, STE, and the temperature and pressure are constant, the ETE can be varied through changes in YO_2 and/or D.O., or, in other words, a change in driving force increases or decreases the mass transfer rate.

The gas phase mass transfer oxygen purity, YO_2, can be increased or decreased through changes in oxygen utilization and/or feed gas mole fraction. When gas phase mass transfer purity is held constant, increases in dissolved oxygen level result in decreases in mass transfer efficiency or increases in system dissolution power in order to satisfy the same oxygen demand. The differential power increases rapidly as utilization efficiency goes up and the corresponding feed oxygen mole fraction decreases. For example, each milligram per liter of dissolved oxygen level elevation can increase dissolution system power from 3 to 6% for feed purities and utilizations of 90% depending on the STE of the aeration devices employed. If air were the aerating gas, oxygen feed gas mole fraction = 0.21, considering optimum oxygen usage, power increases would generally range between 10 to 20% per mg/l of D.O. elevation. Figure 8 presents a qualitative estimation of these effects.

Dissolution power requirements cannot be solely determined without checking to be sure that sufficient energy is supplied with the selected dissolution equipment in order to maintain proper mixing to insure biomass suspension and tank turnover. Oxygen aeration systems, because of their superior mass transfer capabilities, at times

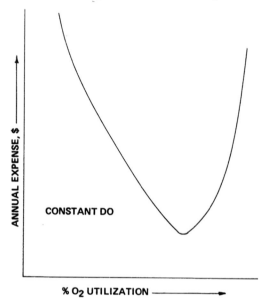

FIGURE 6. Oxygenation system annual operating expense vs. percent oxygen utilization. (Courtesy of Union Carbide Corporation.)

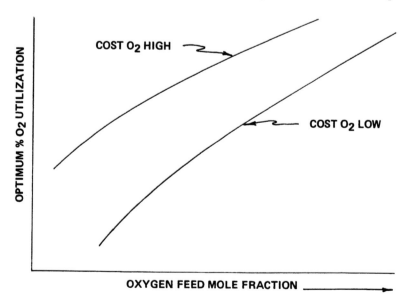

FIGURE 7. Oxygenation system optimum oxygen utilization vs. feed oxygen purity. (Courtesy of Union Carbide Corporation.)

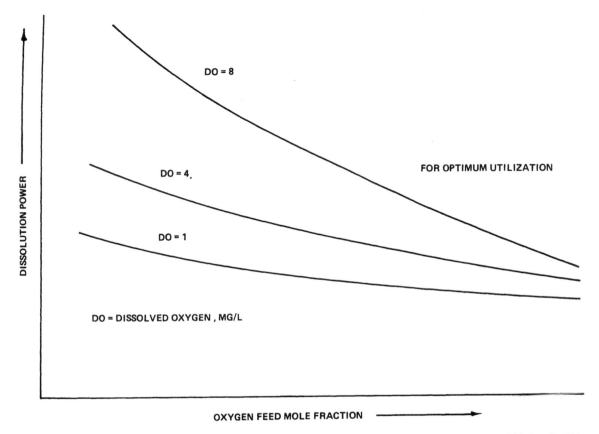

FIGURE 8.   Oxygenation system dissolution power requirement vs. feed oxygen purity. (Courtesy of Union Carbide Corporation.)

become mixing power controlled, especially in the latter stages of multistage systems since oxygen requirements decrease from stage to stage. This occurrence is usually most prevalent when dealing with low strength (BOD$_5$ < 100 mg/l) wastewaters. The design engineer may also use this as a tool in order to maximize matching adequate mixing and mass transfer energy requirements. Systems requiring more power for mixing than mass transfer may decrease mass transfer purities such that oxygen is more efficiently utilized without decreasing dissolution efficiency. As a matter of fact, total system effective energy usage will decrease since less oxygen will have to be generated.

Figure 9 depicts how ETE is affected by a mixing controlled situation where a constant oxygen demand exists. As the volume of the oxygenation tank increases while oxygen dissolution requirements remain constant, the OUR decreases, resulting in higher mixing energy requirements than that required for mass transfer.

After determination of the oxygen supply type and size from the selected feed oxygen mole fraction and utilization, and given the mixing and mass transfer capabilities of the dissolution equipment selected from oxygen consumption requirements, multi-stage gas phase mass transfer purities, dissolved oxygen levels, and temperature and pressure for the design point, the total system may be evaluated for system average power draw and turndown capability. Since invariably most wastewater treatment facilities experience diurnal, weekly, or seasonal load and flow variations, the design point will seldom coincide with the yearly average organic loading conditions. Therefore, power savings over the required design system operating power may be realized. Oxygen generating facilities can respond to the oxygen demand as measured by pressure changes in the gas space of the covered oxygenation tankage. The oxygen generator tracks the system demand, supplying only the required quantity of oxygen, thereby minimizing power consumption or maximizing efficient power usage. This power savings is significant, as generally more than one half of the total installed system power is for oxygen generation. This power savings, or turndown feature, is

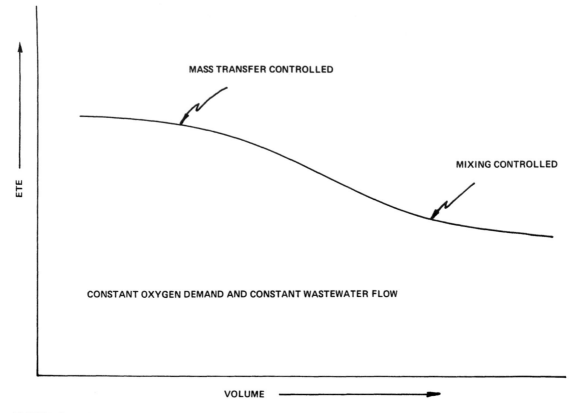

FIGURE 9. Effective transfer efficiency of aeration devices vs. basin volume. (Courtesy of Union Carbide Corporation.)

extremely important, especially for small treatment plants where large variations in oxygen demand are experienced and where turndown is automatic and therefore does not require operator attention. For large plant installations where oxygen demand variations incurred are less, the fractional power savings are less but are of a much greater magnitude and corresponding dollar value. Energy savings can vary from 5 to 20% depending on oxygen generator type and plant size.

Mixing and mass transfer power savings may also be realized in some treatment facilities either through liquid level control, variable speed motors, or gas recirculation compression turndown for sparge turbine systems. Combining the energy reductions for oxygen generation equipment and dissolution equipment can result in overall plant power savings of from 10 to 30%.

These operating expense reductions apply when the treatment plant is at design year. Significant further reductions are possible during the years approaching design year through gearing changes or through removal from service of a fraction of the dissolution and generation equipment available.

VI. OTHER DESIGN CONSIDERATIONS

The biological stability of an activated sludge process may readily be disrupted through severe shifts in environmental conditions. These deleterious changes in environmental conditions may involve increases or decreases in temperature, pH, strength, flow, changes in organic constituents, periods of inadequate nutrient supply, insufficient oxygen transfer capabilty, influent toxic organics or heavy metals, and others. The major concern of the design engineer is not to predict what happens to the effluent quality when environmental disturbances occur, but rather how to best cope with these disturbances and provide for the prevention of deterioration of effluent quality.

A. Shock Loadings
The design engineer may readily take steps to safeguard the biological system from disruptive changes of temperature, pH, changes in organic constituents, insufficient nutrients, and other factors through utilization of pH and temperature control schemes, nutrient addition facilities, equalization basins, etc., when these variations are

known. Once these "knowns" are compensated for, the designer must concentrate on organic and hydraulic surges and oxygen transfer capability and oxygen supply.

Decreases in secondary system removal efficiencies are usually related to biological breakdown, oxygen limitations, or failure of the secondary clarifiers. The biological stability of the process may then be separated into three general areas which may or may not be related. The first area relates to the mass of active biomass in the aeration basin which is available to metabolize the organic pollutants. Sufficient mixed-liquor volatile solids and retention time must be provided for so that the microorganisms can remove the organic material from the solution before entrance into the clarifier. Also, operation must be maintained within a reasonable F/M range so that good settling and compacting of the sludge is achieved since excessive loadings can lead to dispersed growths.

Secondly, the effectiveness of the oxygen transfer equipment in supplying oxygen for peak oxygen uptake rate demands of the biomass must be considered. This is possibly the primary factor leading to the increased stability and reliability of oxygen systems. As previously discussed in Section V.F of this chapter, the ETE, or the oxygen dissolution capacity, may readily be increased through changes in gas phase mass transfer purity and/or the mixed-liquor dissolved oxygen level. The biological mass in an oxygen-activated sludge system is usually functioning in an environment where the dissolved oxygen level is approximately 6 mg/l, or greater prior to an organic shock. UNOX Systems are typically designed to operate at 1 to 2 mg/l D.O. even during shock load periods.

It is obvious that higher D.O. in the mixed liquor is of great value in regard to shock loads not only because of its availability but also due to the fact that when the D.O. decreases, the capability of the mass transfer equipment increases with increased driving force. This combination of high oxygen transfer driving force and automatic oxygen feed control to match system demand favors biological stability of oxygen-activated sludge systems. The ability to maintain adequate dissolved oxygen levels at all times insures biological stability and, therefore, is in essence a shock-resisting aid.

The secondary clarifier provides a relatively quiescent area of solid-liquid separation. It also provides an area and mechanism for thickening of the sludge as well as a convenient vessel from which excess sludge may be wasted. The most common cause of solids washout from secondary settlers is due to hydraulic overloading and/or short circuiting flow. Adequate return sludge capacity and waste sludge control may assist in alleviating this problem provided that bulk upflow velocities are not so great that solids washout cannot be avoided.

It cannot be overemphasized that sufficient mass transfer capability is of utmost importance during peak or shock periods since maintenance of D.O. insures system stability through (1) maintaining the high fraction of active mass, therefore reducing probability of biological breakdown, (2) maintaining aerobic conditions stimulating stability, and (3) maintaining good solids-liquid separation, reducing the probability of mixed liquid solids carryover.

B. Toxicity

Toxic compounds such as heavy metals and some organic compounds will suppress or inhibit the growth of a biological population. Knowledge of the effects of toxic compounds is largely qualitative. Effects vary from system to system and cover ranges of slight inhibition to total failure. The best way to cope with sewage streams with known toxic components is through a source control program and enforcement of industrial waste ordinances.

A major consideration in handling toxic shocks is the effective toxic considerations within the biological system. This effective component concentration is defined as the pounds of toxicant per pound of active mass in the system. Several investigators[34-37] have recently obtained substantial experimental evidence that the controlling parameter in the degree of inhibition that the biomass experiences is this ratio of toxicant applied to viable biomass in the system.

If the toxic component in the waste stream prevails over a long enough period of time such that a steady state condition exists (a long-term shock), all of the biomass will most likely be exposed to the potentially toxic material. If the material collects in the system and associates with the sludge, the concentrations in the sludge will eventually adversely affect the system. If the toxicant does not associate with the sludge but builds up in concentration in the mixed-liquor to equal the influent concentration, a larger volume system will contain a larger mass of the toxicant and will, therefore, most likely suffer somewhat more than a smaller volume system.

When a short-term shock of toxic material occurs, a complete mix system has the ability to dilute the component concentrations to a greater extent than a staged system or a plug flow system. All the biomass in a CMAS is immediately exposed to the toxic material, whereas only a fraction is exposed in a staged system. Unless the dilution effect is sufficient to drop the toxicant concentration below the threshold of toxicity, this system will suffer greatly. If the dilution effect can bring the concentration below the threshold, the CMAS system will very likely not suffer at all. In a liquid staged system, the first or second stages are probably all that will be exposed in a short-term shock, and although the ability to reduce the concentration below the threshold of toxicity is less, the staged system has only a small fraction of the total biomass in the system exposed to the toxic component. Due to the fact that an oxygen system maintains only about half of the system's activated sludge in the aeration tankage and assuming a four-stage oxygen system, only from one eighth to one quarter of the sludge would be exposed to the toxic component.

Because of the wide variety of materials which are inhibiting or toxic to activated sludge, and because of the largely unknown reactions of many of these materials, it is advisable to eliminate these compounds. As previously stated, source control is by far the best. However, if this action is not possible, some form of pretreatment or utilization of a two-sludge system may be effective. A toxicity problem dictates special design considerations.

From the above discussion, it should be apparent that a system's resistance to toxic shocks will depend to a large extent on the nature and duration of the shocks. No one system has an innate advantage over any other in resisting all types of toxic shocks to which activated sludge systems may be exposed. The degree of resistance that a system has to toxic or organic shockloads can really only be determined through experience.

VII. OTHER BIOLOGICAL OXYGENATION SYSTEMS

In addition to the standard flow configuration of the secondary activated sludge process (Figure 10), there are other biological wastewater treatment processes in which oxygen can be advantageously employed. This section briefly presents these other processes in comparison to the standard oxygen-activated sludge process and discusses what advantages application of oxygen in these processes may provide.

A. Contact Stabilization

The first of these processes is the secondary contact stabilization process. The basic purpose of this system is the same as the activated sludge system — to remove soluble and collodial substrate material from wastewater. The operation of the contact stabilization system relies on the fact that the biological substrate removal process is divided into two basic steps. The first, "contact," is an adsorption or absorption process in which the

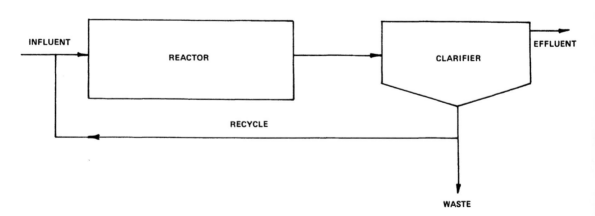

FIGURE 10.   Typical oxygen-activated sludge system.

substrate is attached to or transported into the microorganisms. The second, "stabilization," is the actual aerobic metabolism of the substrate material. The design of the contact stabilization process physically separates these two functions in a system having the flow path illustrated in Figure 11.

The available data show that biologically the contact phase of substrate removal occurs very rapidly compared with metabolism. Thus, the contact basin in an oxygen C-S system has a very short retention time (20 to 40 min). In the effluent from the contact basin, nearly all the soluble substrate has been removed from solution, and what solid substrate there was present in the influent has basically been incorporated into the biological floc. This stream then passes to the clarifier, where the biomass, containing the yet wholly unmetabolized substrate, is separated from the waste stream. The clarifier effluent represents the final process effluent. The clarifier underflow, except for a small portion which is wasted, is recycled to the stabilization basin, where sufficient time is allowed for metabolism to occur. Typical oxygenation system stabilization retention times are in the range of 30 to 90 min. The "stabilized" sludge flow from the stabilization basin is then mixed with the influent waste stream and reenters the contact basin. The concentration of solids in the contact basin is equivalent to that in a normal oxygen-activated sludge system, while the concentration in the stabilization basin is much higher and equivalent to the clarifier underflow concentration. Thus, for a given overall biomass loading, the total volume required in the C-S system is smaller than in the activated sludge system because the average biomass solids concentration is considerably higher.

Both the contact and stabilization zones require oxygen. The contact zone requirement is primarily to maintain aerobic conditions and high dissolved oxygen levels but also to sustain aerobic endogenous activity and whatever metabolism may be occurring. Clearly in the stabilization zone oxygen is required for the high level of metabolism and endogenous respiration occurring and to maintain a high D.O. level. The advantages of using oxygen in this system are basically the same as in the activated sludge system. The smaller volumes needed at a particular biomass loading mean higher volumetric oxygen demands. The ability of the oxygenation system to meet these high volumetric oxygen demands while maintaining the high D.O. level necessary for good floc formation and settling allows the oxygenation system to take better advantage of the potential reactor volume reductions of the C-S system. The better sludge settling characteristics improve the obtainable stabilization reactor solids concentration, thereby further enhancing the potential of the contact

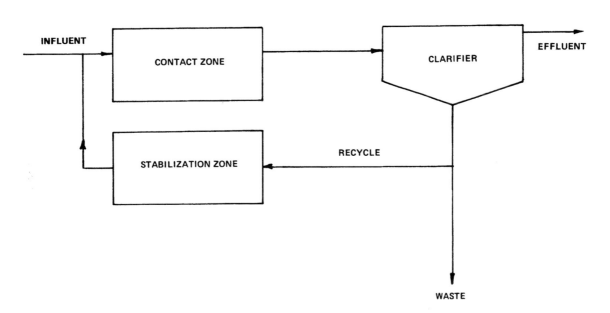

FIGURE 11. Contact stabilization oxygenation system.

stabilization system. Basically, the fact that oxygenation systems do not become mass transfer limited until much higher volumetric oxygen demands are reached makes oxygenation C-S systems much more viable than their more quickly mass transfer limited air aeration counterparts.

B. Nitrification

A second biological process in which oxygen is advantageously used is single sludge nitrification. This process can be thought of as a combination secondary and tertiary treatment system since nitrification is generally referred to as a tertiary treatment process. In a single-step nitrification system the flow pattern is identical to the standard flow oxygen-activated sludge design, but the process is designed to remove carbonaceous substrate and ammonia (NH_3). This is done by providing in the system the proper conditions to cultivate nitrifying bacteria (*Nitrosomanas* and *Nitrobacter*) among the more prevalent carbonaceous bacteria in the biomass. Since the nitrifiers grow much more slowly than the carbonaceous micro-organisms, maintenance of the "proper conditions" consists primarily of assuring that the time which the biomass spends in the reactor system, the steady state sludge retention time (SRT), is at least large enough to provide time for the nitrifiers in the biomass to grow. This, then, is one of the "growth- or kinetic-limited" situations in which the steady state SRT is a particularly important parameter. Recalling the definition of the SRT given previously, its value is increased in a biological system by either increasing the quantity of solids under aeration or by reducing the sum of the effluent solids and sludge wasting rates. For a particular clarifier design, the effluent solids rate is established. It has been shown that the required steady state wasting rate decreases as an inverse function of the biomass loading. Thus, for a particular wastewater and mixed-liquor suspended solids concentration (MLSS), the process SRT is increased by increasing the reactor volume, thus decreasing the biomass loading. Single-step oxygenation nitrification systems will typically have 2.5- to 6.0-hr retention times compared with standard design oxygen-activated sludge systems for carbonaceous removal only having 1.0- to 2.5-hr retention times.

Since nitrification is an aerobic process, oxygen must be supplied to the single-step nitrification system for the oxidation of ammonia as well as for the metabolism of carbonaceous substrate and aerobic endogenous respiration. Since the requirement for oxygen in the nitrification process is high (4.5 lb O_2/lb NH_3 opposed to 0.8 to 1.2 lb O_2/lb BOD_5), the volumetric oxygen demand in a single-step nitrification system is not that greatly reduced from a carbonaceous removal system in spite of the greatly increased retention time. Thus, use of pure oxygen as the aeration gas in such a system is advantageous due to the higher capacity for volumetric oxygen transfer at high D.O. levels. However, the primary advantage of using oxygen in nitrification systems is in the maintenance of high D.O. levels. It has been shown that such high dissolved oxygen concentrations improve the settling and thickening characteristics of the sludge, thus increasing the MLSS which can be maintained in the reactor. It has been additionally illustrated that the high D.O. increases the endogenous activity level of the biomass, resulting in lower net biological sludge production, resulting in a lower required steady state wasting rate. Both of these characteristics of high D.O. oxygenation systems have the effect of decreasing the tank volume required to maintain a particular steady state SRT value.

Thus, single-step nitrification can be accomplished in a much smaller reactor basin for an oxygenation system than for an air aeration system.

A modification of the single-step nitrification system discussed above, called two-step (sludge) nitrification, is depicted in Figure 12. All the foregoing discussion on nitrification applies equally to the two-step system. However, in the two-step, two-sludge system the carbonaceous substrate removal and the ammonia removal are accomplished in separate aeration zones. The first step is a fairly typical high rate activated sludge system in which the bulk of the carbonaceous substrate is removed at high biomass loadings. After a clarification step which confines the first-step sludge to the first step, the waste enters a second aeration zone where nitrification is to occur. Because nearly all the carbonaceous substrate is removed in the first step and the amount of suspended solids entering the second step is also

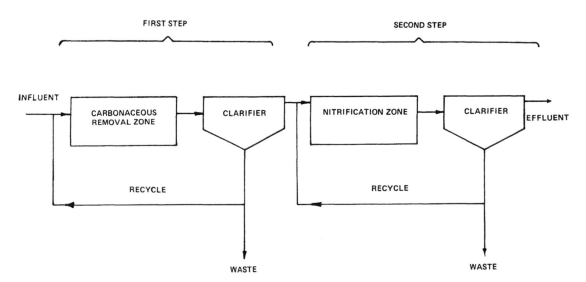

FIGURE 12. Two-step nitrification oxygenation system.

low, the required wasting rate for maintenance of steady state is low. From the definition of SRT, this means that very high SRT values can be maintained in the second step even at low reactor volumes. The net result of designing in this fashion is that the total tank volume required for both steps is typically less than would be required for a similar wastewater using the single-step system. In addition, the high rate first step serves as a buffer zone for the sensitive nitrification step against toxic materials and organic surges which may be associated with the influent wastewater.

The benefits of using oxygen in a two-step nitrification system are essentially the same as discussed for the single step; however, the further reductions in volume necessary to produce a particular effluent quality result in further increases in the volumetric oxygen demand which must be met in order to prevent the system from being mass transfer limited. It has been well established that biological oxygenation systems have significant advantages over air systems in this area.

VIII. EXAMPLE DESIGN FOR OXYGEN-ACTIVATED SLUDGE SYSTEM

The following section gives an example of the use of the equations derived in earlier sections in determination of aeration tankage sizing, oxygen requirements, and expected sludge wasting for an oxygen-activated system to be designed for the municipal application described below. It is assumed that the information required for design has been supplied by the consultant to the municipality and is given below in items 1 through 6. It is further assumed that a comprehensive pilot plant program has been conducted on the waste stream to be treated and that several of the design parameters have been determined from the data from this program. In the general case, this would not be necessary for a waste of this type due to the fact that Union Carbide Corporation has conducted so many pilot programs on a myriad of waste types, and generalized correlations have been developed for use by Union Carbide Corporation engineers.

A. Design Information

1. Wastewater quantity
   Maximum sustained peak — 40 MGD
   Design — 25 MGD
2. Wastewater quality: influent to secondary system
   Design BOD_5 — 200 mg/l — 41,700 lb/day
   BOD_5 at sustained peak — 175 mg/l — 58,300 lb/day
   COD/BOD_5 — 2.0
   Source municipal with 10% light industry

Preceding treatment — screens and primary settling

Design suspended solids	150 mg/l	31,300 lb/day
Design volatile suspended solids	115 mg/l	24,000 lb/day
Suspended solids at sustained peak	135 mg/l	45,000 lb/day
Suspended volatile solids at sustained peak	105 mg/l	35,000 lb/day
Temperature:	Range	10 to 25°C
	Design	20°C
pH:	Range	6.8 to 7.3
	Design	7.0
Alkalinity		250 mg/l as $CaCO_3$
α		0.85
β		0.95
TKN		25 mg/l
NH_3-N		20 mg/l
P		6 mg/l
No known toxic components		

3. Required effluent quality:

BOD_5	20 mg/l
SS	20 mg/l

4. Site limitations — none
5. Solids handling equipment: aerobic digestion, land fill
6. Secondary clarifier specifications:

Overflow rate at design flow	800 gpd/ft²
Overflow rate at sustained peak	1280 gpd/ft²
Depth	12 ft
Feed mechanism	Center well
Take off	Peripheral

B. Tank Volume

From the pilot plant studies, it was determined that the Solid Flux approach (see Chapter 7, Volume I) was applicable and that the initial settling velocity of the mixed-liquor suspended solids could be described by:

$$v_i = ac_i^{-n}$$

where a = 5.0 × 10⁻⁵ ft/lb and n = 2.26. From Equation 3,

$$K = \frac{n}{(n-1)} [180a (n-1)]^{1/n} 10^6 =$$

$$K = \frac{2.26}{1.26} [(180)(5)(10^{-5})(1.26)]^{0.442} (10^6)$$

K = 247,000

From Equation 2,

$$RSS = \frac{247,000}{[OR (R/Q)]^{0.442}}$$

It is now necessary to select a value for the recycle ratio to be used at the design point. In making this selection, the effect on MLSS and RSS

must be considered since the former determines the required aeration tank size at a specified F/M and the latter equals the waste sludge concentration and thus helps determine the flow of waste-activated sludge which must be handled in subsequent operations. As will be shown in Chapter 8, Volume I, the MLSS concentration is not greatly increased once a recycle ratio of from 30 to 60% is exceeded, and since this sludge settles fairly well (a = 5.0 × 10⁻⁵) and the RSS decreases with increasing R/Q for a given clarifier overflow rate, a value of R/Q = 0.30 is selected for design.

$$RSS = \frac{247,000}{[800(0.30)]^{0.422}} = 21,900 \text{ mg/l}$$

By material balance (neglecting influent suspended solids)

$$MLSS = \frac{R/Q}{1 + R/Q}(RSS) = \frac{0.30}{1.30}(21,900) = 5,050 \text{ mg/l}$$

As previously discussed, the system will probably not be biokinetically controlled (this assumption will be checked later), and thus a design point F/M can be selected consistent with attaining good treatment at peak load conditions. For this example, a conservative design point F/M should be about 0.7. This will also be checked later. From Equation 1,

$$F/M = \frac{Q(BOD_s)_A}{V(MLVSS)} = \frac{24(BOD_s)_A}{t_Q(MLVSS)}$$

Assume a VSS/TSS ratio in the middle of the normal range for primary settled municipal wastewater (this is also checked later).

$$\frac{VSS}{TSS} = 0.77$$

$$t_Q = \frac{24(BOD_s)_A}{F/M(MLVSS)} = \frac{24(200)}{(0.7)(0.77)5050} = 1.76 \text{ hr}$$

With this detention time at design, the detention time at the peak will be $\frac{25}{40} \times 1.76 = 1.10$ hr. This is well within a good operating regime as would have been demonstrated in the pilot plant program. The basin volume is $\frac{1.76}{24} * 25 = 1.83$ MG.

Next, check the F/M at the maximum load assuming that the recycle rate is not changed (i.e., OR*R/Q is constant). The RSS value thus remains at 21,900 mg/l and

$$R/Q = 0.3 \frac{25}{40} = 0.188$$

Again by material balance

$$MLSS = \frac{0.188}{1.188} (21,900) = 3,560 \text{ mg/l}$$

F/M at peak $= \frac{24 * 175}{1.10 * 0.77 * 3560} = 1.39$, as can be seen from either of Figures 1, 2, or 3, and, as would have been demonstrated in the pilot plant program, this is well within a good operating regime for an oxygen system and the selection of F/M = 0.7 at design is consistent with a conservative design approach.

C. Oxygen Requirements

The pilot plant program determined that over the range of F/M's from 0.5 to 1.5, A and B values of 0.7 lb O_2/lb BOD_5 removed and 0.16 lb

O_2/day/lb MLVSS were applicable. If it is assumed that all the BOD_5 is removed (this will give slightly conservative answers since usually 95+% is achieved on a total BOD_5 in minus soluble BOD_5 out basis) and that a D.O. level of 6 mg/l must be added into the waste and recycle streams, the oxygen requirement at design is

lbs O_2 dissolved/day = $[A(Q)(BOD_s)_A + B(V)(MLVSS) + (R + Q)\Delta$ DO] 8.34 = [0.7(25)(200) + 0.16(1.83)(0.77)(5050) + 1.3(25)(6)] 8.34
= 40,300 lbs O_2/day

At the sustained peak,

lbs O_2 dissolved/day = 8.34[0.7(40)(175) + 0.16(1.83)(0.77)(3560) + 1.180(40)(6)]
= 49,900 lbs O_2/day

D. Solids Wasting — Design

The parameters utilized for determining the expected solids wasting are determined from pilot plant data, literature studies, and Equations 18 and 26. The unit sludge wasting is then determined from Equation 20. McKinney[33] suggests a value for γ of 0.20 and this is utilized herein. Literature also provides an estimate for Y_c where investigators[30,32] have found that Y_c did not vary significantly with various substrates found in municipal wastewaters. The rationale for this observation was that the major energy consumer in cell synthesis is in converting the intermediate compounds of cell metabolism to cell material and only a minor fraction is utilized in the conversion of the substrate to the intermediate material. From this work, it was determined that a good estimate of the COD-based yield coefficient for soluble wastes in systems of high D.O. (laboratory or oxygen systems) was 0.35 lb of organisms per lb of COD metabolized. It has also been determined that dried organisms are essentially 90% by weight organic material and thus $\delta = 0.90$.

In a typical municipal wastewater, essentially all the influent volatile suspended solids are inactive (that is, they are not viable organisms) and, therefore, f_I is estimated to be unity.

Several of the parameters required for prediction of sludge production are therefore specified ($Y_c = 0.35$, $\gamma = 0.20$, $\delta = 0.9$, $f_I = 1.0$), and the pilot plant data are utilized to define the remaining parameters (f, K_d, ψ) which should be used in the design. The pilot data (in reduced form) are shown below for the two pertinent phases.

Phase	A	B
SRT	1.3	4.4
q	1.46	0.524
$\dfrac{VSS_A}{\Delta COD}$	0.24	0.24
(COD/BOD_s)	2.0	2.0
$M = \dfrac{\% BOD_R}{\% COD_R}$	0.86	0.86
$\left(\dfrac{VSS}{BOD_R}\right)_A$	0.40	0.40
k	0.00125	—
ψ	0.851	0.729
$\epsilon = \left(\dfrac{COD_{NS}}{VSS}\right)_A$	1.6	1.6

Given that $f_I = 1.0$, Equations 18 and 26 may be rewritten as follows:

$$\frac{1}{SRT} = \left[Y\delta\,(1-p) + \frac{f\,VSS_A}{\Delta\,BOD_s} \right] q - K_D \psi\,(\delta - \gamma) \quad (18)$$

$$\psi = \frac{Yq(1-p)}{\dfrac{1}{SRT} + K_D} \quad (26)$$

Since $M = 0.86$ and $COD/BOD_s = 2.0$, and $\delta = 0.9$,

$Y\delta = 0.35(0.9)(2.0)(0.86) = 0.54$ and $Y = 0.60$

It is also a good assumption for all except very short retention time systems that all the soluble substrate is removed biologically and only insoluble material is removed physically. If one assumes the same COD equivalency for the degraded and physically removed solids, a relationship between P and f is established since $f * VSS_A$ is the nondegraded solids, $\left(\dfrac{COD_{NS}}{VSS}\right)_A$ is the COD equivalency of the applied solids, and the product of the two is the COD removed physically.

$$P = f\left(\frac{VSS_A}{\Delta\,COD}\right)\left(\frac{COD_{NS}}{VSS}\right)_A$$

In the case of the pilot plant, therefore, $P = 0.24 * 1.6$ f.

$$P = 0.384\,f$$

Inserting the known parameters in the above equations and substituting the equation for ψ into the SRT equation, we obtain two equations with two unknowns, and the K_D and f values (and therefore P) are determined for the design. It should be noted that in solving the equations, at least four significant figures must be carried in order not to introduce significant inaccuracies. These equations when solved for f and K_D give f = 0.25 and K_D = 0.164. The value for ψ is then determined from Equation 26 above. All required parameters are now determined and the sludge wasting calculation can now proceed. The procedure is to select a value of ψ, determine $\frac{1}{SRT}$ from Equation 18, and check the assumed value of ψ with Equation 26. Once ψ is established, Equation 20 is used to determine the unit sludge wasting.

$$\frac{1}{SRT} = \left[0.54 \left\{1 - 0.384(0.25)\right\} + 0.25\left(\frac{115}{200}\right) \right] q - 0.164(\psi)(0.7) @ q = 0.70 \text{ for design,}$$

$$\frac{1}{SRT} = 0.442 - 0.115\,\psi \quad (18)$$

$$\psi = \frac{0.42[1 - 0.384(0.25)]}{\left(\dfrac{1}{SRT} + 0.164\right)} \quad (26)$$

ψ Assumed	Equation 18 $1/SRT$	Equation 26 ψ
0.80	0.350	0.739
0.75	0.356	0.731
0.72	0.359	0.727
0.727	0.358	0.727

Therefore, the system viability = $\psi\delta = 0.655$ at the design condition of $q = 0.70 \dfrac{\text{lb BOD}_s \text{ removal}}{\text{day} - \text{lb MLVSS}}$ and the expected SRT is 2.79 days. From Equation 20, with $\Delta S = \Delta BOD_5$ and $f_I = 1.0$,

$$\frac{\text{lb SS}_w}{\text{lb BOD}_s \text{ removed}} = \frac{(SS - VSS)_A}{\Delta BOD_s} + f\frac{VSS_A}{\Delta BOD_s} + Y\delta\,(1-P) - \frac{K_D\psi\,(\delta-\gamma)}{q} - \frac{SS_E}{\Delta BOD_s}$$

$$= \frac{(150-115)}{200} + \frac{0.25(115)}{200} + 0.54[1 - 0.384(0.25)] - \frac{0.164(0.727)(0.70)}{0.70} - \frac{20}{200}$$

$$= 0.175 + 0.144 + 0.488 - 0.120 - 0.100 = \underline{0.587}$$

$$\frac{\text{lb SS}_w}{\text{day}} = 25(8.34)(200)(0.587) = 24{,}500 \text{ lb/day @ 2.19\% colids concentration}$$

$$\text{Volume of waste activated sludge/day} = \frac{24{,}500}{0.0219(8.34)} = \underline{134{,}000 \text{ gpd}}$$

E. Checking Assumptions

$$\frac{\dfrac{VSS}{TSS} \text{ Ratio}}{\text{From Equation 24,}} \quad \frac{VSS}{TSS} = \frac{1}{1 + \dfrac{(SS - VSS)_A}{MLVSS(t_Q/SRT)}}$$

since

$$(SS\text{-}VSS)_A = 35 > (S_A - S)\,Y(1-\delta)\,(1-p) = 200(0.6)(0.1)(904) = 11.$$

$$\frac{VSS}{TSS} = \frac{1}{1 + \dfrac{35}{0.77(5050)\left(\dfrac{1.76}{24}\right)\left(\dfrac{1}{2.55}\right)}} = 0.76 \text{ a close check to the assumed 0.77}$$

Kinetics — From the high rate phase of the pilot plant studies, it was determined from Equation 57 that $\psi\delta k = 0.00125$ when $\psi\delta$ was 0.766. Solving Equation 57 for $\psi\delta k$ presumes that the system was biokinetically controlled at this loading. Since the effluent from the plant was still quite good (this we are presuming since it is usually true at this loading; see Figures 1, 2, 3), this assumption will probably give a low value for $\psi\delta k$. However, if the system will not be bio-

kinetically controlled using this $\psi\delta k$ value, then it most certainly will not be biokinetically controlled. In this case,

$$k = 0.00125 \left(\frac{0.655}{0.766}\right) = 0.00107$$

For a four-stage system at the design point, Equation 57 gives

$$\frac{S}{S_A} = \frac{BOD_{5E}}{BOD_{5A}} = \frac{1}{1.3\left[1 + 0.00107(0.77)(5050)\dfrac{(1.76)}{(1.3)(4)}\right]^4 - 0.30} = 0.023$$

$$BOD_{SE} = 0.0230(200) = 4.6 \text{ mg/l}$$

If the effluent SS are 20 mg/l and the BOD_5 associated with them is 7 mg/l (an O_2 system typically has 0.3 to 0.4 mg/l of BOD_5 associated with the effluent SS), expected effluent BOD_5 is 4.6 + 7 = 11.6 mg/l. This is below the required 20 mg/l and

the system should not be biokinetically limited.

At sustained peak — Assume that the peak does not last long enough to attain a steady state and thus the viability does not change from the design case. This is a conservative assumption:

$$\frac{S}{S_A} = \frac{BOD_{5E}}{BOD_{5A}} = \frac{1}{1.19\left[1 + 0.00107(0.77)(3560)\dfrac{(1.1)}{(4)(1.19)}\right]^4 - 0.19} = 0.108$$

$$BOD_{5E} = 0.108(175) = 18.9 \text{ mg/l}$$

Therefore, total effluent BOD_5 will be 26 mg/l during the sustained peak. If the sustained peak lasts 4 hr/day and the design point is maintained for the remaining 20 (a very conservative assumption of the design point was selected properly), the weighted average BOD_5 in the effluent (composite sample) would be

$$\frac{11.6(25)\left(\dfrac{20}{24}\right) + 26(40)\left(\dfrac{4}{24}\right)}{25\left(\dfrac{20}{24}\right) + 40\left(\dfrac{4}{24}\right)} = 15 \text{ mg/l}$$

Therefore, the system is not biokinetically controlled.

NOMENCLATURE

A	Oxygen requirement for synthesis, lb of O_2/lb of BOD_5 removed.
a	Initial settling velocity constant.
a'	Constant in Equation 19.
A_c	Oxygen requirement for synthesis, lb of O_2/lb of COD removed.
A_x	Cross-sectional area for thickening of clarifier.
B	Oxygen requirement for endogenous respiration, lb of O_2/day/lb of MLVSS.
b	Constant in Equation 19.
BOD_5	Five-day biochemical oxygen demand, mg/l.
c_i	Concentration of suspended solids, lb/lb.
COD	Chemical oxygen demand — measured by the dichromate method, mg/l.
COD_{NS}	Chemical oxygen demand of insoluble material.
ΔCOD	COD influent — COD effluent.
D.O.	Dissolved oxygen level in mixed-liquor, mg/l.
$D.O.^*_{T,P}$	Saturation D.O. in equilibrium with air at the operating temperature and atmospheric pressure, mg/l.
$D.O.^*_{20}$	Saturation D.O. in equilibrium with air at 20°C and 1 atm.

ETE	Effective oxygen transfer efficiency, lb O_2/hr/SHP.
f	Fraction of influent volatile solids which are not degraded in the system.
f_I	Fraction of influent volatile solids that is not active biomass.
F/M	Food to biomass ratio, lb of BOD_5 applied/day/lb of MLVSS under aeration.
IOD	Instantaneous oxygen demand, mg/l.
k	Reaction rate constant, lt/mg/hr.
K	Underflow concentration constant, Equation 3.
K_D	Endogenous decay coefficient, lb biomass lysed/day/lb.
K_M	Michaelis-Menten saturation constant, mg/l.
K_s	Monod equation saturation constant, mg/l.
M	Ratio of the percentage COD removal to the percentage BOD_5 removal.
m	Number of stages.
MLSS	Mixed-liquor suspended solids concentration, mg/l.
MLVSS	Mixed-liquor volatile suspended solids concentration, mg/l. 50
n	Exponent in Equation 4 for settling velocity.
OR	Clarifier overflow rate, gpd/ft^2.
P	Fraction of substrate material removed physically.
Q	Influent wastewater flow, millions of gallons per day.
q	Specific removal rate, lb of substrate removed/day/lb of MLVSS under aeration.
R	Recycle flow, millions of gallons per day.
RSS	Return suspended solids concentration, mg/l.
RVSS	Return volatile suspended solids concentration, mg/l.
S	Substrate concentration in mixed liquor or effluent, mg/l.
S_A	Substrate concentration applied to aeration tank, mg/l.
S_B	Substrate concentration in mixed-liquor when complete micromixing prevails, mg/l.
S_{seg}	Substrate concentration in mixed-liquor when complete segregation prevails, mg/l.
SRT	Sludge retention time, days.
SS_A	Suspended solids applied in influent wastewater, mg/l.
SS_E	Effluent suspended solids concentration, mg/l.
STE	Standard transfer efficiency, lb O_2/hr/shp — transferred from air into tap water at zero D.O. and 20°C and under 1 atm of pressure.
t	Time.
t_Q	Hydraulic detention time in aeration tank based on wastewater flow, hr.
t_{Q+R}	Hydraulic detention time in aeration tank based on Q+R, hr.
U	Reaction constant in Michaelis-Menton equation.
V	Volume of aeration tank, MG.
v_i	Zone settling velocity of suspended solids, ft/hr.
VSS_A	Volatile suspended solids applied in influent wastewater, mg/l.
VSS_E	Volatile suspended solids in effluent, mg/l.
VSS/TSS	Fraction of suspended solids under aeration which are volatile.
W	Waste stream flow, mgd.
X	Concentration of viable organisms, mg/l.
Y	Yield coefficient, lb of biomass synthesized per lb of BOD_5 removed.
YO_2	Gas phase oxygen mole (or volume) fraction.
Y_c	Yield coefficient, lb of biomass synthesized per lb of COD removed.
α	Ratio of mass transfer coefficient in the mixed-liquor to that in tap water.
β	Ratio of oxygen saturation concentration in mixed-liquor to that in tap water.
γ	Mass of nondegraded VSS which remains per unit mass of organisms destroyed.
δ	Mass of VSS per unit mass of organisms = 0.9.
ϵ	Lb of oxygen required to oxidize 1 lb of the volatile portion of the biomass completely to carbon dioxide and water.
ϵ'	Lb of oxygen required to oxidize 1 lb of the oxidizable material formed when cells lyse.

θ	Temperature correction factor coefficient.
μ_m	Maximum growth rate constant in Monod equation, in units of reciprocal time.
ϕ	Lb of oxygen required to oxidize 1 lb of BOD_5 completely to CO_2 and H_2O.
ψ	Mass of viable organisms per unit mass of volatile suspended solids.

REFERENCES

1. **Smallwood, C.,** Adsorption and assimilation in activated sludge, *J. Sanit. Eng. Div. Proc. Am. Soc. Civ. Eng.,* Paper 1334, August 1957.
2. **Katz, W. J. and Rohlick, G. A.,** A study of the equilibria and kinetics of adsorption by activated sludge, in *Biological Treatment of Sewage and Industrial Wastes, Vol. I, Aerobic Oxidation,* McCabe, J. and Eckenfelder, W. W., Eds., Reinhold, New York, 1956.
3. **Rich, L. G.,** Activated sludge processes, in *Unit Processes of Sanitary Engineering,* John Wiley & Sons, New York, 1963, 17.
4. **Eckhoff, D. W. and Jenkins, D.,** Transient loading effects in the activated sludge process, *Advances in Water Pollution Research, Part II,* International Association on Water Pollution Research, Munich, 1966.
5. **Eckhoff, D. W. and Jenkins, D.,** Activated Sludge Systems – Kinetics of the Steady and Transient States, Report U.S. Public Health Service Research Grant No. WP-00787, University of California, Berkeley, SERL Report No. 67-12, December 1967.
6. **Ott, C. R. and Bogan, R. H.,** Theoretical analysis of activated sludge dynamics, *J. Sanit. Eng. Div. Proc. Am. Soc. Civ. Eng.,* SA1, February 1971.
7. **Matson, J. V., Characklis, W. G., and Busch, A. W.,** Oxygen supply limitations in full scale biological treatment systems, *Proc. 27th Annu. Ind. Waste Conf.,* Purdue University, Lafayette, Ind., 1972.
8. **Scaccia, C. and Lee, C. K.,** Oxygen Penetration in Activated Sludge Flocs and Effects of Turbulence Intensity and Basin Dimension on Floc Size, paper presented at 50th Annual Water Pollution Control Federation Conference, Philadelphia, October 1977.
9. **Tsai, B. I., Erickson, L. E., and Fan, L. T.,** The effect of micromixing on growth processes, *Biotechnol. Bioeng.,* XI, 181, 1969.
10. **Fan, L. T., Erickson, L. E., Shah, P. S., and Tsai, B. I.,** Effect of mixing on the washout and steady-state performance of continuous cultures, *Biotechnol. Bioeng.,* XII, 1019, 1970.
11. **Sinclair, C. G. and Topiwala, H. H.,** Model for continuous culture which considers the viability concept, *Biotechnol. Bioeng.,* XII, 1069, 1970.
12. **Eckenfelder, W. W.,** Theory of Design of Biological Oxidation Systems for Organic Wastes, Lecture Series, University of Delft, Netherlands, 1964.
13. **Ramanathan, M. and Gaudy, A. F.,** Effect of high substrate concentration and cell feedback on kinetic behavior of heterogeneous populations in completely mixed systems, *Biotechnol. Bioeng.,* XI, 207, 1969.
14. **Atkins, B., Swilley, E. L., Busch, A. W., and Williams, D. A.,** Kinetics, mass transfer and organism growth in a biological film reactor, *Trans. Inst. Chem. Eng.,* 45, 1967.
15. **Herbert, D.,** The chemical composition of micro-organisms as a function of their environment, *Microbial Reaction to Environment,* 11th Symp. Soc. Gen. Microbiol., 1961.
16. **Aiba, S., Humphrey, A. E., and Millis, N. F.,** *Biochemical Engineering,* Academic Press, New York, 1965, 77.
17. **Monod, J.,** La technique de culture continue; theorie et applications, *Ann. Inst. Pasteur,* 73, 390, 1950.
18. **Fair, G. W., Geyer, J. C., and Okun, D. A.,** *Elements of Water Supply and Wastewater Disposal,* John Wiley & Sons, New York, 1971.
19. **Warren, C. E.,** *Biology and Water Pollution Content,* W. B. Saunders, Philadelphia, 1971.
20. **Weston, Roy F., Inc.,** Process Design Manual for Upgrading Existing Wastewater Treatment Plants, EPA Program 17090 GNQ, October 1971.
21. **Gaudy, A. F., Jr. and Gaudy, E. T.,** Biological Concepts for the Design and Operation of the Activated Sludge Process, EPA Project No. 17090 FQJ, September 1971.
22. **Tenny, M. W. and Verhoff, F. H.,** Chemical and autoflocculation of microorganisms in biological wastewater treatment, *Biotechnol. Bioeng.,* XV, 1045, 1973.
23. **Pavani, J. L., Tenny, M. W., and Echelberger, W. F.,** The relationship of algal exocellular polymers to biological flocculation, *Proc. 26th Annu. Ind. Waste Conf.,* Purdue University, Lafayette, Ind., May 4–6, 1971.
24. **Busch, P. L. and Stumm, W.,** Chemical interactions in the aggregation bacteria, *Environ. Sci. Technol.,* 2, 49, 1968.

25. **Argaman, Y. and Kaufman, W. J.,** Turbulence and flocculation, *J. Sanit. Eng. Div. Proc. Am. Soc. Civ. Eng.,* SA2, 223, April 1970.

26. **Thomas, D. G.,** Turbulent disruption of flocs in small particle size suspension, *J. Am. Inst. Chem. Eng.,* 10(4), 517, July 1964.

27. **Matson, J. V., Characklis, W. G., and Busch, A. W.,** Oxygen supply limitations in full scale biological treatment systems, *Proc. 27th Annu. Ind. Waste Conf.,* Purdue University, Lafayette, Ind., 1972.

28. **Kalinske, A. A.,** Effect of dissolved oxygen and substrate concentration on the uptake rate of biological suspension, *J. Water Pollut. Control Fed.,* 43(1), 73, 1971.

29. **Mueller, J. A. et al.,** Oxygen diffusion through zoogloeal flocs, *Biotechnol. Bioeng.,* 10, 331, 1968.

30. **Servizi, J. A. and Bogan, R. H.,** ASCE-Proc. V. 89, *J. Sanit. Eng. Div. Proc. Am. Soc. Civ. Eng.,* n.SA3, Paper 3539, Part 1, June 1963, 17.

31. **Dick, R. I. and Young, K. W.,** Analysis of thickening performance of final settling tank, *Proc. 27th Annu. Ind. Waste Conf.,* Purdue University, Lafayette, Ind., 1972.

32. **McCarty, P. L.,** Thermodynamics of biological synthesis and growth, *2nd Int. Conf. Water Pollut. Res.,* Tokyo, August 1964.

33. **McKinney, P. E.,** Design and operational model for complete mixing activated sludge, *Biotechnol. Bioeng.,* XVI, 703, 1974.

34. **Ghosh, M. M. and Zugger, P. D.,** *J. Water Pollut. Control. Fed.,* 45, 3, March 1973.

35. **Ingols, R. J.,** *Proc. 8th Annu. Ind. Waste Conf.,* Purdue University, Lafayette, Ind., Ext. Ser. 83, 86, 1953.

36. **Directo, L. S. and Moulton, E. Q.,** *Proc. 17th Annu. Ind. Waste Conf.,* Purdue University, Lafayette, Ind., Ext. Ser. 112, 95, 1962.

37. **Salotto, B. V., Barth, E. F., Tolliver, W. E., and Ettinger, M. B.,** *Proc. 19th Annu. Ind. Waste Conf.,* Purdue University, Lafayette, Ind., May 1964.

Chapter 6

OXYGEN-ACTIVATED SLUDGE PILOT PLANT AND TREATABILITY STUDIES

G. P. Breitbach, M. J. Stankewich, Jr., and E. A. Wilcox

TABLE OF CONTENTS

I. INTRODUCTION

During the past several years, Union Carbide Corporation has conducted a comprehensive program investigating UNOX® System performance on a wide variety of wastewaters. After successful full-scale operation of the UNOX System at Batavia, New York, the need to demonstrate the UNOX System on a broad spectrum of wastewaters resulted in one of the most comprehensive pilot plant and treatability programs ever undertaken in the wastewater treatment field. Union Carbide has now completed test programs or is currently conducting demonstrations on the use of the UNOX System at over 200 locations worldwide, treating both municipal and industrial wastewaters. The equivalent of many years of operating experience has been accumulated on wastewaters ranging from ordinary domestic sewage to heavily industrial wastes. Table 1 lists the types of wastewaters treated in UNOX System pilot plants and treatability studies.

Through these bench-scale treatability and field pilot plant programs, considerable experience has been gained demonstrating the performance of the oxygen-activated sludge system and developing design information for the application of this technology on a broad commercial scale. Valuable process knowledge and operational data have been gained from these programs which were conducted under actual plant design operating conditions. Great care was taken to operate the UNOX System bench scale and pilot plant units at the same dissolved oxygen (D.O.) levels, food to biomass ratios, organic loadings, detention times, and diurnal flow patterns that represented practical and economical design conditions for full-scale plants. In addition, a computer data bank has been developed to consolidate these data and to facilitate the development of design parameters for all types of wastewaters.

In this chapter, a listing of Union Carbide's pilot plant and treatability study experience will be presented along with a discussion of the equipment and methods used in conducting these studies. Data on various wastewater types will also be presented to provide a summary of the results obtained from these programs.

TABLE 1

Types of Wastewaters Treated: UNOX® System Pilot Plant and Treatability Studies

Domestic
  Carbonaceous removal
  Carbonaceous removal and nitrification
Domestic and industrial (industrial component identified)
  Pulp and paper
  Petrochemical and allied chemical
  Food processing
  Brewery and distillery
  Textile
  Tanning
  Coke oven waste
  Refinery
  Metal finishing
  Mixed industrial
Pulp and paper
  Kraft process, bleached and unbleached
  Sulfite process, magnesium and ammonia based
  Sulfite and kraft
  Groundwood process
  Chemi-mechanical groundwood
  De-inking/waste paper
Petrochemical and other chemicals
  Acetic and formic acids
  Silicones
  Azo dyes
  Polystyrene and other resins
  Pesticides and insecticide manufacture
  Mixed organic and specialty chemicals
  Phenols
  Nitrocellulose
  Glycols
Food processing
  Citric acid
  Yeast fermentation
  Pickling and relish manufacturing
  Canning
  Grain processing
  Meat packing
  Fish processing
  Lemon processing
Brewery and distillery
  Brewery
  Whiskey distillery
Pharmaceutical
Textile
  Rayon pulping
  Polyvinyl alcohol (PVA)
  Dye wastes
Tanning and glue manufacturing
Steel mill coke oven wastes
Sludge heat treatment supernatant
Film manufacturing and processing
Refinery
Coal gasification

II. PILOT PLANT AND TREATABILITY STUDY EXPERIENCE

A wide variety of wastewater types have been treated in UNOX System pilot plant and treatability studies during the past several years. Table 2 lists programs conducted on wastewaters consisting primarily of domestic sewage. A total of 20 programs, most of which were mobile pilot plants, have been conducted on this type of waste, and these programs have included carbonaceous removal as well as nitrification studies. Most municipal wastewater treatment facilities, however, receive substantial contributions of industrial wastewater from one or more local industries. Table 3 lists over 70 test programs conducted on wastewaters with combined domestic and industrial wastes. These studies cover a broad range of industrial contributors, including pulp and paper, petrochemical, food processing, brewery and distillery, textile, tanning, coke oven ammonia liquor, metal finishing, and refinery wastewaters.

Listed in Table 4 are the pilot plant and treatability studies conducted by Union Carbide worldwide on a variety of totally industrial wastes. Considerable experience has been obtained in the pulp and paper field with over 40 programs completed to date and in the petrochemical industry, with an additional 40 completed programs. In other industrial areas, over 30 studies have been completed. These include food processing, brewery and distillery, pharmaceutical, textile and tanning wastes. Studies have also been conducted on wastewater resulting from coal gasification, refinery operations, photographic film manufacture and processing and coke oven operations producing weak ammonia liquor. The very high strength wastewater resulting from the heat treatment of waste sludges has also been successfully treated in UNOX System studies.

III. GENERAL OBJECTIVES OF PILOT PLANT AND TREATABILITY STUDIES

There are many reasons for conducting an oxygen-activated sludge test program, in addition to the primary objectives of demonstrating the biological treatability of the specific wastewater and evaluating the applicability of the pure oxygen process. Program objectives vary depending on the extent of previous work with the specific wastewater type and the pertinent factors affecting the process selection and system design basis. Test program objectives include the following:

1. Evaluate the biological treatability of the specific wastewater including determination of any toxic or inhibitory characteristics due to heavy metals or organic biotoxins.
2. Develop a design basis for treatment of the wastewater, including determination of oxygen requirements, sludge production, and biomass settling and thickening characteristics.
3. Confirm preliminary process design parameters including the ability to handle shock loads or other nonsteady state factors.

TABLE 2

UNOX® Pilot Plant and Treatability Studies: Domestic Wastes

Location	Type of study
Albstadt-Ebingen, Germany	PP
Amherst, N.Y. (N)	TS, PP
Bartlesville, Okla.	TS
Blue Plains Plant, Washington, D.C.	PP
Denver, Colo.	PP
Grand Island, N.Y.	PP
Harrisburg, Pa.	PP
Humber Plant, Toronto, Ontario, Canada	PP
Indianapolis, Ind. (N)	PP
Kempton Park, South Africa	PP
Miami, Fla.	PP
Newtown Creek Plant, New York	TS, PP
North San Mateo County, Cal.	TS
Perth, Australia	PP
Phoenix, Ariz. (N)	PP
Seattle, Wash.	PP
Spokane, Wash.	PP
Ward's Island Plant, New York	TS, PP
Wayne County, Mich.	PP
Wuppertal, Germany	PP

Note: TS, treatability study; (N), nitrification; and PP, pilot plant.

TABLE 3

UNOX® Pilot Plant and Treatability Studies: Combined Domestic and Industrial Wastes

Location	Major industrial component	Type of study
Pulp and Paper		
Duluth, Minn.	Pulp and paper	PP
Goppingen, Germany	Pulp and paper, dye, and leather-works	PP
Hopewell, Va.	Kraft pulp and paper, agricultural chemicals	PP
Luke, Md.	Bleached kraft	TS
North Tonawanda, N.Y.	Pulp and paper, plastics	PP
Petrochemical and Allied Chemical		
Baltimore, Md.	Chemical production	PP
Basel, Switzerland	Petrochemical	PP
Wuppertal-Buchenhofen, Germany	Petrochemical	PP
Cincinnati, O.	Dye, paint, detergent, brewery	TS, PP
Middlesex, N.J.	Chemical, plastics, food, pulp and paper	PP
Ocean County, N.J.	Organic dye production	PP
Pensacola, Fla. (N)	Chemical	PP
South Charleston, W. Va.	Petrochemical	PP
Food Processing		
Asahi Kasei, Nobeoka, Japan	Food processing	TS
Batavia, N.Y.	Dairy	PP
Bremerhaven, Germany	Fish processing	PP
Camden, N.J.	Tomato canning	TS
Cedar Rapids, Ia.	Grain wastes, meat packing	TS, PP
Dubuque, Ia.	Meat packing	PP
Holland, Mich.	Food processing	TS
Kansas City, Kans.	Meat packing	PP
Muscatine, Ia.	Grain wastes	PP
New Orleans, La.	Seafood processing, poultry, brewery	PP
Oakland, Cal.	Canning	PP
Sacramento, Cal.	Canning	PP
Salem, Ore.	Canning	PP
Totnes, England	Dairy	PP
Turlock, Cal.	Poultry, cannery	TS, PP
Brewery and Distillery		
Copenhagen, Denmark	Brewery, chemical	PP
Frankfurt-Nierrad, Germany	Brewery	TS
Jacksonville, Fla.	Brewery	PP
Kobe Higashinada, Japan	Brewery	PP
Louisville, Ky.	Distilleries, dairy, chemical, slaughterhouse	PP
Munchen Grosslappen, Germany	Brewery, slaughterhouse	PP
Philadelphia SE, Pa.	Brewery	PP
St. Louis, Mo.	Brewery, dairy, food, chemical	PP
Tampa, Fla.	Brewery	PP
Textile		
Asahi Kasei, Nobeoka, Japan	Textile	TS

TABLE 3 (continued)

UNOX® Pilot Plant and Treatability Studies: Combined Domestic and Industrial Wastes

Location	Major industrial component	Type of study
Chattanooga, Tenn.	Textile, chemical	TS
Concord, N.C.	Textile	TS
Danville, Va.	Textile, polyvinyl alcohol (PVA)	TS
Kyoto Kisshoin, Japan	Textile (dyes)	PP
Monroeville, Alabama	Textile (dyes)	TS
Morganton, N.C.	Textile (dyes), and poultry	PP
Newton, N.C.	Textile	TS
Prato, Italy	Textile	PP
Roanoke Rapids, N.C.	Textile	TS
Rocky Mount. N.C.	Textile	TS

Tanning

Location	Major industrial component	Type of study
Brockton, Mass. (N)	Tannery	TS, PP

Coke Oven — Weak Ammonia Liquor

Location	Major industrial component	Type of study
Motherwell, England	Coke oven	PP

Metal Finishing

Location	Major industrial component	Type of study
Huntington, W. Va.	Chrome, alloys, dyes	PP

Mixed Industrial

Location	Major industrial component	Type of study
Barceloneta, Puerto Rico		TS
Detroit, Mich.		PP
Euclid, O.		PP
Fall River, Mass.		PP
Kalamazoo, Mich.		PP
Kawasaki Iriezaki, Japan		PP
Knoxville, Tenn.		PP
Norfolk, Virginia (N)		PP
Passaic Valley, N.J.		PP
Ponthir, England		PP
Philadelphia NE, Pa.		PP
Philadelphia SW, Pa.		PP
Rochester, Minn.(N)		PP
San Francisco, Cal.		PP
Springfield, Mo.(N)		TS
Taunton, Mass.		TS
Winston-Salem, N.C.		PP
Wyandotte, Mich.		PP

Refinery

Location	Major industrial component	Type of study
Hyperion Plant, Los Angeles, Cal.	Refineries	PP
Los Angeles County, Carson, Cal.	Refineries	PP
Newark, N.J.	Refineries	PP
Tonawanda, N.Y.	Refineries, chemical producers, coke producers	PP

Note: TS, treatability study; (N), nitrification; and PP, pilot plant.

TABLE 4

UNOX® Pilot Plant and Treatability Studies: Industrial Wastes

Location	Waste components	Type of study
Pulp and Paper		
Alaska Lumber and Pulp, Sitka, Alas.	Magnesium-based bisulfite	PP
Alton Boxboard, Jacksonville, Fla.	Unbleached kraft	TS
Appleton Papers, Combined Locks, Wisc.	Chemi-mechanical groundwood	TS
Bergstrom Papers, Neenah, Wisc.	De-inking	TS
Boise Cascade Paper, International Falls, Minn.	Bleached kraft	TS
Boise Cascade Insulite, International Falls, Minn.	Construction board and paneling	TS
Chesapeake Corp., West Point, Va.	Unbleached kraft	PP
Container Corp. of America, Fernandina Beach, Fla.	Unbleached kraft	TS
Crown Zellerbach Corp., Wauna, Ore.	Bleached kraft	PP
Daishowa Fuji, Japan	Pulp and paper	TS
Denpo Pulp, Japan	Unbleached kraft	TS
Fibreboard Corp., Antioch, Cal.	Bleached and unbleached kraft	PP
Finch Pruyn and Co., Glens Falls, N.Y.	Ammonia-based bisulfite	PP
Georgia Kraft Corp., Rome, Ga.	Unbleached kraft	PP
Great Lakes Paper Co., Thunder Bay, Ontario, Canada	Kraft	PP
Gulf States Paper Co., Tuscaloosa, Ala.	Bleached kraft	
Hokuetsu, Japan	Pulp and paper	TS
Honshu Paper Co., Fuji, Japan	De-inking	TS
ITT Rayonier Corp., Hoguiam, Wash.	Ammonia-based sulfite	TS
Jujo Akita, Japan	Unbleached kraft	PP
Jujo Kimbery, Japan	De-inking	TS
Jujo Kushiro, Japan	Chemi-mechanical groundwood	TS
Kojin Sacki, Japan	Unbleached kraft	TS
Kojin Toyama, Japan	Unbleached kraft	TS
Longview Fibre Co., Longview, Wash.	Unbleached kraft	PP
Mitsubishi Pulp and Paper, Hachinoe Plant, Japan	Kraft	TS
Mosinee Paper Co., Mosinee, Wisc.	Pulp and paper	TS
Nekoosa-Edwards Paper Co., Nekoosa, Wisc.	Kraft and sulfite	TS
Nova Scotia Forest Industries, Port Hawkesbury, Nova Scotia, Canada	Bleached sulfite	PP
Oji Kasugai, Japan	Kraft	TS
Oji Tomakoru Paper Co., Japan	Pulp and paper	TS
Sanyo Kokusaku Pulp Co., Japan	Kraft	TS
Simpson Timber, Shelton, Wash.	Groundwood	TS
St. Regis Paper Co., Tacoma, Wash.	Unbleached kraft	PP
Thilmany Pulp and Paper, Thilmany, Wisc.	Groundwood	TS
Tokai Pulp Co., Shimada, Japan	Unbleached kraft	TS
Tokiwa Sangyo, Japan	Pulp and paper	TS
Toyo Pulp Co., Kure, Japan	Kraft	TS
Vetrni Pulp and Paper, C.S.S.R.	Sulfite	PP
S. D. Warren Co., Westbrook, Me.	Kraft	TS
Weyerhaeuser Corp., Everett, Wash.	Pulp and paper	TS
Weyerhaeuser Corp., Longview, Wash.	Kraft and sulfite	TS
Weyerhaeuser Corp., Tacoma, Wash.	Pulp and paper	TS

TABLE 4 (continued)

UNOX® Pilot Plant and Treatability Studies: Industrial Wastes

Location	Waste components	Type of study
Petrochemical and Allied Chemical		
BASF Wyandotte, Jamesbury, N.J.	Polystyrene	TS
BASF Wyandotte, S. Kearny, N.J.	Azo dyes	TS
Bayer, Elberfeld, Germany	Petrochemical	PP
Bayer, Leverkusen, Germany	Petrochemical	PP
Borg-Warner Chemicals	Phenols	TS
Chemagro, Kansas City, Mo.	Pesticides, insecticides	PP
Ciba Geigy, Basel, Switzerland	Petrochemical	TS, PP
Ciba Geigy, Cranston, R.I.	300 specialty chemicals	TS, PP
Denka Chiba, Japan	Petrochemical	TS
Dow Chemical, Plaquemine, La.	High salt petrochemical	TS
Dow Chemical, Freeport, Tex.	High salt propylene glycol	TS
Dow Chemical, Sarnia, Ontario, Canada	Ethylene Glycol	TS
DuPont, Deepwater, N.J.	Dyes, nitrocellulose, 700 other batch products	PP
EXXON Chemical, Baton Rouge, La.	Low molecular weight acids and alcohols, other organic products	PP
Futamura Kagaku, Japan	Chemical compounds	TS
General Electric, Mount Vernon, Ind.	Plastics processing	TS
Hercules, Inc., Wilmington, N.C.	Dimethyl terrphthalate (DMT), acetic and formic acids	PP
Hoechst, Germany	Petrochemical	TS
Kasei Mizushima, Japan	Petrochemical	TS
Kyowa Yuka Yokkaichi, Japan	Petrochemical	TS
Lin Yuan, Taiwan	Multiple plant consortium	TS
Missho Kayaku, Japan	Petrochemical	TS
Mitsubishi Kasei, Mizushima, Japan	Petrochemical	TS
Mitsui Toatsu, Omuda, Japan	Agricultural chemicals	TS
Mitsui Toatsu, Osaka, Japan	Petrochemical	TS
Naphtachimie, France	Petrochemical	TS
Nissokasei, Chiba, Japan	Chemical	TS
Sandoz, Switzerland	Petrochemical	TS
Shell-Berre, France	Petrochemical	PP
Shell-Norco, Louisiana	Acrolein, petrochemical	TS
Shell-Rotterdam, Holland	Petrochemical	TS
Showa-Neoprene, Japan	Petrochemical	TS
Sumitomo Chiba Chemical, Japan	Petrochemical	TS
Sumitomo Nihama, Japan	Chemical	TS
Tennaco Chemicals, Pasadena, Tex.	Methanol, ammonia	PP
Union Carbide Corp., Brownsville, Tex.	Acetic and formic acids	TS
Union Carbide Corp., Institute, W. Va.	Petrochemicals	PP
Union Carbide Corp., Marietta, O.	Polystyrenes, polysulfone, phenolic and epoxy resins	TS
Union Carbide Corp., Sistersville, W. Va.	Silicones	TS
Union Carbide Corp., Taft, La.	Acrolein, glycol, amines, aromatics, olefins	PP
Union Carbide, Antwerp, Belgium	Petrochemical	TS
Union Carbide, Montreal East, Quebec, Canada	Glycol, amines, aromatics, olefins	PP

TABLE 4 (continued)

UNOX® Pilot Plant and Treatability Studies: Industrial Wastes

Location	Waste components	Type of study
Brewery and Distillery		
Nikka Whiskey, Japan	Distillery	TS
Sapporo, Japan	Brewery	TS
Whitbread, England	Brewery	PP
Pharmaceutical		
Knoll, Minden, Germany	Pharmaceutical	PP
Lederle Labs, Pearl River, N.Y.	Fermentation	TS
Textile		
Deering Millikin, Blacksburg, S.C.	Textile, polyvinyl alcohol (PVA)	TS
Kojin Co., Japan	Rayon pulping	TS
Teijin Hercules Ehime, Japan	Textile	TS
Tanning		
Himeji Takagi, Japan	Tanning	PP
Peter Cooper Corp., Gowanda, N.Y.	Glue from animal hides	TS
Tatsuno, Japan	Tanning hides	TS
Coke Oven Weak Ammonia Liquor		
Jones and Laughlin Steel, Aliquippa, Pa.	Coke oven weak ammonia liquor	PP
Kawatetsu Kugaku Mizushima	Steel manufacturing, coke oven	TS
U.S. Steel, Clairton, Pa.	Coke oven, phenols	TS
Coal Gasification and Liquefaction		
Panhandle Eastern Pipe Line, Houston, Tex.	Coal gas liquor	TS
Sasol, South Africa	Coal gas liquor, coal liquefaction liquor	PP
Sludge heat treatment		
Cincinnati, O.	Zimpro® supernatant	TS
Fujisawa City, Japan	Heat treatment liquor	TS
Levittown, Pa.	Zimpro supernatant	TS
Rothchild, Wisc.	Zimpro supernatant	TS
Food processing		
Ajinomoto Kawasaki, Japan	Food processing	PP
Frito Lay, Dallas, Tex.	Corn and potato processing	TS
H. J. Heinz, Holland, Mich.	Canning wastes, pickles	TS
Lachema, C.S.S.R.	Citric acid	TS
Oriental Yeast, Japan	Yeast processing	TS
Shiogama, Japan	Fish processing	TS
Standard Brands, Peekskill, N.Y.	Yeast processing	TS
Standard Brands, Peoria, Ill.	Food processing	TS
Sunkist Growers, Corona, Cal.	Lemon processing	TS, PP

TABLE 4 (continued)

UNOX® Pilot Plant and Treatability Studies: Industrial Wastes

Location	Waste components	Type of study
	Miscellaneous	
Fuji Film, Japan	Film manufacture	TS
Osaka Gas, Japan	Refinery	TS

Note: TS, treatability study; PP, pilot plant.

4. Optimize a preliminary design for better treatment efficiency and reduced operating costs.
5. Demonstrate the capabilities and characteristics of high-purity oxygen systems, including the degree of biochemical oxygen demand (BOD_5), chemical oxygen demand (COD), and total suspended solids (TSS) removal.
6. Compare the oxygen system performance with other processes to arrive at a process selection.
7. Characterize the influent wastewater by providing for daily collection and analysis.
8. Provide a facility for operator training prior to full-scale plant start-up.
9. Generate secondary sludge or effluent for further testing, such as solids dewatering, aerobic digestion, ozonation, etc.
10. Determine pH neutralization and nutrient addition requirements, if any.

Depending on the specific program objectives, timing requirements, and economic considerations, there are various testing methods possible. The two most common are bench scale treatability studies and on-site field pilot plants. In some cases, such as industrial wastes known to be difficult to treat or with wastes containing potential toxins, it may be desirable to precede field pilot studies with a bench scale treatability program to initially determine wastewater treatability at a scale providing greater process flexibility and lower program costs.

IV. UNOX FIELD PILOT PLANT PROGRAMS

An extensive pilot plant demonstration capability is maintained by Union Carbide for providing comprehensive services to consultants, municipalities, and industries interested in investigating the use of the UNOX System for treatment of their wastewater. Eight mobile pilot plant units, such as that shown in Figure 1, are used to collect the necessary process information and operational data to provide direct scale-up to actual full-scale plant design conditions.

A. Scope of Pilot Plant Programs

A typical UNOX pilot plant program can be expected to be 3 to 4 months in duration, with almost all studies being less than 6 months in length. Programs generally consist of a start-up phase plus three operating phases. Two weeks is generally sufficient for pilot plant start-up and biomass acclimation, followed by 3 to 4 weeks at each operating phase. The first operating phase usually investigates system operation at the proposed design conditions, with the second phase evaluating performance at peak organic loading conditions. The third phase typically investigates system response to diurnal loading variations or shock loads. Specific objectives of the program are defined by the client and can include any or all of those mentioned previously. Typical objectives are:

1. Establish UNOX System process design parameters for treatment of the specific wastewater under investigation.
2. Determine the ability of the UNOX System to achieve the required degree of removal of BOD_5, COD, suspended solids, and any other constituents of concern.
3. Determine oxygen requirements of the system.
4. Determine sludge production of the system.
5. Determine biomass settling and thickening characteristics.

FIGURE 1. UNOX® System mobile pilot plant. (Courtesy of Union Carbide Corporation.)

6. Assess the effect of diurnal or shock loads on the system.

Table 5 is a sample System Performance Summary for a pilot plant demonstration program showing the type of information collected and included in the final report. Oxygen consumption, sludge production and biomass set tling characteristics are summarized for each program phase along with organic and suspended solids removals and operating conditions. The data presented not only permit effective design of a UNOX System but also provide information required for the design of sludge handling and disposal systems.

Many intangible benefits are derived from conducting a 3 to 6 month pilot plant study at the plant site. A UNOX pilot plant study provides the customer's technical personnel and operators with an opportunity to evaluate and operate a small-scale system in advance of the construction of a full-scale plant. For industrial applications, a continuously operating pilot plant system facilitates the identification and isolation of process streams which could cause difficulties during full-scale plant treatment. Thus, industrial production modifications might be accomplished before full-scale problems are encountered. Finally, a field pilot plant study, because it is conducted on a representative wastewater stream, assures that the UNOX System will meet effluent standards while operating with typical wastewater varia tions.

B. Mobile Pilot Plant Equipment

A schematic diagram of a UNOX Mobile Pilot Plant and its specifications are shown in Figure 2. The UNOX System and all auxiliary equipment, except the final clarifier, are housed in a 40 ft warehouse van, as shown in Figure 3. The UNOX System utilizes a covered, staged aeration basin for contact of oxygen gas and mixed-liquor. High-purity oxygen (90 to 99.8% by volume) enters the first stage of the system and flows cocurrently with the wastewater being treated through the aeration basin. The gastight mobile pilot plant reactor consists of four cocurrent gas-liquid stages with a total ca-

TABLE 5

UNOX® Pilot Plant System Performance Summary

Parameters	Effluent concentrations
Phase	Biochemical oxygen demand
Duration (weeks)	Total (mg/l)
Oxygenation time	Soluble (mg/l)
Q (hr)	Chemical oxygen demand
Q + R (hr)	Total (mg/l)
Recycle fraction (R/Q), %	Soluble (mg/l)
Biomass loading	Suspended solids
lb BOD$_s$/day/lb MLVSS	Total (mg/l)
lb COD/day/lb MLVSS	Volatile (mg/l)
Organic loading	Removals
lb BOD$_s$/day/1000 ft.^{3}	Biochemical oxygen demand
lb COD/day/1000 ft.^{3}	Total (%)
Mixed-liquor suspended solids (mg/l)	Soluble (%)
Mixed-liquor volatile SS (mg/l)	Chemical oxygen demand
Recycle SS (mg/l)	Total (%)
Clarifier overflow rate (gal/day/ft.^{2})	Soluble (%)
Clarifier mass loading (lb SS/day/ft.^{2})	Suspended solids
Influent concentrations	Total (%)
Biochemical oxygen demand	Volatile (%)
Total (mg/l)	Oyxgen consumption
Soluble (mg/l)	lb oxygen/lb BOD$_s$ removed
Chemical oxygen demand	lb oxygen/ lb COD removed
Total (mg/l)	Oxygen utilization (%)
Soluble (mg/l)	Sludge production
Suspended solids	lb TSS wasted/lb BOD$_s$ removed[a]
Total (mg/l)	lb VSS wasted/lb BOD$_s$ removed[a]
Volatile (mg/l)	lb TSS wasted/lb BOD$_s$ removed[b]
	lh VSS wasted/lb BOD$_s$ removed[b]
	Sludge age (lb MLVSS/lb VSS wasted/day)
	Sludge volume index (ml/g)
	Initial settling velocity (stirred) (ft/hr)

[a] Corrected for inventory shift
[b] Corrected for inventory shift and effluent

pacity of 1600 gal, as shown in Figure 2. Each stage contains a separate sparger/impeller contacting unit, which consists of a rotating sparger and a marine-type propeller used as the primary liquid pumping device. These units are powered by a gear box and motor drive, as shown in Figure 2.

The oxygen gas fed into the first stage gas space is maintained at a pressure of about 2 in of water. This pressure is sufficient to maintain control and, through a slight pressure drop, prevent backmixing of the gas from stage to stage. This minimization of backmixing in the system is important, since the cocurrent flow of liquid wastewater and oxygen gas effectively matches the oxygen demand to the oxygen purity. The highest oxygen purity is at the influent end of the tank, where the oxygen uptake rate is greatest, thereby allowing for efficient oxygen usage with low power requirements. As the oxygen-enriched gas passes through the system, small diaphragm compressors in each stage pump the gas down the hollow mixer shafts and through the rotating spargers. Stage one has three compressors to handle shock loads that may be encountered during a study. The rate at which oxygen gas is fed to the first stage is automatically controlled by the first stage gas pressure. A simple pressure sensor is installed in the first stage tank cover which detects changes in gas pressure resulting from a decrease or increase in oxygen uptake rate as wastewater flow or strength varies. A signal is relayed to a flow control valve on the inlet oxygen line which adjusts oxygen flow to maintain the desired gas pressure set point under the first stage cover. The unit acts like a respirometer; as the oxygen uptake rate increases, the first stage pressure decreases, which in turn causes the rate of oxygen feed to increase.

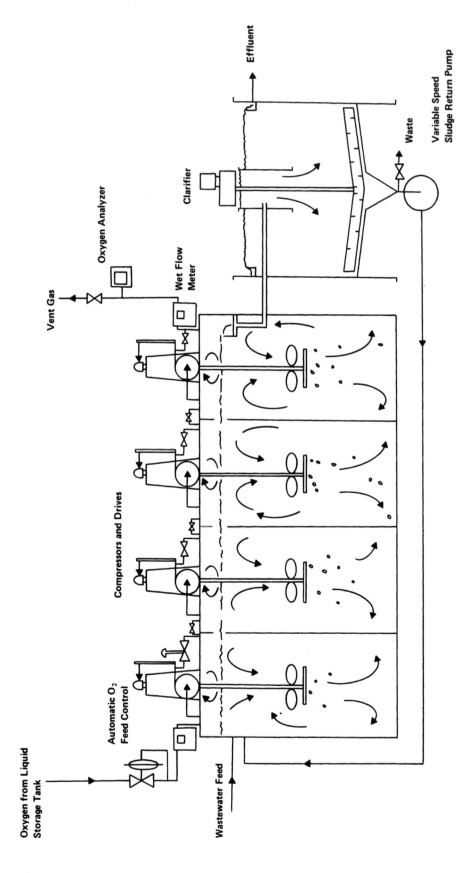

FIGURE 2. UNOX® System pilot plant schematic. (Courtesy of Union Carbide Corporation.)

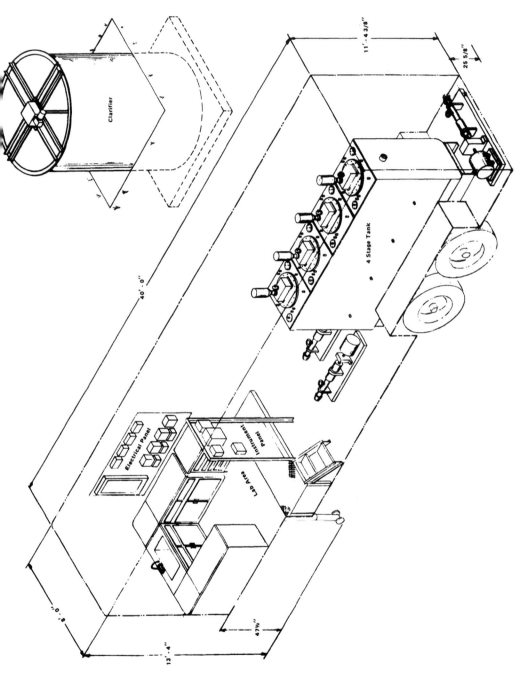

FIGURE 3. UNOX® System mobile pilot plant van. (Courtesy of Union Carbide Corporation.)

When less oxygen is taken up by the biomass, the oxygen feed rate decreases. Before entering the reactors, the oxygen gas passes through a dry test meter. The gas then flows through the gas space openings between the stages, and the waste gas finally passes through a vent gas meter and is vented to the atmosphere. Both the oxygen feed rate and vent gas rate are monitored continuously. All compressors have manual discharge control valves to control the dissolved oxygen level in each stage. The composition of the oxygen gas in each stage is also monitored periodically with a Servomex® paramagnetic oxygen analyzer.

The influent wastewater is pumped into stage one by a variable speed Moyno® pump (1 to 30 GPM) so that a range of feed rates and, therefore, a comparable range of liquid retention times can be simulated. Mixed-liquor from the biological reactor flows to a 7 ft diameter center-feed clarifier, which contains a peripheral effluent weir. The clarifier has a surface area of 34.3 ft² and a thickening area of 38.0 ft². This clarifier contains a variable speed plow-type scraper with center takeoff for settled solids. Clarifier underflow solids are withdrawn with a variable speed Moyno pump for recycling. A separate Moyno pump is used in conjunction with an automatic timer for sludge wasting. There is a provision for manual wasting if desired.

By the substitution of subwall assemblies in the clarifier shell, clarifier surface area can be varied to suit any range of flows. It is also possible to step feed any stage of the reactor or run as a two or three stage system with the first one or two stages of the reactor isolated. In other words, the UNOX Pilot Plant can be customized to fit the requirements of a specific study. Each UNOX Pilot Plant has the necessary auxiliary equipment needed for carefully monitoring and controlling the system to give accurate design information. Dissolved oxygen and pH meters are provided to measure these parameters in the influent wastewater, reactor stages, or clarifier effluent. Instrumentation for measuring the first and fourth stage gas space pressures, the temperature of the gas and mixed-liquor in stage one, the flow of gas being fed to each sparger, and the power consumption and revolutions per minute of each sparger makes monitoring and control of the system

very simple. A continuously operating combustible gas analyzer with automatic alarm system insures completely safe operation even in the event of a spill of volatile, combustible material into the feed to the UNOX System.

The laboratory area in the pilot plant van is equipped with a portable pH meter, a portable dissolved oxygen meter for making oxygen uptake rate (OUR) measurements, one-liter graduated cylinders with stirring devices for running stirred and unstirred settling tests and sludge volume index (SVI) determinations, a magnetic stirrer, and a microscope for periodic examinations of the biomass.

V. UNOX TREATABILITY STUDIES

In addition to pilot plant facilities, Union Carbide also has the capability of demonstrating UNOX System performance with bench-scale treatability studies. The overall objective of these studies is to demonstrate the biological treatability of a specific wastewater utilizing the oxygen-activated sludge process. The data collected are used in selecting a design basis, realizing that this data is subject to the normal limitations of bench-scale data. These bench-scale treatability studies can be conducted either at the site where the wastewater is generated or in an off-site laboratory with wastewater collected and shipped to the lab periodically.

A. Scope of Treatability Studies

Because of the physical limitations of the size of the treatability study and the need to ship wastewater samples for off-site studies, several restrictions are placed on the operation of the study. The logistics of sample handling generally limit the size of samples available for testing; therefore, smaller volume reactors are used. The limited reactor size is restrictive primarily in the data relating to sludge production and handling. Effects which need to be considered when using the smaller reactors include the following:

1. Physical factors, such as attached biomass on reactor surfaces or solids deposited as scum above the liquid, may affect system performance.
2. Clarifier performance may be somewhat poorer than in full-scale systems due to in-

creased wall effects and liquid surging in small scale clarifiers.

3. The limited quantity of waste solids available will limit examination of sludge dewatering characteristics.

Limitations also arise due to the method of wastewater collection used in treatability studies. In off-site studies, samples are often collected as grab samples or short-term composites and then shipped to a laboratory. The method of collection, as well as the need to ship the samples, introduces a degree of uncertainty into the study results. The use of relatively few grab samples or composites taken during the course of the treatability study is not likely to be truly representative of the wastewater variability. Additionally, the wastewater characteristics may change during the period of sample collection and shipment. Within the limitations discussed above, treatability studies do provide positive useful information in the following areas:

1. Determination of the biological treatability of the wastewater
2. Determination of design parameters, including biological stabilization rate, oxygen requirements, sludge production, and biomass settling and thickening characteristics

Because the treatability study utilizes an external small-scale clarifier, limited information is available directly from observation of clarifier effluent and underflow concentrations. As in pilot plant programs, separate settling tests are used as the primary source of information on solids-liquid separation characteristics.

Wastewater for treatability studies may require pretreatment before being exposed to the biological population in the UNOX System. If the wastewater is toxic, pretreatment may include precipitation of certain metals or reactive alteration of specific organic materials. When required, the wastewater stream may be chemically pretreated by pH adjustment or nutrient addition to insure a proper environment for biological activity.

B. Sample Collection and Shipment

The method of collecting wastewater samples for treatability studies should be chosen with great care and with specific objectives in mind. If wastewater variability is known to be minor, the time of collection of samples may be unimportant, and either a grab or composite sample will be equally representative. If significant wastewater variability is anticipated, it may be desirable to collect grab samples at times that are known to represent specific wastewater conditions. In industrial studies specifically, the critical wastewater composition which may determine the success or failure of biological treatment may occur for only brief periods each day or shift. Composited samples of a variable wastewater stream representing average conditions should only be utilized where averaged results are acceptable or where equalization of the wastewater stream is anticipated.

Where practical, wastewater samples should be collected for immediate shipment. If immediate shipment is not practical or if long-term compositing is planned, the collected wastewater sample should be refrigerated. Control of biological activity prior to shipment using techniques such as pH adjustment is not recommended because of the potential for alteration of the wastewater characteristics. Wastewater samples should reach the location of the treatability studies within 24 hr of the time of collection. In most instances, shipment by air express with direct delivery to the wastewater treatment laboratory will be necessary.

In order to determine if any changes in wastewater characteristics have occurred during shipment, a one quart sample of the shipment should be quick frozen, packed in dry ice, and shipped to the treatability laboratory. If this is not feasible, pH adjustment of the 1 qt sample to a pH of 2 is acceptable. Analysis of the preserved sample compared to analysis of the unstabilized samples as received provides information on wastewater alterations which may have occurred during shipment.

The above discussion applies primarily to off-site treatability studies. However, the same sample collection procedure should be followed for on-site treatability studies, unless a continuous representative wastewater supply is available. Due to plant limitations, some on-site studies may not have fresh wastewater continuously available.

FIGURE 4. UNOX® System treatability study open-top reactor. (Courtesy of Union Carbide Corporation.)

C. Treatability Study Equipment

The equipment for bench-scale treatability studies may range from very simple open-top vessels, as shown in Figure 4, to more complex systems, such as the laboratory fermenter system shown in Figure 5 and the covered multistage reactor shown in Figure 6. The program objectives will dictate which approach is more appropriate. For a basic determination of the biological treatability of a wastewater, the simple open top apparatus is usually adequate. However, this type of apparatus cannot be used with wastewaters containing a significant concentration of volatile organic materials. For

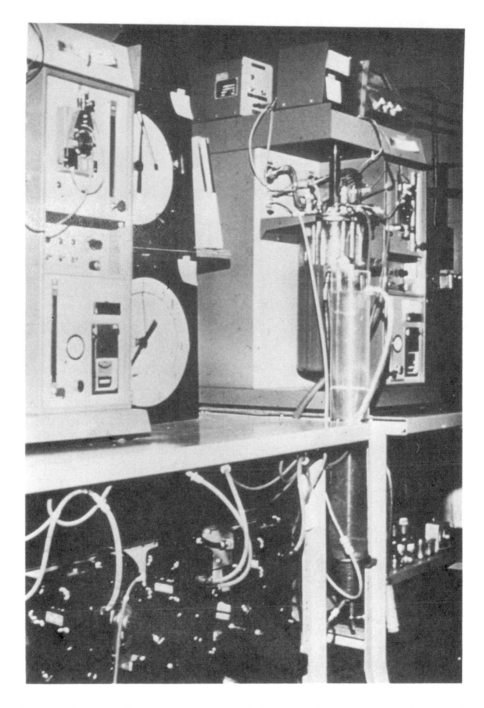

FIGURE 5. UNOX® System treatability study laboratory fermenter reactor. (Courtesy of Union Carbide Corporation.)

wastewaters of this type, closed reactors must be utilized to prevent the stripping of organic materials to the atmosphere. In addition to the above case, the more sophisticated enclosed reactor systems are also employed where the objective is to demonstrate the performance of the oxygen-activated sludge process but timing or financial restraints preclude using a field pilot plant. These systems are also employed in situations where insufficient wastewater is available for larger scale testing or where significant information is already available concerning ox-

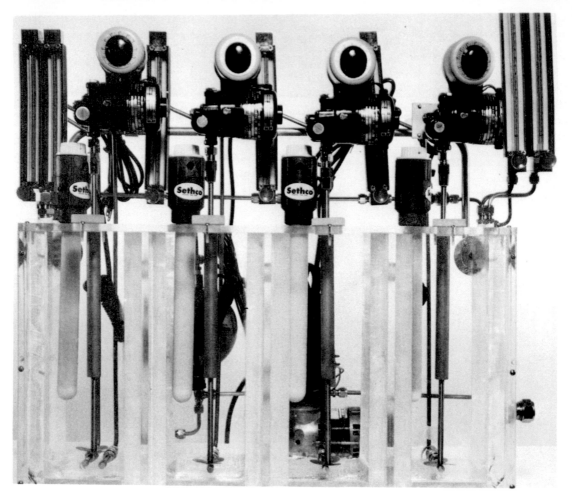

FIGURE 6. UNOX® System treatability study covered multi-stage reactor. (Courtesy of Union Carbide Corporation.)

ygen-activated sludge treatment of the waste-water, and the principal objective is, therefore, design point verification of the process.

1. Open-Top Treatability Reactor

A photograph of this type of treatability reactor is shown in Figure 4. Typically, one or more 2-l single-stage reactors are used. Each reactor utilizes oxygen gas, which is transferred through a sparger-impeller contacting unit. The open-top reactors are sparged with a predetermined flow of CO_2 gas to approximate the operation of a full-scale enclosed UNOX System. The quantity of CO_2 added is determined by the type and organic strength of the wastewater.

These treatability reactors are equipped with variable speed finger pumps (positive flow) for both influent flow and recycle solids flow. The systems have external clarifiers with the capability of variable sludge recycle rates. A pri-

mary clarifier can also be employed prior to the reactor, if necessary.

Where temperature control is necessary to simulate operation at actual industrial plant discharge temperatures or to evaluate summer and winter operation for municipal waste-waters, a water bath is employed. The treatability reactors are placed within the water bath, which is either heated with immersion heaters or cooled with an auxiliary refrigeration system to the predetermined temperature. Chemicals for pH control or nutrient addition can be added to the reactor feed drum or directly into the reactor by positive flow peristaltic pumps.

2. Laboratory Fermenter System

The laboratory fermenter system provides for more sophisticated control of reactor conditions. However, unlike the open-top and covered multi-stage reactors, it does not lend itself readily to transport for on-site studies; therefore, wastewater samples must be shipped to the testing laboratory. Traditionally this type of fermenter equipment is used for batch operations. For oxygen-activated sludge treatability studies, however, they are modified for continuous operation by drilling the glass kettle for effluent discharge to a secondary clarifier. Return sludge from the clarifier and feed to the unit are introduced through the top of the reactor. Instrumentation associated with this equipment allows for precise control of pH, temperature, and dissolved oxygen concentrations. This equipment is available for purchase from several manufacturers including Virtis, Gardener, New York and New Brunswick Scientific, New Brunswick, New Jersey. Typically, a 14-l glass kettle is utilized providing a single-stage closed-top reactor with a 10-l liquid volume. Other volumes can be employed, and these units can also be arranged in series to provide a multi-staged treatability apparatus.

These units can be operated either with once-through gas flow or with gas recycle. If strippable organics are present in the wastewater being tested, gas recycle is employed. In this case, pure oxygen is supplied into the gas space of the fermenter jar. A vent line regulated by a drum meter controls the volume of vent gas. Provisions can be made to maintain the oxygen purity of the gas space, if precise oxygen utilization control is desired. A small recirculation compressor with a by-pass to the gas space is utilized for gas recirculation. Regulation of the amount of by-pass gas will provide variable control of the quantity of gas which is sparged into the wastewater. A mechanical impeller system is used for mixing. The quantity of gas sparged to the wastewater is adjusted to maintain a particular dissolved oxygen concentration. If once-through gas flow is employed, a mixture of pure oxygen and CO_2, as described in the discussion of open-top systems, is fed to the unit.

Instrument probes immersed in the liquid are connected to automatic control equipment. With this equipment the pH, D.O., and temperature levels can be automatically controlled on a continuous basis. The reactor is equipped with a heat exchange coil which can be connected to a water heater or water refrigeration system to automatically control reactor temperature. The system has its own pumps for maintaining influent waste and recycle sludge flow rates. The system also uses an external secondary clarifier, which can be temperature controlled by heating tapes or a cooling water bath.

3. Covered Multi-stage Treatability Reactor

A bench-scale model of the multistage UNOX System has been designed and fabricated for treatability studies. A schematic diagram of this unit and a list of its specifications are shown in Figure 7. The reactor consists of either three or four aeration stages, each with its own heater and stirrer. There is a single gas recirculation compressor for all stages. Influent and sludge recirculation pumps and an external clarifier are also provided. The materials of construction are plexiglass for the reactor and clarifier and stainless steel for the hardware. The pump is a finger type, which uses Tygon® tubing for the wetted surface. The gas recirculation system is composed of individual gas spargers and rotameters for each stage, as well as feed flow and vent flow rotameters. The gas recirculation line also includes a by-pass valve for improved flow control, a surge tank, and a drain valve for condensate.

The operating mode of this system is very nearly identical to that of a full-scale or pilot-scale UNOX System. A covered, staged aeration basin is used for contact of oxygen gas and mixed-liquor. High-purity oxygen enters the first stage of the system and flows cocurrently with the wastewater being treated. Each stage of the aeration basin is a completely mixed unit containing a sparger/impeller unit for gas dissolution and liquid mixing. The only operational difference between the covered multi-stage treatability unit and pilot plant units is the method of gas recirculation. To minimize equipment complexity in the treatability unit, a single gas recirculation compressor is used to supply the recirculated gas to the spargers in all stages. As a result, the gas above the last-stage liquid is used as the sparging gas for all stages. Since the UNOX System vent gas normally contains approximately 50% oxygen, maintenance of elevated D.O. with low power input is still readily achievable. This gas flow scheme differs from the pilot plant unit, where individual compressors on each stage recirculate the gas from that stage back into the liquid in the same stage. The liquid flow scheme in this system is identical to pilot plant and full-scale systems. Wastewater is pumped into Stage 1 by a variable

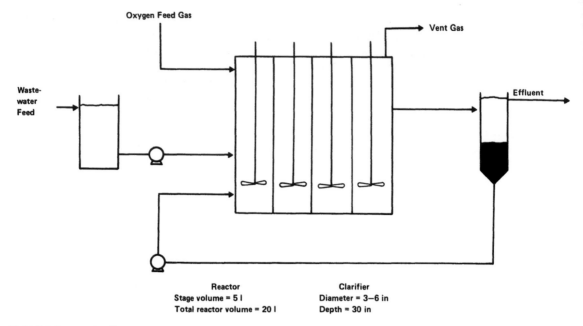

FIGURE 7. UNOX® System treatability study covered multi-stage reactor schematic. (Courtesy of Union Carbide Corporation.)

speed finger pump, so that a variety of feed rates and liquid retention times are possible. Mixed-liquor from the biological reactor flows to a clarifier containing a peripheral effluent weir. Clarifiers used with this system vary in size according to the liquid flow rate in the particular study but generally are 3 to 6 in in diameter. Clarifier underflow solids are withdrawn with a variable speed pump for recycling. Sludge wasting is performed on a batch basis either from the recycle line or the last-stage mixed-liquor. Continuous wasting, usually with an automatic timer, can also be performed from the recycle line.

VI. OPERATIONAL PROCEDURES AND PROCESS MONITORING

A. Biomass Acclimation

An acclimated biomass may be developed by adapting an existing biomass to a new wastewater or by developing a new biomass from a seed source. Adapting an existing biomass is often preferred because of the shorter time period required. If an existing biomass is to be utilized, the best source is usually a biomass grown on domestic wastewater, since it is likely to have a large variety of microorganisms present. If a new biomass is to be developed, the seed source is usually a sample of domestic wastewater that is known to be biologically active.

Acclimation by adapting an existing culture of microorganisms is accomplished by continuous feeding with a waste stream composed of the wastewater normally fed to the biomass mixed with the new wastewater. The concentration of the new wastewater in the mixture is gradually increased until a complete changeover has been achieved. Response of the biomass to introduction of the new wastewater may be examined by monitoring the oxygen uptake rate of the biomass and the COD reduction across the system. Decreases in oxygen uptake rate below the level which would be expected at that particular organic loading can mean that acclimation has not been completely achieved.

Initially, the development of a new biomass from a seed source is achieved most practically by a fill and draw procedure. Before introducing new feed to the aeration chamber, the surfaces are cleaned of slime and the resulting suspended solids are permitted to settle. A portion of the supernatant is removed and replaced by wastewater feed. The fill and draw procedure is followed until a suitable concentration of suspended solids have developed which have settling characteristics making continuous flow operation feasible. Biological stabilization of organic material can be observed by calculating the COD at the time of feeding based on the COD of the supernatant plus the COD of the feed and comparing the result with the COD of the supernatant removed at the time of the next feeding. A decrease in COD over the aeration period indicates that organic material is being biologically stabilized. If volatile organic materials are present in the wastewater, potential removals of COD by stripping should be considered. Measurements of total organic carbon (TOC) or total oxygen demand (TOD) may also be utilized.

Observation of the acclimation of a new biomass grown from seed can be achieved both microscopically and macroscopically. Microscopic examination provides visual inspection of the developing biomass. The presence of a mixed culture, including higher forms, can be ascertained. With experience, changes in the numbers of higher forms or the degree of activity of higher forms can be used as an indicator of stress on the biological system. Generally, when higher forms decrease in numbers or become sluggish, the biomass is being stressed. Macroscopic observations can include the development of slime growth on surfaces within the system and, eventually, the development of suspended solids. The macroscopic observations can be supported by oxygen uptake rate measurements.

B. Oxygen Supply

Oxygen for pilot plant programs is generally supplied from liquid oxygen storage vessels. Liquid oxygen must be vaporized to provide a gaseous oxygen supply to the pilot plant. Treatability studies require smaller amounts of oxygen and the use of high-pressure cylinders is recommended.

In the enclosed reactors such as the pilot plant or covered treatability units, the oxygen supply is fed through the gas space of the reactor and the feed rate is adjusted to provide a suitable purity of oxygen in the vent gas, typically about 50%. This gas flow rate will result in concentrations of oxygen, carbon dioxide, and other materials in the gas phase similar to those obtained in a full-scale UNOX System.

In open-top treatability reactors, consideration must be given to the amount of carbon dioxide in the system. Contact with the atmosphere permits the carbon dioxide to be stripped and results in lower concentrations of carbon dioxide dissolved in the mixed-liquor than would be obtained in an enclosed system. Dissolved carbon dioxide, in the form of carbonic acid, represents acidity and results in a slight depression of the mixed-liquor pH level. This may be important when treating high-strength wastes and must be considered. A suitable level of carbon dioxide dissolved in the mixed-liquor of an open-top reactor can be achieved by mixing carbon dioxide with the oxygen feed gas. The concentration of carbon dioxide in the feed gas stream should be equal to that which would be present in the gas phase of an enclosed reactor. It is recommended that carbon dioxide be mixed with the oxygen supply in the approximate amounts listed below. As can be seen, the amount of CO_2 added is dependent on the BOD_5 content of the wastewater.

Feed BOD_5 concentration (mg/l)	CO_2 concentration in feed gas (%)
300	5—15
300—1000	15—25
1000—5000	25—50

C. Biomass Loading

For most types of waste waters, a reasonable estimate of the kinetic rate of biological stabilization is available from past experience. Based on this experience, a narrow range of reactor loadings and detention times expected to achieve the desired objectives can be selected. In such cases, the objective of the pilot plant or treatability program is to confirm the selected operating conditions and to provide additional data which may be used to select and optimize the process design basis.

With bench-scale programs, in order to reduce the time required to examine the kinetic rates of biological stabilization, it is sometimes desirable to operate two or more units in parallel. With field pilot plant programs, various loading rates are investigated in sequential phases of operation.

D. Process Monitoring

The process monitoring for a UNOX System pilot plant or treatability study includes both analytical and operational evaluations. The analytical sampling is carried out by the plant operators or laboratory technicians and consists

TABLE 6

UNOX® System Process Monitoring: Analytical Testing Schedule

Source and analysis	Day of week						
	S	M	T	W	T	F	S
Influent wastewater							
BOD$_s$ total	X	X	X	X	X	X	X
BOD$_s$ soluble	X	X	X	X	X	X	X
COD total	X	X	X	X	X	X	X
COD soluble	X	X	X	X	X	X	X
TSS and VSS	X	X	X	X	X	X	X
NH$_3$-N		X		X		X	
PO$_4$-P		X		X		X	
Effluent							
BOD$_s$ total	X	X	X	X	X	X	X
BOD$_s$ soluble	X	X	X	X	X	X	X
COD total	X	X	X	X	X	X	X
COD soluble	X	X	X	X	X	X	X
TSS and VSS	X	X	X	X	X	X	X
NH$_3$-N		X		X		X	
PO$_4$-P		X		X		X	
Mixed-liquor							
TSS and VSS	X	X	X	X	X	X	X
Recycle sludge							
TSS and VSS	X	X	X	X	X	X	X
COD total	X	X	X	X	X	X	X

Note; All samples are 24 hr composites.

of daily 24-hr composite samples for all analytical tests. These samples include wastewater feed, clarifier effluent, mixed-liquor, and recycle sludge. A typical analytical testing schedule for a carbonaceous removal study is shown in Table 6. All laboratory tests are performed as indicated in *Standard Methods.**

Operational monitoring includes the determination of oxygen uptake rates, biomass settling rates, and the sludge volume index. The oxygen uptake rates are usually determined one to three times per day on mixed-liquor samples from each stage. The settling tests, both stirred and unstirred, are conducted each day using samples of mixed-liquor from the last stage. Settling tests are also conducted using various mixtures of mixed-liquor and recycle sludge, in order to evaluate the biomass settling character-

istics over a broad range of solids concentrations.

Operational monitoring also includes measuring and calibrating influent and recycle flow rates, as well as monitoring the amount of sludge being wasted daily. Determination of pH and temperature in the feed, effluent, and first and last stages of the reactor is performed routinely during the day. The sludge blanket depth is monitored frequently to determine whether or not sludge should be wasted. Dissolved oxygen measurements in all stages and in the secondary clarifier effluent are performed regularly. In the pilot plant units, monitoring also includes the feed gas and vent gas flow rates, gas composition in each stage, gas space pressure, hydrocarbon readings in Stages 1 and 4, mixer motor revolutions per minute, and power consumption. Clarifier scraper revolutions per

* *Standard Methods for the Examination of Water and Wastewater,* 14th ed., American Public Health Association, Washington, D.C., 1976.

hour are recorded in all studies. Routine microscopic examinations are also made to determine the presence of any filamentous organisms and free-swimming or stalked ciliates and to observe general biomass characteristics. Table 7 lists the type and method of operational monitoring performed on UNOX System pilot plant and treatability studies.

VII. DATA ANALYSIS

A UNOX System pilot plant or treatability study will provide information relating to four significant parameters of biological wastewater treatment systems: substrate removal, oxygen consumption, sludge production, and solids-liquid separation. With careful analysis of the data obtained in the study, a large amount of information useful in the full-scale design effort can be obtained.

A. Substrate Removal

In any wastewater treatment system, one of the most important indicators of process performance is final effluent quality. Therefore, in an activated sludge system, the degree of BOD_5 and COD removal achieved in the biological reactor is of great importance. Process performance is evaluated by a plot of food to biomass ratio (F/M) removed (pound BOD_5 or COD removed per day per pound MLVSS) vs. applied biomass loading (lb BOD_5 or COD applied per day per pound MLVSS). The F/M removed is calculated based on the pounds of influent total substrate minus the pounds of effluent soluble substrate. This procedure allows for a more accurate analysis by separating reactor performance from clarifier performance. Most of the insoluble substrate in the effluent is due to biological solids which are not removed by the secondary clarifier.

The slope of the line on the plot mentioned above is equal to the percent BOD_5 or COD removal. A linear relationship indicates a constant percent removal regardless of biomass loading and further indicates that the system is not kinetically limited at any loading applied during the program. A reduced slope at higher loadings can be taken as an indication of biomass overloading, resulting in higher effluent substrate concentrations. Figure 8 illustrates the two types of curves which may be obtained from pilot plant or treatability study results.

TABLE 7

UNOX® System Process Monitoring: Operational Monitoring

Parameter	Method of Measurement
Oxygen uptake rate	Weston and Stack Dissolved Oxygen Analyzer
Settling rate and sludge volume index	One-liter graduated cylinder, stirred (at 10 rph) and unstirred
Liquid flow rate	Calibration drum
Gas flowrate	Dry test meter for feed gas, wet test meter for vent gas
pH	Leeds and Northrup pH meter
Temperature	Thermometer
Sludge blanket depth	Photocell blanket depth indicator (pilot plant), visual observation (treatability study)
Dissolved oxygen concentration	Weston and Stack Dissolved Oyxgen Analyzer
Gas composition	Servomex® Paramagnetic Oyxgen Analyzer
Gas pressure	Magnihelic pressure gauge
Hydrocarbon level	Johnson and Williams combustible gas analyzer
Mixer Motor RPM	Tachometer
Mixer Motor Power Consumption	Wattmeter
Clarifier scraper revolutions per hour	Visual observation
Microscopic examination	Microscope (150 ×)

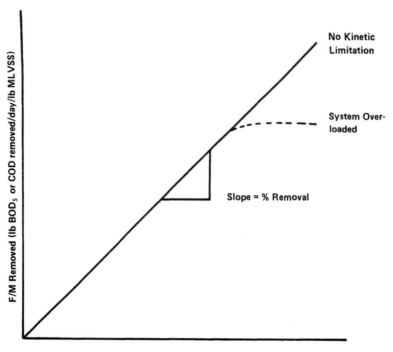

FIGURE 8.   F/M applied vs. F/M removed.

B. Oxygen Consumption

The oxygen requirement is one of the most important aspects in the design of a UNOX System or any biological treatment system. For this reason, considerable effort is made to accurately determine the amount of oxygen required by the biomass for cell synthesis and endogenous respiration at both design and peak operating conditions. A UNOX pilot plant or treatability program permits accurate determination of the oxygen requirement by providing up to three separate methods of determining the amount of oxygen being consumed in the process.

The first method is the gas-phase oxygen mass balance. Since the feed rate of pure oxygen and the vent gas flow rate and oxygen purity are continuously monitored during a pilot plant study, the oxygen utilized by the biomass can be calculated by a simple balance around the closed system. This balance can be further refined by deducting the amount of oxygen used to increase the dissolved oxygen concentration in the system. Influent D.O. can be measured along with effluent D.O. to determine the change which, when multiplied by the flow rate, gives the amount of oxygen used for

D.O. elevation. This method is accurate only if the system is kept gastight throughout the study.

The second method of determining oxygen consumption is by periodically measuring the oxygen uptake rate (OUR, expressed in milligrams per liter per hour) of the biomass. OUR measurements are made one to three times per day on mixed-liquor from each of the stages of the UNOX reactor. The dissolved oxygen level in a mixed-liquor sample withdrawn from the reactor is monitored to depletion by using a polarographic oxygen analyzer designed for *in situ* determinations. The slope of the straight line section of the curve of D.O. vs. time is the OUR. Because of the response time of the D.O. probe and the inherent time lag caused by removing samples from the system, OUR measurements have been found to yield consistently low oxygen requirements, particularly in a system with very high biomass concentrations and, therefore, high OURs. However, OUR measurements are important as a means of determining the relative amounts of oxygen consumed in each of the stages of the UNOX System reactor. This information permits proper sizing of oxygen transfer equipment for a full-scale UNOX System. OUR measurements are also very useful as indicators of process stability, since sudden decreases in OURs not attributable to biomass loading changes generally indicate significant process problems.

The most accurate method of determining oxygen consumption is by an overall COD balance. Since, by definition, a pound of oxygen is required to consume a pound of COD, oxygen consumption is directly related to a complete inventory of COD around the system. If a system is at steady state, there is one "source" and two "sinks" of COD in the system:

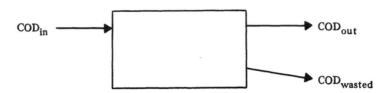

However, under nonsteady state conditions, a third COD "sink" exists, i.e., accumulation in the mixed-liquor and clarifier solids inventory (COD_{accum}):

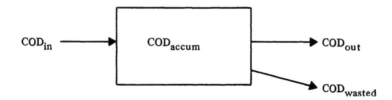

The following equation, therefore, represents a complete COD balance around the system:

$$COD_{in} - COD_{out} - COD_{wasted} - COD_{accum} = \text{oxygen consumed}$$

COD_{in} is the COD content of the influent wastewater; COD_{out} refers to the COD of the secondary clarifier effluent; COD_{wasted} is the COD content of the intentionally wasted sludge; COD_{accum} is the COD gained or lost due to changes in the mixed-liquor or clarifier solids inventory. Note that due to the scale of the treatability study equipment, COD_{wasted} must include the COD content of mixed-liquor and recycle sludge samples taken for analysis. During UNOX pilot plant and treatability studies, total COD is determined on daily composite samples of influent, effluent, and recycle sludge so that an accurate COD balance can be made.

Due to analytical variability in COD determinations performed on recycle sludge samples and short-term errors in accounting for system solids inventory changes, it is recommended that the COD balance be performed over no shorter a time period than a weekly average. Additionally, a significant portion of the sludge inventory of the system is found in the secondary clarifier. In order to minimize the effect of the accumulation term in the COD balance, the clarifier sludge blanket level should be kept as constant as is practical.

By using the COD balance, the amount of oxygen consumed per day can be calculated. By using this information along with the pounds of BOD_5 or COD removed per day (based on total influent substrate minus soluble effluent substrate), the oxygen consumption ratio, expressed as pounds of oxygen consumed per pound of substrate removed, can be determined. A plot of oxygen consumption ratio vs. F/M removed, using data obtained at different biomass loadings, will result in a curve with the general shape shown in Figure 9. Due to the increasingly significant effect of endogenous respiration at low biomass loadings, the oxygen consumption ratio increases dramatically as F/M decreases.

C. Sludge Production

The quantity of excess biological solids produced in a secondary wastewater treatment system is an important parameter needed for the design of sludge handling and disposal equipment. Because the cost of excess sludge disposal can reach 50% of the total cost of secondary wastewater treatment, it is important that accurate sludge production data be collected dur-

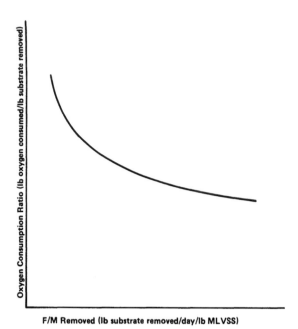

FIGURE 9.   Oxygen consumption ratio vs. F/M removed.

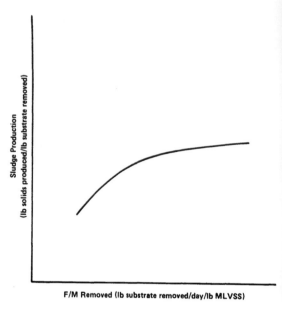

FIGURE 10.   Sludge production vs. F/M removed.

ing all secondary wastewater treatment studies in order to preclude overdesign or underdesign of the sludge disposal system. Excess sludge production is a function of the degree of endogenous respiration occurring in the biological reactor, which is governed primarily by the biomass loading (F/M). Since the UNOX System employs a multi-stage gas-liquid contacting system with all streams entering the first stage, the biomass loading decreases rapidly from stage to stage. Therefore, a high degree of endogenous respiration occurs in the latter stages, resulting in decreased sludge production. It has also been determined that the high dissolved oxygen environment results in a more aerobic biological floc and contributes to lower sludge production. This occurs since all parts of the biomass are in a working mode and, when placed in a food-limiting situation, the sludge will undergo a higher degree of endogenous respiration or auto-oxidation than sludge floc in a low D.O. environment.

The measurement of sludge production in any biological system requires careful control of wasting schedules and system sludge inventory levels, as well as careful analytical monitoring of the system. Sludge production data are collected during all UNOX pilot plant and treatability studies and include total suspended solids (TSS) and volatile suspended solids (VSS) analyses on daily composite samples of influent, effluent, mixed-liquor, and recycle sludge. The method of calculating sludge production is similar to the method used to calculate oxygen consumption. As with the COD balance, it is recommended that data representing a weekly or longer average be used to determine sludge production. "Total solids produced" includes solids leaving the system in the effluent and waste streams, as well as solids accumulated in the mixed-liquor and clarifier inventory. "Total solids requiring disposal" includes only the waste and accumulation terms and is the figure which should be used in the design of sludge handling facilities. Both total solids produced and total solids requiring disposal are usually expressed as pounds of solids per pound of substrate removed.

A plot of pounds of solids produced per pound substrate removed vs. F/M removed is shown in Figure 10. The effect of biomass loading on sludge production is apparent from the figure. Sludge production in oxygen-activated sludge systems is discussed in depth in Chapter 2, Volume II.

In addition to the determination of sludge production amounts, the quality of the excess sludge and its dewatering characteristics are often of concern. Limited information about the dewatering characteristics of biological solids is available from treatability studies because of the limited quantity of sludge available. If sludge dewatering is an important objective in a study, the use of a mobile pilot plant is recommended. With treatability study equipment, dewatering studies are generally limited to simple filter leaf tests to establish filtration rates and cake characteristics employing selected conditioning agents. The dewatering characteristics of oxygen-activated sludges are discussed in Chapter 3, Volume II.

D. Solids-liquid Separation

Small secondary clarifiers do not always provide meaningful information on sludge settling and thickening characteristics. This is particularly true for treatability study clarifiers. However, a characterization of sludge settling properties can be made by some relatively simple additional testing. A one-liter graduated cylinder fitted with a stirring device is used to determine initial settling velocities over a range of sludge solids concentrations. The stirrer, ro-

tated at approximately 10 RPH, is used to reduce wall effects, which would otherwise be disproportionately large compared to full-scale clarifiers. The settling test is performed by placing a sludge sample of known solids concentration in the graduated cylinder and recording the height of the solids-liquid interface as a function of time. A settling curve, such as that illustrated in Figure 11, is generated. The initial settling velocity is defined as the slope of the steepest portion of the curve. This procedure is repeated over a range of initial sludge concentrations (various combinations of mixed-liquor, recycle, and effluent streams may be used to produce sludges with a range of initial solids concentrations), and the results are plotted on log-log paper as initial settling velocity vs. initial solids concentration, as shown in Figure 12.

Over the range of solids concentrations normally encountered in the activated sludge system, this plot will yield a straight line relating initial solids concentration, C_i to initial settling velocity, V_i, by the following relationship:

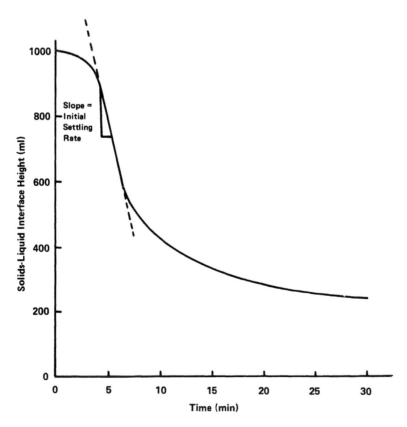

FIGURE 11.   Determination of initial settling rate.

$$V_i = a\, C_i^{-n}$$

where the constants a and n are determined from the log-log plot. This relationship may be used to predict performance in full-scale clarifiers through use of the limiting solids flux theory as discussed in Chapter 7, Volume I.

VIII. PILOT PLANT DATA BANK

To assist in the reduction of raw pilot plant data into design correlations, Union Carbide has developed an extensive wastewater treatment pilot plant data bank. This wastewater treatment data bank is a comprehensive information storage and retrieval system which provides an efficient means of assessing and analyzing data. Input data, the source of which are the results obtained from pilot plant operations, are stored in the system via coded variable identification assignments. Input data are divided into groups, which can be further subdivided into runs; e.g., a group might consist of all data from a pilot plant program, which

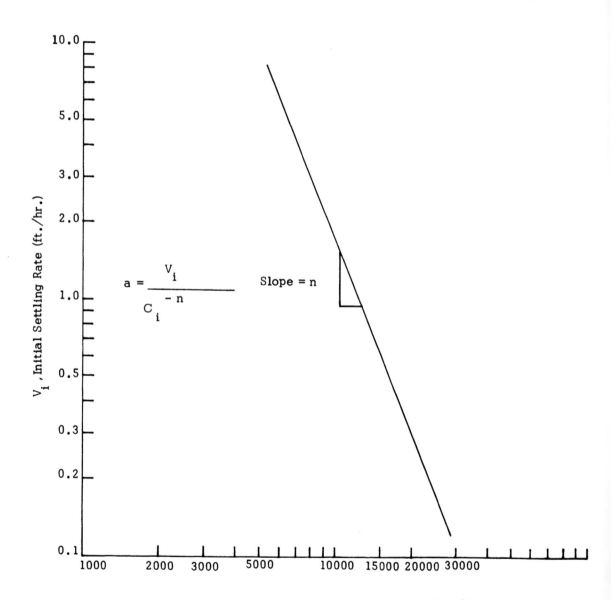

FIGURE 12.   Determination of biomass settling constants.

might then be divided into runs that corresponded to the phases of operation. Data retrieval is accomplished through the use of the coded variable assignments together with a series of system commands that identify which data are to be assessed and the format in which they are to be presented. Data for a particular variable or set of variables can be retrieved from any or all groups and runs.

The use of the data bank as a central storage facility precludes the possibility of data becoming lost or inaccessible. It also eliminates a great deal of the manpower otherwise required to assimilate the data. The capabilities of the data bank are not limited to simple data storage and retrieval, however. Even a minimal knowledge of computer languages and programming techniques is sufficient to allow the user to direct the system to retrieve selected information, perform a linear regression analysis, and plot the results. This feature is particularly valuable in fields such as wastewater treatment, where a significant amount of design information is obtained from empirical correlation of data, as it allows the updating of design correlations with relatively minor effort.

The value of the data bank increases as the amount of data within it increases. Extensive data generation is useless if the data cannot be readily obtained and used. The data bank serves the necessary function of providing convenient and flexible data processing, while eliminating the time-consuming and often frustrating aspects typically associated with it.

IX. OPERATING EXPERIENCE

As a result of conducting numerous pilot plant and treatability studies over the past several years, Union Carbide has amassed a considerable amount of information on the treatment of many types of wastewaters. Actual performance data from over 50 of these pilot plant and treatability studies are presented below. Each wastewater type is discussed in terms of the general conclusions that can be made regarding wastewater treatability and design basis.

A. Municipal Wastewaters

Table 8 lists UNOX System performance on four entirely domestic wastewaters. As shown in the data, efficient treatment occurred at biomass loadings of 0.55 to 0.65 lb BOD_5/day/lb MLVSS and retention times of 1.3 to 2.0 hr. Clarifier underflow concentrations of 1.5 to 3.0% resulting in mixed-liquor suspended solids levels of 4000 to 6000 mg/l were achieved at clarifier overflow rates of 460 to 900 GPD/ft². These values represent the typical range of design values for full-scale UNOX Systems treating domestic wastes.

Table 9 contains a listing of several nitrification programs conducted on municipal wastes. The generally low biomass loadings and long system sludge retention time (SRT) requirements characteristic of the nitrification process can be seen in the data. Retention times of 0.8 to 2.3 hr and mixed-liquor suspended solids concentrations of 2300 to 6400 mg/l were achieved in the nitrification portion of these studies. Ammonia removals of 91 to 97% were seen providing final effluent ammonia levels of 0.3 to 1.7 mg/l.

A large percentage of UNOX System pilot plant and treatability studies have been conducted on combined domestic and industrial wastes. Tables 10 to 16 list data from more than 20 of these studies. Table 10 shows data demonstrating effective treatment of combined domestic and pulp and paper wastewater. The effect of the industrial fraction of the waste is indicated by the influent COD/BOD_5 ratio of 2.7 to 3.1. System performance is not significantly different from that seen with domestic wastes. Biomass loadings of 0.65 to 1.05, retention times of 1.0 to 1.5 hr, recycle sludge solids concentrations of 2.0 to 2.2%, and MLSS levels of 5500 to 5800 mg/l are all comparable to results obtained in municipal UNOX System applications.

Data obtained on combined domestic and petrochemical wastes are shown in Table 11. With the exception of the Wuppertal-Buchenhofen data on retention time and biomass loading, the data indicate typical domestic wastewater design parameters. The influent wastewater in the Wuppertal-Buchenhofen study had a very large industrial contribution, as shown by the influent COD concentration of 767 mg/l. Despite this contribution, 95% BOD_5 removal was achieved.

Listed in Table 12 are data obtained in nine pilot plant and treatability studies conducted on combined domestic and food processing wastes. These wastes include contributions from grain processing, seafood processing, meat packing, canning and dairy operations. The influent wastewaters in these studies had BOD$_s$ concentrations varying from 72 to 660 mg/l with COD values of 170 to 1200 mg/l. In all cases, effluent BOD$_s$ concentrations were 23 mg/l or less. Biomass loadings in these studies ranged from 0.4 to 1.4 lb BOD$_s$/day/lb MLVSS, requiring retention times of 0.7 to 4.1 hr. Clarifier underflow solids concentrations of 1.3 to 3.2% and MLSS levels of 3700 to 7400 mg/l were achieved at clarifier overflow rates usually between 500 and 1500 GPD/ft². Lower overflow rates resulted from equipment limitations in two studies.

Table 13 contains data obtained on combined domestic and brewery and distillery wastewaters. These wastes are characterized by rela-

tively high organic concentrations (100 to 1000 mg/l BOD$_s$, 230 to 1800 mg/l COD) but can also be effectively treated at high biomass loadings. F/M levels of 0.5 to 1.9 were demonstrated in these studies, resulting in effluent BOD$_s$ concentrations less than 27 mg/l in all cases. Only the Jacksonville study, which had an influent BOD$_s$ of 900 to 1000 mg/l and COD of 1500 to 1800 mg/l, required retention times greater than typical UNOX domestic wastewater treatment systems. Clarifier underflow solids concentrations of 1.2 to 3.1% and MLSS concentrations of 2600 to 7800 mg/l were obtained at clarifier overflow rates of 380 to 1300 GPD/ft².

The results obtained from treatment of combined domestic and textile wastes are listed in Table 14. These wastewaters are generally high in organic constituents and/or suspended sol-

TABLE 8

UNOX® System Operating Data: Domestic Wastes

Parameters Phase	Spokane, Wash.	Wuppertal, Germany	Miami, Fla. I	Miami, Fla. II	Denver, Colo.
Retention time, RT$_Q$ (hr)	1.5	2.0	1.7	1.3	1.7
Recycle ratio, R/Q (%)	20	—	27	26	—
Wastewater temperature (°C)	18	—	31	31	—
MLSS (mg/l)	3700	5100	6200	5900	4700
MLVSS (mg/l)	2800	4200	2700	2900	3200
Biomass loading (lb BOD$_s$/lb MLVSS/day)	0.57	0.60	0.55	0.64	0.61
Biomass loading (lb COD/lb MLVSS/day)	1.12	1.47	1.14	1.50	1.41
Influent characteristics (mg/l)					
BOD$_s$	102	211	184	170	137
COD	200	517	380	398	316
SS	64	89	183	172	174
Effluent characteristics (mg/l)					
BOD$_s$	16	31	6	5	13
COD	47	219	67	63	60
SS	21	40	17	13	16
Removals (%)					
BOD$_s$	84	85	97	97	90
COD	77	58	82	84	80
SS	67	55	91	92	90
Clarifier overflow rate (gal/day/ft²)	460	—	730	890	650
Clarifier underflow concentration (%)	1.9	1.6	2.7	2.3	2.0
Sludge production (lb TSS/lb BOD$_{sR}$)	0.59	0.40	0.31	0.36	0.84

TABLE 9

UNOX® System Operating Data: Nitrification Studies

Parameters	Amherst, N.Y. Step 1 Carbonaceous	Amherst, N.Y. Step 2 Nitrification	Indianapolis, Ind.	Phoenix, Ariz.
Retention time, RT_Q (hr)	0.45	1.37	2.32	0.78
Recycle ratio, R/Q (%)	29	53	66	50
Wastewater temperature (°C)	17	20	24	31
MLSS (mg/l)	6800	6400	5900	2300
MLVSS (mg/l)	5300	4600	4500	1800
Biomass loading (lb BOD_s/lb MLVSS/ day)	1.48	0.16	0.37	—
Biomass loading (lb COD/lb MLVSS/ day)	2.76	0.37	0.74	0.86
Influent characteristics (mg/l)				
$\quad BOD_s$	147	43	164	—
$\quad COD$	274	100	330	46
$\quad SS$	158	45	105	15
$\quad NH_3$-N	22	18	22	28
$\quad TKN$	29	23	31	—
Effluent characteristics (mg/l)				
$\quad BOD_s$	39	14	3	—
$\quad COD$	108	64	36	42
$\quad SS$	49	34	9	14
$\quad NH_3$-N	18	1.7	0.3	0.9
$\quad TKN$	23	5	2.7	—
$\quad NO_2$-N	—	0.2	0.1	0.1
$\quad NO_3$-N	—	15.3	16.5	26
Removals (%)				
$\quad BOD_s$	73	66	98	—
$\quad COD$	61	36	89	9
$\quad SS$	69	25	91	6
$\quad NH_3$-N	18	91	97	97
$\quad TKN$	20	78	91	—
Clarifier overflow rate (gal/day/ft²)	770	770	500	1140
Clarifier underflow concentration (%)	3.0	1.6	1.6	0.7
Sludge production (lb TSS/lb BOD_{sR})	0.75	0.00	0.48	—
Sludge retention time, SRT (days)	1.5	11.1	9.5	8.4

ids. The wastes treated in the Morganton and Danville studies had COD concentrations of 420 to 690 mg/l and suspended solids levels as high as 410 mg/l. Effluent quality of 8 to 35 mg/l BOD_s and 9 to 33 mg/l suspended solids (SS) was achieved at biomass loadings of 0.2 to 0.7. The low biomass loading at Danville is due to the extremely high COD/BOD_s ratio (4.0 to 4.5), indicating the large and difficult to degrade industrial contribution. Recycle sludge concentrations (2.8 to 2.9%) and mixed-liquor levels (3200 to 5900 mg/l) were within typical operating ranges.

Treatment of combined domestic and refinery wastewater is illustrated by two Los Angeles area studies which are listed in Table 15. At biomass loadings of 1.0 to 1.1, influent BOD_s concentrations of 170 to 220 mg/l were reduced to 12 to 20 mg/l, with effluent suspended solids of 20 to 22 mg/l. The systems were operated at retention times of 1.4 to 2.0 hr.

Table 16 lists data from two studies conducted on combined domestic and mixed industrial wastes. Biomass loadings and retention times were similar to studies conducted with domestic wastewaters. Recycle sludge and mixed-

TABLE 10

UNOX® System Operating Data: Combined Domestic and Pulp and Paper Wastes

Parameters Phase	Location Duluth, Minn.	
	I	II
Retention time RT_Q (hr)	1.5	1.04
Recycle ratio, R/Q (%)	30	30
Wastewater temperature (°C)	14	14
MLSS (mg/l)	5800	5500
MLVSS (mg/l)	5000	4600
Biomass loading (lbs BOD_5/lb MLVSS/day)	0.65	1.05
Biomass loading (lb COD/lb MLVSS/day)	2.02	2.76
Influent characteristics (mg/l)		
BOD$_5$	202	201
COD	633	544
SS	145	132
Effluent characteristics (mg/l)		
BOD$_5$	31	38
COD	247	223
SS	52	51
Removals (%)		
BOD$_5$	85	81
COD	61	59
SS	63	63
Clarifier overflow rate (gal/day/ft²)	590	840
Clarifier underflow concentration (%)	2.2	2.0
Sludge production (lb TSS/lb BOD$_{5R}$)	0.76	0.90

TABLE 11

UNOX® System Operating Data: Combined Domestic and Petrochemical Wastes

Parameters	Location	
	Baltimore, Md.	Wuppertal-Buchenhofen, Germany
Retention time, RT_Q (hr)	1.8	6.2
Recycle ratio, R/Q (%)	36	—
Wastewater temperature (°C)	21	—
MLSS (mg/l)	4200	5200
MLVSS (mg/l)	3300	4300
Biomass loading (lb BOD_5/lb MLVSS/day)	0.72	0.32
Biomass loading (lb COD/lb MLVSS/day)	1.31	0.70
Influent characteristics (mg/l)		
BOD$_5$	177	348
COD	321	767
SS	90	114
Effluent characteristics (mg/l)		
BOD$_5$	13	18
COD	60	176
SS	25	36
Removals (%)		
BOD$_5$	93	95
COD	84	77
SS	71	68
Clarifier overflow rate (gal/day/ft²)	650	—
Clarifier underflow concentration (%)	1.5	2.0
Sludge production (lb TSS/lb BOD$_{5R}$)	0.49	0.52

TABLE 12
UNOX® System Operating Data: Combined Domestic and Food Processing Wastes

Parameters / Phase	Cedar Rapids, Ia.		Sacramento, Cal.		New Orleans, La.		Oakland, Cal.
	I	II	I	II	I	II	
Retention time, RT_Q (hr)	1.8	1.9	1.5	0.7	1.8	1.8	2.2
Recycle ratio R/Q (%)	29	42	25	25	25	26	31
Wastewater temperature (°C)	18	18	23	18	21	26	19
MLSS (mg/l)	4700	4200	4000	4200	5600	7400	6200
MLVSS (mg/l)	4000	3600	2900	3200	2700	5400	4800
Biomass loading (lb BOD_5/lb MLVSS/day)	1.40	1.21	0.67	1.44	0.76	0.56	0.48
Biomass loading (lb COD/lb MLVSS/day)	2.80	3.00	1.31	4.03	1.11	0.92	1.04
Influent characteristics (mg/l)							
$\quad BOD_5$	415	331	133	143	210	229	209
$\quad$ COD	826	825	260	400	308	377	454
$\quad$ SS	180	386	160	136	191	236	126
Effluent characteristics (mg/l)							
$\quad BOD_5$	12	23	11	15	12	12	11
$\quad$ COD	99	132	57	95	64	66	90
$\quad$ SS	18	42	19	32	18	28	30
Removals (%)							
$\quad BOD_5$	97	93	92	89	94	95	95
$\quad$ COD	88	84	78	76	69	82	80
$\quad$ SS	90	89	88	76	91	88	76
Clarifier overflow rate (gal/day/ft²)	520	530	590	780	650	650	530
Clarifier underflow concentration (%)	1.5	1.3	1.9	1.8	2.5	3.2	2.3
Sludge production (lb TSS/lb BOD_{5R})	0.64	1.08	0.46	0.86	0.45	0.27	0.50

TABLE 12 (continued)
UNOX® System Operating Data: Combined Domestic and Food Processing Wastes

Parameters Phase	Batavia, N.Y. I	II	III	Kansas City, Kans. I	II	Salem, Oreg. I	II	Camden, N.J. I	II	Bremerhaven, Germany I	II
Retention time, RT_Q (hr)	4.1	2.8	1.5	1.9	2.5	1.2	2.5	3.3	2.0	3.9	3.4
Recycle ratio R/Q (%)	24	45	34	23	34	37	30	—	—	—	—
Wastewater temperature (°C)	15	21	19	29	28	22	18	—	—	—	—
MLSS (mg/l)	3700	5900	7000	5700	6300	4400	5200	5400	6600	6300	4100
MLVSS (mg/l)	2200	4300	4500	4300	4600	3300	4400	3900	4800	5600	3600
Biomass loading (lb BOD_5/lb MLVSS/day)	0.41	0.55	0.79	0.73	0.50	0.42	0.53	0.38	0.58	0.66	0.79
Biomass loading (lb COD/lb MLVSS/day)	0.93	1.15	1.17	1.59	1.48	0.99	1.03	0.78	1.22	1.21	1.40
Influent characteristic (mg/l)											
BOD_5	159	262	220	249	244	72	242	205	230	658	418
COD	352	578	325	541	711	169	469	405	485	1207	739
SS	221	430	174	161	257	118	314	156	212	212	166
Effluent characteristics (mg/l)											
BOD_5	11	14	23	9	8	8	17	15	10	20	17
COD	73	89	97	160	123	49	70	57	47	165	194
SS	9	12	19	8	12	22	22	18	9	15	15
Removals (%)											
BOD_5	92	94	90	96	97	89	93	93	96	97	96
COD	80	84	71	70	83	71	85	96	90	86	74
SS	96	97	89	95	95	81	93	88	96	93	91
Clarifier overflow rate (gal/day/ft²)	1520	1130	110	620	460	590	290	—	—	130	150
Clarifier underflow concentration (%)	1.9	1.9	3.0	3.1	2.8	1.4	2.0	—	—	1.7	1.3
Sludge production (lb TSS/lb BOD_{sr})	0.81	0.20	0.66	0.42	0.34	1.27	1.56	0.35	0.37	0.65	0.57

TABLE 13

UNOX® System Operating Data: Combined Domestic and Brewery and Distillery Wastes

Parameters	St. Louis, Mo.		Louisville, Ky.		Jacksonville, Fla.			Philadelphia SE, Pa.
Phase	I	II	I	II	I	II	III	
Retention time, RT_Q (hr)	1.8	2.0	1.8	1.6	6.1	6.0	5.0	0.7
Recycle Ratio, R/Q (%)	37	30	28	28	58	57	49	16
Wastewater temperature (°C)	29	27	14	19	36	34	36	28
MLSS (mg/l)	5100	7000	6100	4000	7800	4400	3500	2600
MLVSS (mg/l)	3800	5400	3800	3000	6100	3800	2800	2100
Biomass loading (lb BOD_s/lb MLVSS/day)	0.83	0.50	0.66	1.02	0.66	0.96	1.59	1.87
Biomass loading (lb COD/lb MLVSS/day)	1.64	1.03	1.34	1.86	1.17	1.72	2.59	5.11
Influent characteristics (mg/l)								
BOD_s	238	227	185	200	1010	915	918	100
COD	471	467	375	365	1793	1640	1493	227
SS	83	76	180	135	428	465	357	83
Effluent characteristics (mg/l)								
BOD_s	10	8	21	17	17	25	27	7
COD	106	81	88	72	95	84	78	37
SS	27	15	20	16	25	65	34	7
Removals (%)								
BOD_s	95	96	89	92	98	97	97	93
COD	76	82	77	80	95	95	95	83
SS	64	80	89	88	94	86	90	91
Clarifier overflow rate (gal/day/ft²)	650	580	660	710	380	380	460	1300
Clarifier underflow concentration (%)	1.9	3.1	2.6	1.7	1.9	1.3	1.2	1.9
Sludge production (lb TSS/lb BOD_{sR})	0.33	0.27	0.64	0.51	0.25	0.32	0.45	1.01

TABLE 14

UNOX® System Operating Data: Combined Domestic and Textile Wastes

Parameters	Morganton, N.C.			Danville, Va.	
Phase	I	II	III	I	II
Retention time RT_Q (hr)	3.5	2.0	3.5	6.0	6.5
Recycle ratio, R/Q (%)	25	25	25	—	—
Wastewater temperature (°C)	26	25	25	—	—
MLSS (mg/l)	5900	5900	4400	4200	3200
MLVSS (mg/l)	4400	4400	3400	3400	2400
Biomass loading (lb BOD_s/lb MLVSS/day)	0.54	0.70	0.48	0.18	0.20
Biomass loading (lb COD/lb MLVSS/day)	1.09	1.36	0.85	0.81	0.81

TABLE 14 (continued)

UNOX® System Operating Data: Combined Domestic and Textile Wastes

	Location				
	Morganton, N. C.			Danville, Va.	
Parameters Phase	I	II	III	I	II
Influent characteristics (mg/l)					
BOD$_s$	343	270	239	137	132
COD	690	525	425	616	534
SS	317	411	366	54	26
Effluent characteristics (mg/l)					
BOD$_s$	31	35	15	8	13
COD	123	117	28	319	272
SS	22	33	15	12	9
Removals (%)					
BOD$_s$	91	87	94	94	90
COD	82	78	93	48	49
SS	93	92	96	78	65
Clarifier overflow rate (gal/day/ft²)	330	580	330	—	—
Clarifier underflow concentration (%)	2.9	2.8	2.9	—	—
Sludge production (lb TSS/lb BOD$_{sR}$)	—	0.23	0.40	0.10	0.10

TABLE 15

UNOX® System Operating Data: Combined Domestic and Refinery Wastes

	Location	
Parameters	Hyperion Plant, Los Angeles, Cal.	Los Angeles County, Carson, Cal.
Retention time, RT$_Q$ (hr)	1.4	2.0
Recycle ratio, R/Q (%)	44	25
Wastewater temperature (°C)	—	27
MLSS (mg/l)	—	3600
MLVSS (mg/l)	2800	2500
Biomass loading (lb BOD$_s$/lb MLVSS/day)	1.10	1.04
Biomass loading (lb COD/lb MLVSS/day)	2.10	2.11
Influent characteristics (mg/l)		
BOD$_s$	170	216
COD	322	490
SS	105	168
Effluent characteristics (mg/l)		
BOD$_s$	12	20
COD	70	90
SS	22	20
Removals (%)		
BOD$_s$	93	91
COD	78	82
SS	79	88
Clarifier overflow rate (gal/day/ft²)	420	580
Clarifier underflow concentration (%)	0.8	1.8
Sludge production (lb TSS/lb BOD$_{sR}$)	1.05	—

TABLE 16

UNOX® System Operating Data: Combined Domestic and Mixed Industrial Wastes

Parameters	Location	
	San Francisco, Cal.	Philadelphia SW, Pa.
Retention Time, RT_Q (hr)	2.0	1.0
Recycle ratio, R/Q (%)	31	13
Wastewater temperature (°C)	22	15
MLSS (mg/l)	5600	4200
MLVSS (mg/l)	4500	2800
Biomass loading (lb BOD_5/lb MLVSS/day)	0.39	0.94
Biomass loading (lb COD/lb MLVSS/day)	1.42	2.23
Influent characteristics (mg/l)		
$\quad BOD_5$	127	105
$\quad COD$	480	269
$\quad SS$	154	128
Effluent characteristics (mg/l)		
$\quad BOD_5$	9	15
$\quad COD$	135	93
$\quad SS$	29	20
Removals (%)		
$\quad BOD_5$	93	86
$\quad COD$	72	65
$\quad SS$	81	84
Clarifier overflow rate (gal/day/ft²)	590	870
Clarifier underflow concentration (%)	2.4	3.2
Sludge production (lb TSS/lb BOD_{sR})	0.31	0.84

liquor concentrations were 2.4 to 3.2% and 4200 to 5600 mg/l, respectively. Effluent quality of 9 to 15 mg/l BOD and 20 to 29 mg/l SS was achieved at clarifier overflow rates of 600 to 900 GPD/ft².

B. Industrial Wastewaters

A listing of operating data obtained in eight pilot plant and treatability studies conducted on pulp and paper wastewaters is contained in Table 17. These wastes show a large variation in influent characteristics with COD values of 300 to 5400 mg/l and BOD_5 values of 90 to 3800 mg/l. Effective biological treatment providing effluent BOD_5 concentrations of 4 to 40 mg/l was achieved at biomass loadings of 0.1 to 1.7 with most studies conducted at F/Ms of 0.3 to 1.0. Retention times in these studies ranged from 1.8 to 14 hr, with typical values of 2 to 6

hr. Clarifier underflow concentrations of 2 to 3% were achieved at clarifier overflow rates of 400 to 650 GPD/ft². MLSS concentrations were generally high, averaging 7800 mg/l for all studies listed, with a range of 3800 to 14,800 mg/l.

Union Carbide has also conducted many studies on wastewaters from the petrochemical industry. A listing of data from eight of these studies is contained in Table 18. As would be expected with wastes from the chemical industry, a wide variation in wastewater characteristics is evident with influent BOD_5 values of 180 to 5370 mg/l and COD values of 500 to 7500 mg/l. In most cases, the influent suspended solids concentration was much lower than the BOD_5 or COD levels. With the exception of the Union Carbide-Brownsville study, generally lower biomass loadings were required to treat

TABLE 17

UNOX® System Operating Data: Pulp and Paper Wastes

Parameters		Chesapeake Corp., West Point, Va.		Fibreboard Corp., Antioch, Cal.		Gulf States Paper, Tuscaloosa, Ala.		Honshu Paper, Fuji, Japan	
Phase		I	II	I	II	I	II	I	II
Retention time, RT_O (hr)		1.8	2.2	2.0	3.0	2.6	2.8	4.3	3.5
Recycle ratio, R/Q (%)		33	33	27	30	—	—	—	—
Wastewater temperature (°C)		35	34	36	36	—	—	28	21
MLSS (mg/l)		5,300	6,600	5,700	5,400	4,400	4,400	6,400	8,600
MLVSS (mg/l)		4,700	5,900	5,100	4,800	3,400	3,400	5,100	6,200
Biomass loading (lbs BOD_5/lb MLVSS/day)		0.89	0.54	0.69	0.44	0.36	0.34	0.10	1.06
Biomass loading (lb COD/lb MLVSS/day)		2.90	1.63	2.47	1.56	1.02	0.95	0.69	1.23
Influent characteristics (mg/l)									
	BOD_5	274	294	291	256	126	131	88	956
	COD	893	888	1,040	932	358	366	610	1,110
	SS	84	79	103	86	44	37	166	10
Effluent characteristics (mg/l)									
	BOD_5	22	27	31	24	13	17	6	15
	COD	425	357	579	540	149	152	185	107
	SS	49	46	72	63	30	29	11	3
Removals (%)									
	BOD_5	92	90	90	91	90	87	93	98
	COD	52	60	44	42	58	58	70	90
	SS	42	40	42	37	32	24	93	70
Clarifier overflow rate (gal/day/ft²)		650	520	580	410	—	—	—	—
Clarifier underflow concentration (%)		2.0	2.7	2.9	2.5	—	—	—	—
Sludge production (lb TSS/lb BOD_{sR})		0.20	0.26	0.49	0.39	0.40	0.29	—	—

TABLE 17 (continued)

UNOX® System Operating Data: Pulp and Paper Wastes

	Jujo Kimberly, Japan			Hokuetsu, Japan	Sanyo Kokusaku Pulp, Gotsu, Japan					Toyo Pulp, Kure, Japan
Parameters / Phase	I	II	III		I	II	III	IV	V	
Retention time, RT_Q (hr)	3.5	5.0	3.0	1.9	5.5	6.0	6.0	7.5	14.0	2.3
Recycle ratio, R/Q (%)	—	—	—	—	—	—	—	—	—	—
Wastewater temperature (°C)	20	—	—	23	—	—	—	—	25	23
MLSS (mg/l)	6,500	5,900	13,500	3,800	13,700	5,600	14,800	11,500	11,700	5,200
MLVSS (mg/l)	3,800	3,800	5,900	2,700	5,500	4,200	9,100	6,000	6,300	4,000
Biomass loading (lbs BOD_5/lb MLVSS/day)	0.59	0.38	0.27	1.17	0.79	1.71	0.70	0.50	1.04	0.43
Biomass loading (lb COD/lb MLVSS/day)	0.94	0.54	0.38	3.04	1.11	2.47	1.09	0.74	1.47	1.42
Influent characteristics (mg/l)										
BOD$_5$	325	310	200	250	1,000	1,800	1,600	940	3,800	165
COD	520	440	280	650	1,400	2,600	2,500	1,400	5,370	545
SS	130	290	130	142	60	30	280	30	1,060	70
Effluent characteristics (mg/l)										
BOD$_5$	11	6	4	7	20	39	38	40	7	8
COD	114	53	37	370	170	246	156	178	463	216
SS	5	4	6	6	9	12	20	6	18	11
Removals (%)										
BOD$_5$	97	98	98	97	98	98	98	96	99	95
COD	78	88	87	43	88	91	94	87	91	60
SS	96	99	95	96	85	60	93	80	98	84
Clarifier overflow rate (gal/day/ft²)	—	—	—	—	—	—	—	—	—	—
Clarifier underflow concentration (%)	—	—	—	—	—	—	—	—	—	—
Sludge production (lb TSS/lb BOD$_{5R}$)	—	—	—	—	—	—	—	—	—	—

Location

TABLE 18

UNOX® System Operating Data: Petrochemical Wastes

Parameters Phase	Hercules, Wilmington, N.C. I	II	III	IV	Tenneco Chemicals, Pasadena, Tex. I	II	Union Carbide, Brownsville, Tex. I	II
Retention time, RT_Q (hr)	24.2	16.0	16.0	16.0	2.7	1.8	15.9	11.2
Recycle ratio, R/Q (%)	200	100	78	37	33	33	—	—
Wastewater temperature (°C)	—	—	—	—	—	—	39	39
MLSS (mg/l)	24,300	20,300	18,400	7,300	6,500	9,600	8,000	9,600
MLVSS (mg/l)	12,900	11,400	11,000	8,900	4,500	4,800	5,500	6,800
Biomass loading (lb BOD_s/lb MLVSS/day)	0.11	0.21	0.16	0.31	0.44	0.64	1.34	1.70
Biomass loading (lb COD/lb MLVSS/day)	0.19	0.30	0.28	0.43	1.19	1.39	—	—
Influent characteristics (mg/l)								
BOD$_s$	1,458	1,576	1,200	1,751	180	226	4,880	5,370
COD	2,507	2,250	2,128	2,439	486	492	—	—
SS	—	—	—	—	38	73	—	—
Effluent characteristics (mg/l)								
BOD$_s$	—	29	22	68	27	48	54	30
COD	192	239	158	371	126	122	—	—
SS	—	—	—	—	31	29	63	50
Removals (%)								
BOD$_s$	—	98	98	96	85	79	99	99
COD	92	89	93	85	74	75	—	—
SS	—	—	—	—	18	60	—	—
Clarifier overflow rate (gal/day/ft²)	95	150	150	150	430	650	—	—
Clarifier underflow concentration (%)	3.4	3.6	4.0	4.6	2.5	2.4	3.0	3.5
Sludge production (lb TSS/lb BOD_{sR})	—	0.21	0.30	0.30	0.61	0.83	0.21	0.21

TABLE 18 (continued)

UNOX® System Operating Data: Petrochemical Wastes

Parameters Phase	Union Carbide Antwerp, Belgium	Union Carbide Montreal, Canada	BASF-Wyandotte, Jamesbury, N.J.	E. I. DuPont, Deepwater, N.J.	Shell Chemical Norco, La.	Mitsui Toatsu, Osaka, Japan	Denka Chiba, Japan I	Denka Chiba, Japan II
Retention time, RT_Q (hr)	2.6	5.8	3.6	4.0	3.6	3.6	5.0	7.0
Recycle Ratio, R/Q (%)	—	—	—	—	—	—	—	—
Wastewater temperature (°C)	—	—	—	—	—	17	25	26
MLSS (mg/l)	7,800	—	4,100	12,300	12,700	5,600	6,200	6,000
MLVSS (mg/l)	6,500	5,000	3,600	6,000	5,000	3,200	5,200	5,100
Biomass loading (lb BOD_s/lb MLVSS/day)	0.17	0.35	0.45	0.29	0.53	0.46	0.50	0.51
Biomass loading (lb COD/lb MLVSS/day)	0.53	0.79	0.83	0.57	1.03	1.15	0.74	0.69
Influent characteristics (mg/l)								
BOD_s	2,400	397	303	294	461	220	560	750
COD	7,500	893	851	576	899	550	830	1,010
SS	20	30	179	59	146	110	40	100
Effluent characteristics (mg/l)								
BOD_s	15	16	35	18	16	29	3	10
COD	600	186	459	176	193	321	37	67
SS	80	19	20	14	—	15	12	13
Removals (%)								
BOD_s	99	96	87	94	97	87	99	99
COD	92	79	46	69	79	42	96	93
SS	—	37	89	76	—	86	70	87
Clarifier overflow rate (gal/day/ft²)	300	200	160	290	—	—	—	—
Clarifier underflow concentration (%)	1.0	2.8	—	3.1	2.1	—	—	—
Sludge production (lb TSS/lb BOD_{sR})	—	0.41	—	0.52	—	—	—	—

these wastes, with biomass loadings of 0.1 to 0.65 listed. Retention times varied from 1.8 to 24 hr, as a result of the variety of waste strengths encountered. Sludge thickening qualities were excellent in these studies, as indicated by clarifier underflow concentrations averaging 3.0%. The lower clarifier overflow rates seen in these studies do not result in significantly higher clarifier costs, since most petrochemical waste treatment plants handle relatively small liquid flows, compared to municipal plants. Therefore, even at overflow rates of 100 to 400 GPD/ft², clarifier requirements are reasonable. Mixed-liquor solids concentrations in these studies were extremely high, ranging from 4100 to 24,300 mg/l and averaging 10,500 mg/l.

Table 19 lists data obtained in studies conducted on food processing wastes. Biomass loadings of 0.3 to 0.7 resulted in reductions in BOD_s concentration from influent levels of 280 to 1300 mg/l to effluent values of 10 to 15 mg/l. Effluent suspended solids ranged from 10 to 12 mg/l. Required retention times were 2.3 to 17 hr at mixed-liquor solids levels of 5400 to 9000 mg/l.

Treatment of brewery and distillery wastes is shown by the data listed in Table 20. BOD_s was almost completely removed from these wastes, decreasing from influent levels of 710 to 2100 mg/l to effluent concentrations of 7 to 15 mg/l. Biomass loadings of 0.5 to 0.7 were achieved with retention times of 5 to 11 hr and mixed-

TABLE 19

UNOX® System Operating Data: Food Processing Wastes

Parameters Phase	H. J. Heinz, Holland, Mich.	Shiogama, Japan I	II	III
Retention time, RT_Q (hr)	2.3	1.7	9.5	6.5
Recycle ratio, R/Q (%)	—	—	—	—
Wastewater temperature (°C)	—	6	14	14
MLSS (mg/l)	5400	7000	8000	9000
MLVSS (mg/l)	4200	5900	7200	8000
Biomass loading (lb BOD_s/lb MLVSS/day)	0.70	0.32	0.42	0.60
Biomass loading (lb COD/lb MLVSS/day)	1.11	0.34	0.48	0.63
Influent characteristics (mg/l)				
BOD$_s$	282	1330	1200	1300
COD	446	1420	1360	1370
SS	185	200	170	130
Effluent characteristics (mg/l)				
BOD$_s$	14	15	12	10
COD	80	65	55	60
SS	12	10	10	12
Removals (%)				
BOD$_s$	95	99	99	99
COD	82	95	96	96
SS	94	95	94	91
Clarifier overflow rate (gal/day/ft²)	—	—	—	—
Clarifier underflow concentration (%)	—	—	—	—
Sludge production (lb TSS/lb BOD_{sR})	0.39	—	—	—

TABLE 20

UNOX® System Operating Data: Brewery and Distillery Wastes

Parameters	Sapporo Brewery, Japan	Nikka Whiskey, Japan
Retention time, RT_Q (hr)	4.8	11.0
MLSS (mg/l)	6,600	10,500
MLVSS (mg/l)	5,100	9,200
Biomass loading, (lb BOD_s/lb MLVSS/day)	0.70	0.50
Biomass loading (lb COD/lb MLVSS/day)	0.75	0.85
Influent characteristics (mg/l)		
BOD$_s$	714	2,100
COD	765	3,600
SS	38	500
Effluent characteristics (mg/l)		
BOD$_s$	7	15
COD	43	129
SS	14	—

TABLE 20 (continued)

UNOX® System Operating Data: Brewery and Distillery
Wastes

Parameters	Location	
	Sapporo Brewery, Japan	Nikka Whiskey, Japan
Removals (%)		
BOD_s	99	99
COD	94	96
SS	63	—

TABLE 21

UNOX® System Operating Data: Tannery Wastes

Parameters	Location		
	Tatsuno, Japan		Himeji Takagi, Japan
Phase	I	II	
Retention time, RT_Q (hr)	4.5	2.3	6.0
Recycle ratio, R/Q (%)	—	—	—
Wastewater temperature (°C)	27	23	23
MLSS (mg/l)	8500	7500	6900
MLVSS (mg/l)	6000	4900	6000
Biomass loading (lb BOD_s/lb MLVSS/day)	0.26	0.79	0.53
Biomass loading (lb COD/lb MLVSS/day)	0.62	1.86	0.81
Influent characteristics (mg/l)			
BOD_s	290	370	788
COD	690	870	1200
SS	570	670	245
Effluent characteristics (mg/l)			
BOD_s	7	5	19
COD	74	95	115
SS	4	18	12
Removals (%)			
BOD_s	98	99	98
COD	89	89	90
SS	99	97	95
Clarifier overflow rate (gal/day/ft²)	—	—	180
Clarifier underflow concentration (%)	—	—	2.5
Sludge production (lb TSS/lb BOD_{sR})	—	—	0.35

liquor solids concentrations of 6600 to 10,500 mg/l.

Table 21 lists data from two studies conducted on tannery wastes. Results indicate 98 to 99% BOD_s removal at biomass loadings of 0.3 to 0.8 and retention times of 2.3 to 6.0 hr. Suspended solids removal was also excellent with 95 to 99% removal. Final effluent values were 5 to 19 mg/l BOD_s and 4 to 18 mg/l SS.

MLSS concentrations of 6900 to 8500 mg/l were recorded.

A study conducted at Fuji Film, in Japan, indicates the level of treatment attainable on film manufacturing and processing wastes. As listed in Table 22, the influent characteristics of this waste were 300 to 500 mg/l COD and 160 to 320 mg/l BOD_s. Biomass loadings of 0.4 to 1.9 lb BOD_s/day/lb MLVSS achieved final ef-

TABLE 22

UNOX® System Operating Data: Film Manufacturing and Processing Wastes

| Parameters | Fuji Film, Fuji, Japan | | |
Phase	I	II	III
Retention time, RT_Q (hr)	4	3	1.4
MLSS (mg/l)	3200	4200	3800
MLVSS (mg/l)	2400	3200	2900
Biomass loading (lb BOD_s/lb MLVSS/day)	0.40	0.80	1.90
Biomass loading (lb COD/lb MLVSS/day)	0.75	1.57	2.97
Influent characteristics (mg/l)			
$\quad BOD_s$	160	320	320
$\quad$COD	300	500	500
$\quad$SS	80	50	50
Effluent characteristics (mg/l)			
$\quad BOD_s$	15	14	23
$\quad$COD	100	85	120
$\quad$SS	17	10	6
Removals (%)			
$\quad BOD_s$	91	96	93
$\quad$COD	67	83	76
$\quad$SS	79	80	88

fluent BOD_s concentrations of 14 to 23 mg/l and 6 to 17 mg/l SS. The retention time and mixed-liquor solids levels were 1.4 to 4.0 hr and 3200 to 4200 mg/l, respectively.

Chapter 7
SOLID-LIQUID SEPARATION AND CLARIFIER DESIGN

D. W. Gay and K. W. Young

TABLE OF CONTENTS

I. INTRODUCTION

The effectiveness of a secondary wastewater treatment facility is generally determined by its ability (or inability) to maintain low levels of effluent suspended solids (SS_E) and effluent BOD_5. Therefore, the importance of proper final settling tank performance on the overall efficiency of the activated sludge process is well known. The ineffective removal of biological solids causes increases in all of the effluent parameters critical to overall plant performance. As the magnitude of the effluent suspended solids increases, the effluent BOD_5 and COD also increase causing deterioration of plant performance which in some cases is not attributable to any biological causes.

In addition to solids capture (or clarification), a secondary clarifier must effectively consolidate (or thicken) the biological solids. Ineffective thickening of the biological solids prior to recycle to the aeration tanks can result in inefficient utilization of the secondary system. Thickening capabilities less than anticipated result in higher secondary aeration system loadings; this in turn could potentially decrease sludge settleability in a cycle which results in further deterioration of both biological system operation and solid-liquid separation. Although it is not the intent of this chapter to discuss in detail the interactions of the biological system with the clarifier, it is apparent that the design or operation of either unit can significantly affect the design or operation of the other unit.

The thickening capabilities of the clarifier also have a significant effect on the economics of waste sludge handling and disposal, an interrelationship that is sometimes not adequately considered by the design engineer. Dewatering and disposal facilities for handling the sludge from an activated sludge system can represent 35 to 40% of the total annual capital, operating, and maintenance costs of a typical secondary sewage treatment plant. The ability to quantitatively predict the thickening capabilities of the secondary clarifier, therefore, can have a significant effect on overall plant economics.

Classical wastewater treatment plant design practices have recognized that performance of the biological phase of secondary treatment is within the control of the design engineer. However, the capability to exercise similar design or operational control of the secondary clarifiers has oftentimes not been fully recognized. Instead, many established design procedures[1,2] are based on empirisms, using parameters such as hydraulic loading, retention time, or solids loading to establish "safe" solids separation unit design. The use of such techniques is a concession on the part of the designer that he is powerless to quantitatively determine the relative degree to which thickening proceeds.

In those cases where the individual settling characteristics have been considered, the sludge volume index (SVI) has generally been used as the design basis. Empirical correlations relating the thickening capabilities of secondary clarifiers to the SVI have been proposed by many investigators[3-5] as quantitative design approaches which are sensitive to the uniqueness of sludge settleability. However, the SVI is highly dependent on the initial solids concentration and has been shown to have a significant dependency on the geometry of the laboratory equipment used to determine its magnitude.[6] Use of an estimated return sludge concentration determined from SVI relationships[3,6] also denies the designer the opportunity of considering alternative final settling tank designs to accomplish the optimum degree of thickening dictated by overall plant economics.

Since final settling tank design can affect process performance as well as overall plant economics, both the clarification and thickening functions must be taken into account in the design of this unit operation. In general, the rate limiting design function, i.e., the step that controls the size of the clarification unit, is the thickening function. Therefore, in this chapter, primary attention is given to the thickening step in final clarifier design.

The purpose of this chapter is to present a rational approach to clarifier design based on sound theoretical and engineering principles. Graphical design techniques as well as a rigorous mathematical model are developed, and illustrative design examples are presented for each method. The design equations are also modified so the model can be used to estimate the performance of operating final settling tanks. The predicted performance is then compared with the actual performance of full- and pilot-scale operating clarifiers.

II. THEORETICAL ANALYSIS OF SETTLING TANK PERFORMANCE

The basis of theoretical thickener performance has been established by Kynch[7] and others.[8,9] The fundamental approach has been applied to the analysis of activated sludge final settling tanks by Dick[10] and Dick and Young.[11] The following discussion of theory is a recapitulation of these researchers' analysis.

The total downward solids flux in a continuous thickener is the sum of the solids flux due to gravity subsidence of the solids through the liquid and the solids flux due to the bulk downward movement of the liquid/solid mixture because of underflow withdrawal. The total thickener solids flux is mathematically represented as follows

$$G_t = G_g + G_u \tag{1}$$

where G_t is the total flux; G_g is the flux due to gravity; and G_u is the flux due to underflow, all in pounds per square foot per day.

At any infinitesimal, cross-sectional element in a settling tank where solids exist at concentration c_i, the solids flux due to gravity is

$$G_g = 24 \rho c_i v_i \tag{2}$$

where ρ is the density of the sludge; C_i is the solids concentration; and V_i is the settling velocity. Similarly, the flux through that element due to recycle sludge or underflow withdrawal is

$$G_u = 8.34 c_i u \tag{3}$$

where u (ft/hr) is the downward liquid/solid mixture velocity in a settling tank of area A_x (ft^2) caused by removal of sludge at flow rate Q_u (gal/day) or

$$u = \frac{Q_u}{A_x} \tag{4}$$

In activated sludge processes in which excess biological and/or inert solids are wasted from the final settling tank underflow, the sludge removal flow rate is the sum of the recycle flow rate (R) and the waste sludge flow rate (W). Generally, the waste sludge flow rate has a much smaller magnitude than the recycle sludge flow rate. Therefore, to facilitate the preparation of a simple, functional settling tank model, the flux due to waste sludge withdrawal is ignored for this analysis. As the following mathematical evolution shows, however, the waste sludge flow rate could easily be included in the derivations. By neglecting W, Equation 4 becomes

$$u = \frac{R}{A_x} 10^6 \qquad (5)$$

where R is the return sludge rate in MGD.

Substitution of Equations 2 and 3 into Equation 1 yields an expression for the total solids flux for each concentration of solids that might exist in the settling tank

$$G_t = 24 \rho c_i v_i + 8.34 c_i u \qquad (6)$$

Equation 6 is graphically shown in Figure 1, where it can be seen that the solids flux due to underflow withdrawal varies linearly with concentration and depends only on the underflow removal magnitude. However, the value of the gravity flux depends on both the concentration and settling characteristics of the solids. Low solids concentrations result in a small gravity flux (G_g) because c_i is small, and similarly high concentrations also result in small fluxes because v_i approaches zero. At intermediate concentrations, the gravity flux (G_g) passes through a maximum.

Analysis of the total flux curve yields valuable insight into the design and operation of thickening and clarification equipment. Figure 2 illustrates the total flux curve. If the clarifier influent solids concentration (c_o) is greater than c_m (which it should always be since operation at less than c_m would be very uneconomical) and it is desired to thicken the solids to concentration c_u prior to withdrawal, then of all of the sludge concentrations that may exist in the sludge blanket, there is one (c_l) that provides the lowest possible or limiting solids flux (G_l). To assure proper clarifier thickening performance, sufficient settling tank area must be provided so that the operating solids loading is less than or equal to the limiting solids handling capacity or flux. If negligible solids are lost in the effluent and there is negligible solids accumulation, then the area required for thickening is

$$A_x = \frac{8.34 \, c_o \, (Q + R)}{G_l} \qquad (7)$$

Figure 3 illustrates the effect of the basic operating variable (u) or R/A_x. For a given clarifier area, larger solids handling capacities (G_l) can be obtained by increasing R as the dashed lines indicate on Figure 3. However, the increased solids handling capacity results in a sacrifice in sludge thickening capability. Therefore, as the recycle rate increases, the solids handling capacity increases but the underflow solids concentration decreases. In design as well as in operation, the engineer has to determine the optimum trade-off between these two parameters based on overall plant requirements and economics.

Before proceeding further, the assumptions inherent in this theoretical analysis should be summarized. The preceding relationships assume that the overall effects of solids in the aeration tank influent, solids generated by biosynthesis, and solids lost in the effluent can be ignored. Also, as previously stated, the effect of sludge wasting on the downward liquid velocity in the clarifier is ignored. The efficacy of not considering these parameters will be seen later. In addition to the above, the three basic assumptions underlying this analytical procedure are contained in Equations 4

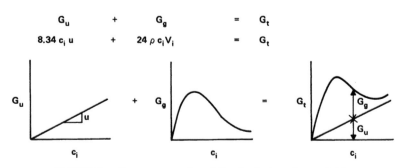

FIGURE 1.   Components of total flux in final settling tank.[11]

and 7. These assumptions are that the settling velocity is a function of the solids concentration only; that the removal of sludge produces uniform downward velocities throughout the tank cross-sectional area; and that the solids are uniformly applied over the tank cross-sectional area, respectively. The validity of these tacit assumptions depends on the clarifier hydraulic and equipment

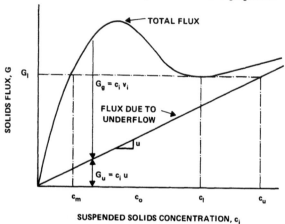

FIGURE 2. Total flux curve.[11]

design. It has previously been reported[11] that all of the above assumptions are valid for properly designed and operated pilot-scale equipment.

III. DEVELOPMENT OF INFORMATION ON SETTLING CHARACTERISTICS

The above theoretical analysis shows that in order to evaluate or design a final settling tank it is necessary to know the relationship between the initial settling velocity and the suspended solids concentration over the entire range of solids concentrations encountered in the final settling tank.

The initial settling velocity of sludges is usually determined by use of laboratory batch settling tests in transparent columns. The procedure for conducting sludge settling tests has been documented by Dick[34] and Vesilind.[30] The transparent column is filled with sludge at a predetermined inital solids concentration and mixed to distribute the solids uniformly. At time zero, the mixing is stopped, and the solids are allowed to settle. Following an initial period of particle agglomera-

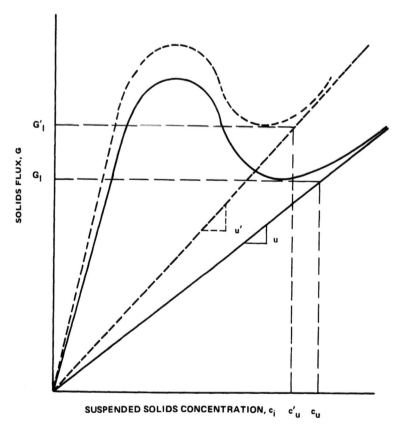

FIGURE 3. The effects of the operating variable U (underflow rate) on clarifier operation. (Courtesy of Union Carbide Corporation.)

tion and possibly channel formation, the solids-liquid interface subsides at a uniform rate in the zone settling regime particularly for sludges with sufficiently high initial solids concentration, greater than 1500 mg/l of total solids (TSS). Zone settling, as pointed out by Vesilind,[30] means that the solids will settle as a blanket without any interparticle movement. This creates a very distinct solids-liquid interface, the height of which is measured as a function of time. Then layers of higher solids concentration are formed after which the solids-liquid interface subsides at a slower rate. The solids-liquid interface position is plotted as a function of time as shown in Figure 4. The slope of the straight line portion of this curve is the initial or zone settling velocity characteristic of the sludge at the initial solids concentration. This procedure is then repeated at different initial solids concentration. Typically six to eight such tests judiciously distributed over the concentration range of interest, e.g., 2000 to 25,000 mg/l for oxygenated activated sludge, would be sufficient to determine the relationship between initial settling velocity and the initial solids concentration.

In spite of the apparent simplicity of the batch settling test procedure described, serious error can still result if proper precaution is not taken to insure or at least approach the conditions of the sludge as it exits in the actual final settling tanks.

In filling the settling column initially with sludge, turbulence that is too intense or not intense enough can produce nonuniform distribution of solids at the beginning of the settling test.[34] It is found with data reported in this chapter that gently shaking the column and its contents several times for smaller settling columns will insure initial uniform distribution of solids in the settling column. For larger columns, it is necessary to "rock" the column and its contents or fill the column rapidly and then use the stirrer as a plunger to stir up the sludge. Use of air aeration as a means to distribute the sludge uniformly in the settling column prior to the settling test is not recommended, as this will alter the D.O. of the sludge. As discussed later, mixed-liquor D.O. has a significant influence on the settling velocity of the sludge.

The effect of wall friction in small laboratory settling columns can cause settling velocities to be either greater or less than those in full-scale settling tanks depending on the sludge concentration.[30] To minimize the wall effect, the diameter of laboratory settling columns should be as large as possible (and usually not less than 2 or 3 in., preferably 5 to 6 in.).[11] Also, slow stirring of the sludge during the settling test is essential. Slow rotating agitators constructed of heavy gauge wire and driven by clock motors are often used for this purpose.[14] The speed of stirring should be in the range of 10 rph, as too fast a speed would tend to destroy any natural bridging occurring in the sludge sample, thereby artificially giving higher settling velocity in the test as compared to that in the full-scale settling tank. This is especially true for poor settling sludges, e.g., filamentous sludge.

Initial sludge depth also influences settling velocity,[31] especially for poor settling sludges, whereas for sludges that settle very well, the initial height has an influence only at very shallow initial depths. The age of the sludge also affects the results of settling tests. Settling tests should be run with "fresh sludge" samples immediately after they have been taken from the aeration tank or the clarifier and, as mentioned previously, should not be reaerated before running the test. Another variable is temperature, because of its effect on the fluid viscosity. Studies have shown that for very diluted sludges, the temperature can have a significant effect on the settling velocity.[32] For more concentrated sludges, the temperature seems to have a negligible effect on the settling velocity.[33]

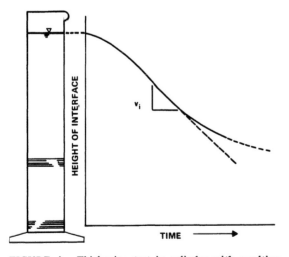

FIGURE 4. Thickening test in cylinder with resulting interface height vs. time curve. (Courtesy of Union Carbide Corporation.)

IV. DESIGN OF
FINAL SETTLING TANKS

Section II of this chapter presented the theoretical basis for a rational clarifier design procedure. It should again be emphasized that the applicability of this design procedure is premised upon the assumptions inherent in the development of the basic theory. There are two approaches (each with its particular advantages) the engineer may choose in the design and/or evaluation of final clarifiers: (1) the graphical technique and (2) the mathematical technique.

To facilitate the presentation of each approach, clarifier surface areas for a plant receiving an average of 10 MGD of wastewater will be calculated. Furthermore, because of sludge handling facility constraints, it is assumed that the design clarifier underflow concentration must fall within the range of from 16,000 to 24,000 mg/l and that the design recycle ratio cannot exceed 50%. Since the purpose of this chapter is not to integrate the design of the clarifier and aeration basin, no attempt will be made to determine the optimum secondary clarifier design from the above information. Rather, the calculational procedures germane to clarifier design will be presented for inclusion in Chapter 8, Volume I, dealing with design optimization of the clarifier and aeration basin.

A. Graphical Clarifier Design Procedure

There are two graphical procedures. The first, described in detail by Dick and Young,[11] involves the graphical addition of the two components (G_g + G_u) contributing to the bulk downward flux in a final clarifier. Figure 1 shows the total flux (G_t) as the sum of the solids flux due to gravity (G_g) and the solids flux due to sludge recycle (G_u). The graphical representation of G_t is then utilized to determine the operational characteristics of the final clarifier.

For the designer, however, this procedure is somewhat cumbersome since for each recycle rate (R) a separate G_t plot must be constructed. The second graphical design procedure circumvents this problem by utilizing only the batch flux curve. As noted by Yoshioka et al.[8] and shown in Figure 5, a tangent to the batch flux curve intersects the ordinate axis at the value of the limiting flux (G_l) and intersects the abscissa at the corresponding underflow concentration (c_u). Furthermore, the absolute value of the slope of this tangent is the downward flow velocity (u).

Pilot plant, field, or empirical data are collected in accordance with the procedures outlined in Section III of this chapter. To facilitate this presentation, typical data collected during an oxygen activated sludge pilot plant program will be used and are shown in Table 1. From these data the batch flux curve is constructed as presented in Figure 6. That section of the curve corresponding to low initial solids concentrations has been included as a broken line to complete the classical batch flux-type curve. The broken portion of the curve has not been verified with data and, therefore, only indicates relative position. Tangents are then added, which correspond to the desired design (or operating) recycle sludge concentrations (RSS).

Three typical design clarifier underflow concentrations for the example problem are shown in Figure 6. The tangents associated with each recycle sludge concentration are labeled as A, B, and C corresponding to recycle sludge concentrations of 16,000, 20,000, and 24,000 mg/l, respectively. The magnitude of the limiting solids flux for each tangent is also indicated. Four potential recycle ratios (r = 0.2, 0.3, 0.4, and 0.5) which are less than the maximum (r ⩾ 0.5) specified in the example will be investigated for each clarifier underflow concentration.

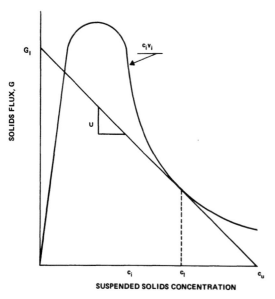

FIGURE 5. Thickening characteristics of a final clarifier as determined from the batch flux curve. (Courtesy of Union Carbide Corporation.)

TABLE 1

Oxygen-activated Sludge Pilot Plant Data

Solids concentration (c_i)			Initial settling velocity (v_i) (ft/hr)	Solids flux $G_g = c_i v_i$ (lb/ft² /day)
(mg/l)	(lb/lb)	(lb/ft³)*		
2800	0.0028	0.175	14	58.8
3800	0.0038	0.237	11.5	65.4
5100	0.0051	0.318	6.2	47.3
5400	0.0054	0.334	7.8	62.5
5400	0.0054	0.337	7.0	56.6
5500	0.0055	0.343	6.2	51.0
10200	0.0102	0.636	1.2	18.3
11000	0.011	0.686	1.05	17.3
13500	0.0135	0.842	0.54	10.9
14000	0.014	0.874	0.84	17.6
15000	0.015	0.936	0.32	7.2
16000	0.016	0.998	0.40	9.6
19000	0.019	1.186	0.29	8.3
20500	0.0205	1.279	0.17	5.2
22000	0.022	1.373	0.15	4.9
24500	0.0245	1.529	0.14	5.1

*Assuming sludge density equals density of water, i.e., 62.4 lb/ft³ .

Data courtesy of Union Carbide Corporation.

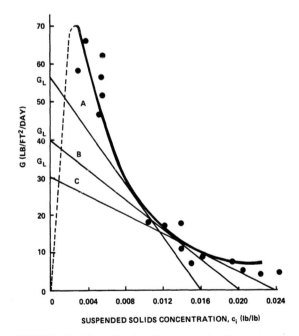

FIGURE 6. Use of batch flux curve for the design of final clarifiers. (Courtesy of Union Carbide Corporation.)

With the design information specified above, the mixed-liquor suspended solids concentration (MLSS) is obtained from a mass balance around the aeration tank.

$$MLSS = \left(\frac{r}{1+r}\right) RSS \qquad (8)$$

This equation assumes that the suspended solids contribution of the waste flowing to the aeration tank, the bacterial synthesis in the aeration tank, and the solids lost with the effluent are all negligible as compared to the solids in the recycle. The solids load to the final clarifier can be determined from the MLSS concentration, and hence, the area required for final clarification is obtained from

$$A_x = \frac{8.34 \, (MLSS) \, Q \, (1+r)}{G_l} \qquad (9)$$

The results of the example problem are presented in Table 2. From these results, the clarifier area is selected. Once the design point clarifier area has been determined, Figure 6 and Equations 8 and 9 can be utilized to determine off design point conditions (i.e., RSS and MLSS).

In classical clarifier design, a required depth was ordinarily specified in addition to the area re-

TABLE 2

Results of Graphical Clarifier Design Procedure

Recycle sludge concentration (RSS)		Limiting solids flux (G_l)	Recycle ratio (r)	Suspended solids concentration (SS)		Clarifier area (A_x)*
(mg/l)	(lb/lb)	(lb SS/ft²/day)	(% of Q)	(mg/l)	(lb/lb)	(ft²)
16000	0.016	56.0	0.2	2667	0.0027	4820
			0.3	3690	0.0037	7160
			0.4	4570	0.0046	9590
			0.5	5330	0.0053	11840
20000	0.02	39.5	0.2	3333	0.0033	8360
			0.3	4615	0.0046	12630
			0.4	5715	0.0057	16850
			0.5	6667	0.0067	21220
24000	0.024	30.0	0.2	4000	0.004	13340
			0.3	5538	0.0055	19880
			0.4	6857	0.0069	24940
			0.5	8000	0.008	33360

*Rounded off to the nearest even unit of ten.

Data courtesy of Union Carbide Corporation.

quired for thickening. The depth requirement has been commonly associated with the desire to provide time to accomplish compression of the solids. Analysis of the volume requirement for thickening, however, has not been developed on a rational basis. Therefore, the depth requirement for a clarifier is that which assures containment of the sludge blanket during periods of operation at transient conditions. A more detailed presentation of blanket level variations and depth requirement is given in the following chapter. Additionally, sufficient depth must be provided to allow for the accumulation of solids during any period when the rate at which solids enter the clarifier exceeds the rate of removal (i.e., when there is no wasting). Also, ample height above the blanket (> 4 ft) must be allowed to account for hydraulic imperfections and to minimize solids carry over into the effluent.

B. Mathematical Clarifier Design Procedure

The development of a mathematical expression for predicting final settling tank performance requires a relationship for solids transport due to gravity subsidence (G_g) and solids transport due to recycle sludge withdrawal (G_u). The mathematical expression for the flux caused by sludge withdrawal has previously been developed and is shown as Equation 3. The relationship for the solids transport due to gravity subsidence, however, is not so clearly defined. This is because the amount of solids flux due to gravity subsidence is dependent upon the experimentally determined relationship between suspended solids concentration and settling velocity. Therefore, to develop a mathematical description of the final settling tank performance, the concentration-velocity relationship must be expressed as an equation.

Many researchers[12-13] have attempted to express the settling velocity of the sludge solids as a function of the concentration. The activated sludge data presented here, however, were found to fit the following empirical relationship reasonably well

$$v_i = ac_i^{-n} \qquad (10)$$

where the constants a and n depend on the physical characteristics of the sludge. This relationship is not valid at low solids concentrations, but it has been found to closely describe sludge settling data collected by others.[11] This empirism has, therefore, been adopted as a reasonable representation of settling behavior within the practical concentration ranges encountered in most operating sewage treatment plants. However, it should be emphasized that any expression relating the settling velocity to the solids concentration can be utilized in the following analytical approach although the mathematics may become considerably more cumbersome. The substitution of Equations 5 and 10 into Equation 6 yields

$$G_t = 24 \, a\rho \, c_i^{(1-n)} + 8.34 \times 10^6 \, \frac{R}{A_x} \, c_i \qquad (11)$$

As noted in Figure 2, the value of G_t that corresponds to the limiting concentration c_1 is the limiting solids flux G_1. As this figure indicates, c_i attains the value of c_1 when

$$\frac{dG_t}{dC_i} = 0 \text{ and } \frac{d^2 G_t}{dc_i^2} > 0 \qquad (12)$$

Therefore, the differentiation of Equation 11 gives

$$\frac{dG_t}{dc_i} = 24 \, a \, \rho \, (1-n) \, c_i^{-n} + 8.34 \times 10^6 \, \frac{R}{A_x} \qquad (13)$$

By substituting

$$R = rQ \qquad (14)$$

(where r is the recycle sludge ratio) and setting Equation 13 equal to zero the following relationship is obtained:

$$c_1 = \left[\frac{24 \, a\rho(n-1)}{(8.34 \times 10^6)r} \left(\frac{A_x}{Q} \right) \right]^{1/n} \qquad (15)$$

Differentiation of Equation 13 yields

$$\frac{d^2 G_t}{dc_i^2} = \frac{24 \, a \, \rho \, (n-1)}{c_i^{(1+n)}} \qquad (16)$$

Since a, ρ, and c_i are all positive, the value of Equation 16 is positive, as long as $n > 1.0$, indicating that the c_1 expressed in Equation 15 is the limiting concentration, corresponding to the flux G_1.*

Substitution of Equations 14 and 15 into

Equation 11 gives an expression for the limiting solids flux as a function of the clarifier overflow rate (Q/A_x), the recycle ratio (r), and the settling properties of the sludge.

$$G_1 = \frac{(8.34 \times 10^6)n}{n-1} \left[2.88 \times 10^{-6} \, (a)\rho(n-1) \right]^{1/n}$$
$$\left[\frac{Qr}{A_x} \right]^{\frac{n-1}{n}} \qquad (17)$$

A mass balance around a clarifier loaded at an applied flux of G_1 indicates that the underflow from that clarifier will equal

$$RSS = \left(\frac{G_1 A_x}{8.34 \, rQ} \right) \qquad (18)$$

Therefore, by substituting Equation 17 into Equation 18 and solving for A_x, the expression for the clarifier area as a function of influent flow (Q), recycle ratio (r), desired underflow concentration (c_u), and the settling properties of the sludge is obtained.

$$A_x = \frac{Qr \, (10^6)}{2.88 \, a \, \rho \, (n-1)} \left[\frac{RSS \, (n-1)10^{-6}}{n} \right]^n \qquad (19)$$

The pilot plant data collected for the illustrative example are shown on Figure 7. These data correlate reasonably well with the form presented as Equation 10, with empirical values for the constants a and n of 1.28×10^{-5} and 2.52, respectively. With these constants, the differences between the results of the example shown in Tables 2 and 3 are generally less than 1% and can be attributed to the rounded off assumptions used to facilitate each calculational procedure.

This mathematical procedure also can be utilized in the evaluation of an operating clarifier. Proper mathematical manipulation of Equation 19 yields an expression for the dependency of RSS on the overflow rate (OR), recycle ratio, and physical characteristics of the sludge (a and n).

$$RSS = \frac{K}{[(OR) \, (r)]^{1/n}} \qquad (20)$$

where

---

*With these variable constraints, Equation 16 cannot have a negative value indicating the failure of this mathematical model to predict the maximum point of Figure 2. This failure can be related to the inability of Equation 10 to adequately predict velocities at low concentrations. This anomaly is of little consequence, however, since the normal magnitudes of c_1 are usually within the range of concentrations, for which Equation 10 does satisfactorily describe the settling velocity.

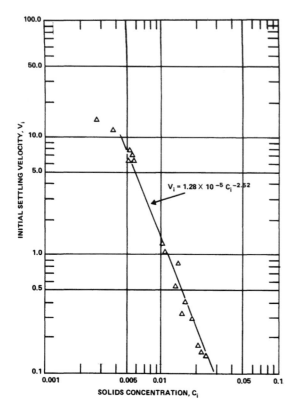

$$K = \frac{10^6 \, n}{n - 1} \left[2.88 \, a \, \rho \, (n - 1) \right]^{1/n} \qquad (21)$$

Whether used for design or evaluation, the mathematical approach is ideal for computerization.

V. FIELD PERFORMANCE DATA

The applicability and utility of the solids flux model have previously been shown by Dick and Young.[11] In their paper, the mathematical representation of the solids flux theories was used to predict the clarifier performance at two oxygen activated sludge pilot plants. Solids settling tests were conducted in either one-liter graduated cylinders or a tall (8.5 ft) 5.5-in. diameter plexiglass column, and the data collected were found to correlate well with Equation 10. From the sludge settling characteristics and the operating and physical characteristics of the pilot facilities, the mathematical model was utilized to predict clarifier underflow and MLSS concentrations for comparison with those actually observed.

The results of the Dick and Young[11] analysis are shown in Figures 8 and 9. Error lines encompassing most of the data are shown on each figure to assist with the interpretation of the results. In

FIGURE 7. Typical solids concentration-initial settling velocity relationship for oxygen-activated sludge. (Courtest of Union Carbide Corporation.)

TABLE 3

Results of Mathematical Clarifier Design Procedure

Recycle sludge concentration (RSS)		Recycle ratio (r) (% of Q)	Mixed-liquor suspended solids concentration (MLSS)		Clarifier area (A_x)* (ft²)
(mg/l)	(lb/lb)		(mg/l)	(lb/lb)	
16000	0.016	0.2	2667	0.0027	4770
		0.3	3690	0.0037	7150
		0.4	4570	0.0046	9540
		0.5	5330	0.0053	11920
20000	0.02	0.2	3333	0.0033	8370
		0.3	4615	0.0046	12550
		0.4	5715	0.0057	16740
		0.5	6667	0.0067	20920
24000	0.024	0.2	4000	0.004	13250
		0.3	5538	0.0055	19880
		0.4	6857	0.0069	26500
		0.5	8000	0.008	33130

Note: Density (ρ) in Equation 19 has been assigned a value of 62.4 lb/ft³.

*Rounded off to nearest even unit of ten.

Data courtesy of Union Carbide Corporation.

general, the agreement shown between the predicted and observed underflow concentrations was considered to be excellent in view of the many potential sources of error.

The data presented by Dick and Young were collected as part of the very comprehensive pilot plant program initiated by Union Carbide Corporation at the conclusion of the government funded

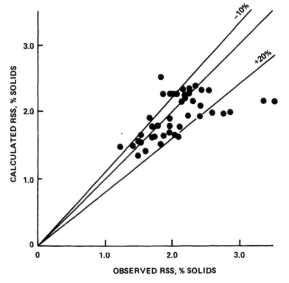

FIGURE 8. Middlesex County Pilot Plant calculated vs. observed returned sludge concentration.[11]

oxygen system demonstration work at Batavia, N.Y.[16] Utilizing a total of approximately ten mobile pilot plant units, programs have been conducted in more than 60 locations throughout the U.S. and Canada to obtain design data and operating information on a wide variety of waste streams. Furthermore, a number of Japanese, European, and Australian programs have been carried out. Extensive process performance data are noramlly collected during all of these pilot plant programs; however, the data presented in this section will be confined to those relating to solids-liquid separation.

A. Data Collection and Measurement

The settling properties reported in this section were determined by observing the subsidence rate of the solids-liquid interface during settling tests. Two types of transparent, plexiglass settling columns were used for these tests: either a large 5.5-in. diameter column 8.5 ft tall or a standard one-liter graduated cylinder. Each column was equipped with a variable speed stirring mechanism that had rotational speeds from 0 to 60 rph. Activated sludge samples of various MLSS concentrations were formed by combining plant effluent, mixed-liquor, and clarifier underflow in different

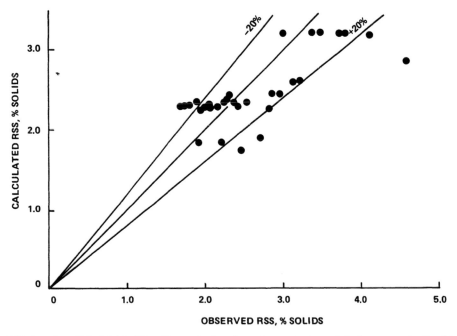

FIGURE 9. Philadelphia Pilot Plant calculated vs. observed returned sludge concentration.[11]

proportions. For example, low concentration samples were obtained by diluting mixed liquor from the last stage of the plant with clarifier effluent, while intermediate concentrations were obtained by blending varying fractions of mixed-liquor and clarifier underflow. The highest concentrations were samples taken directly from the clarifier underflow. Both columns were filled with samples by pouring from the top, a method which proved to be the most convenient field procedure. The data presented by Dick and Young[11] and those contained here indicate that the anomalies created by this filling procedure are generally insufficient to significantly affect the resultant settling velocities. The solids in the large column were also gently redistributed after the column was filled by plunging the stirring device up and down. All of the settling tests were conducted at the temperature and dissolved oxygen concentration of the mixed-liquor, and unless otherwise indicated, all were performed immediately after the sludge sample was withdrawn from the system.

B. Pilot Plant Data

Most of the pilot plant data (with the exception of Detroit, Mich., which will be described later) presented in this section were obtained from the high-purity oxygen mobile pilot plant units described in Chapter 6, Volume I. These mobile, continuous flow (~43,000 gpd capacity) pilot units have circular center feed, peripheral withdrawal clarifiers with sloped bottoms (~13° ptich) and rotating sludge scrapers. The diameter of these clarifiers is nominally 7 ft with a liquid depth of approximately 9 ft. The center feed well has a diameter of 26 in. extending 6 ft below the liquid surface and is equipped with a baffle plate at the bottom end to prevent disruption of the clarifier sludge blanket by the clarifier influent.

The data presented below are representative of the data collected and analyzed. The settling data from the following five pilot plant locations have been included: the Kaw Point Plant, Kansas City, Kansas;[17] the Southwest Water Pollution Control Plant, Philadelphia, Pennsylvania;[18] the Southeast Water Pollution Control Plant, Philadelphia, Pennsylvania;[19] the East Bay Municipal Utilities District Wastewater Treatment Plant, Oakland, California;[20] and the Detroit Metro Wastewater Plant, Detroit, Michigan. With the exception

of Detroit (which will be discussed in detail later) the duration of all of the above programs was from 3 to 5 months.

Pertinent operating data (including operating temperature, pH, F/M, etc.) from each program are presented in Table 4. Although the system loadings and mixed-liquor temperatures varied significantly during many of these programs, no attempt was made to determine the settling characteristics for specific system loadings or temperatures.

As indicated earlier, data from the Detroit pilot facility were not collected during a typical mobile pilot plant program. This permanent Midwestern pilot facility is approximately three times larger than the mobile units and was on-line continuously for approximately 3 years. It operates with two center feed, peripheral take-off clarifiers. Each clarifier is 9 ft in diameter with a nominal liquid depth of 9 ft. The center feed well is 4 ft in diameter and extends 5 ft below the liquid surface. The clarifier bottoms are gently sloped (~13°) with sludge scrapers moving the sludge blanket to the center withdrawal sump.

The settling information presented here was selected from the operating data collected during 1973 and is representative of the data collected prior and subsequent to that period. During this period, the system was fed municipal-industrial sewage from existing full-scale primary clarifiers. It was operated through weekly average organic loading and biomass loading ranges of 86 to 181 lb BOD_5/1000 ft^3/day and 0.27 to 1.42 lb BOD_5/lb MLVSS/day, respectively. The volatile fraction of the mixed-liquor solids averaged 55%, and the mixed liquor D.O. concentration ranged from 2 to 10 mg/l. The mixed-liquor temperature and pH ranges were 60 to 78°F and 6.3 to 7.5, respectively.

Figures 10 and 11 show the measured relationship between the solids concentration and settling velocity for the above pilot plant tests. As shown, the data fit the relationship presented in Equation 10 very well. Table 5 presents the magnitudes for the exponent n and the coefficient a determined from the data for all the pilot programs. Young[21] has reported that for oxygen activated sludge plants treating primary effluent, the exponent n has a relatively constant magnitude of approxi-

TABLE 4

Pilot Plant Operating Data

Location	Type of pre-treatment	Organic loading (lb BOD$_5$/1000 ft^3/day)		Biomass loading (lb BOD$_5$/lb MLVSS/day)		Mixed-liquor characteristics						VSS/TSS (%)
						Temperature (°F)		pH		D.O.		
		Minimum	Maximum	Minimum	Maximum	Minimum	Maximum	Minimum	Maximum	Minimum	Maximum	
Kansas City, Kansas	Primary	38	329	0.14	1.27	72	87	5.7	7.4	6	15	72
Philadelphia, Southwest, Pennsylvania	Primary	90	250	0.40	1.72	53	66	6.2	6.9	5	12	70
Philadelphia, Southeast, Pennsylvania	Primary	112	276	0.67	2.61	70	86	6.2	6.9	6	–	70
Oakland, California	Primary	112	200	0.37	1.06	61	77	5.8	6.8	6	12	80
Detroit, Michigan	Primary	86	181	0.27	1.42	60	78	6.3	7.5	2	10	55

Data courtesy of Union Carbide Corporation.

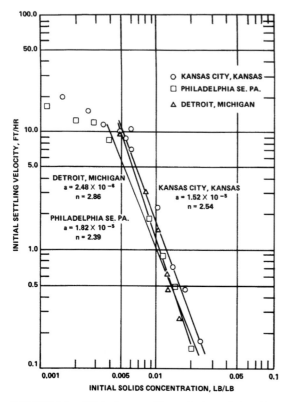

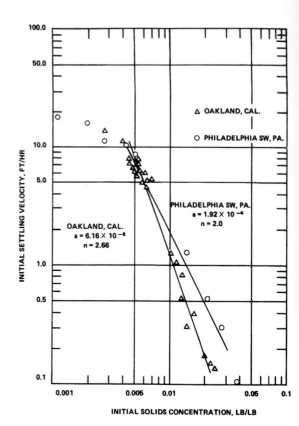

FIGURE 10. Pilot plant settling velocity relationships. (Courtesy of Union Carbide Corporation.)

FIGURE 11. Pilot plant settling velocity relationships.

mately 2.3, while the coefficient a varies as a function of the sewage characteristics.

As these data indicate, there are limitations to this generalization, and, in the absence of specific data, the selection of settling constant values (as outlined in Section VI) should be tempered by the above data. Using Equation 20, the settling constants presented for each pilot plant in Table 5, and obtaining the pilot plant flow rate (Q), recycle ratio (r), and clarifier area (A_x) from plant operating records, the predicted performance of the pilot plant clarifiers was calculated and compared with the observed performance. Tabulations of the observed and calculated data along with recycle ratios and overflow rates for each pilot program are shown in Tables 6, 7, 8, 9, and 10. As shown, all results represent weekly average data.

Comparison of the observed and predicted values for the MLSS and RSS concentrations are shown for Kansas City, the two Philadelphia plants, and Oakland on Figures 12 and 13, respectively. Because of the large number of data points, the observed vs. predicted comparisons for Detroit are shown separately in Figures 14 and 15.

As the figures indicate, most of the data fall within the ±20% error lines added to each figure to assist with the interpretation of the results. The agreement of the theory with the operational data exhibited in these figures is considered excellent since the settling characteristics determined at specific operating conditions (i.e., system loading, temperature, etc.) were applied generally for all system operating conditions. In addition to the errors which may have been imparted because of the general use of the settling characteristics, the tables indicate that in many cases the mass balance relating the MLSS concentration to the RSS concentration (Equation 8) was violated when the measured data were used. Therefore, some of the error indicated on the figures can surely be attributed to experimental inaccuracies associated with the measurement of solids concentrations. The results of these pilot plant studies indicate that allowances for aeration tank influent solids, biological synthesis, loss of solids in the effluent, and the influence of sludge wastage on the downward velocity in the clarifier can be ignored without sacrificing accuracy.

TABLE 5

Pilot Plant Sludge Settling Constants

Location	Type of pretreatment	a (ft/hr)	n
Kansas City, Kansas	Primary	1.52×10^{-5}	2.54
Philadelphia, Southwest, Pennsylvania	Primary	1.92×10^{-4}	2.00
Philadelphia, Southeast Pennsylvania	Primary	1.82×10^{-5}	2.39
Detroit, Michigan	Primary	2.48×10^{-6}	2.86
Oakland, California	Primary	6.16×10^{-6}	2.66

Data courtesy of Union Carbide Corporation.

TABLE 6

Pilot Plant Data, Oakland, California

Week of	Recycle fraction (R/Q)	Clarifier overflow rate (gal/day/ft²)	Mixed-liquor suspended solids (mg/l) Observed	Calculated	Recycle suspended solids (%) Observed	Calculated
10/3/71	0.25	577	4520	4630	1.82	2.31
10/10/71	0.28	577	5570	4850	2.04	2.22
10/17/71	0.33	577	5960	5200	1.87	2.10
10/24/71	0.32	577	4480	5110	1.61	2.11
10/31/71	0.31	575	4790	5060	1.78	2.14
11/7/71	0.31	580	5260	5040	1.65	2.13
11/14/71	0.31	577	5350	5050	1.89	2.13
11/21/71	0.31	530	5200	5220	2.08	2.20
11/28/71	0.31	512	5940	5280	2.42	2.23
12/5/71	0.31	499	6430	5330	2.14	2.26
12/12/71	0.31	502	6270	5320	2.20	2.25
12/19/71	0.31	510	6350	5290	2.45	2.24
12/26/71	0.31	502	6070	5320	2.15	2.25
1/2/72	0.31	485	6640	5390	2.32	2.28
1/9/72	0.31	502	6390	5320	2.41	2.25
1/16/72	0.31	502	7527	5320	2.69	2.25
1/23/72	0.31	502	6770	5320	2.47	2.25
1/30/72	0.31	530	6190	5220	2.38	2.20
2/6/72	0.31	530	6810	5220	2.70	2.20

Data courtesy of Union Carbide Corporation.

TABLE 7

Pilot Plant Data, Southeast Philadelphia, Pennsylvania

Week of	Recycle fraction (R/Q)	Clarifier overflow rate (gal/day/ft²)	Mixed-liquor suspended solids (mg/l)		Recycle suspended solids (%)	
			Observed	Calculated	Observed	Calculated
5/21/72	0.34	737	3420	4540	1.31	1.79
5/28/72	0.40	737	3190	4770	1.12	1.67
6/4/72	0.40	737	3270	4770	1.14	1.67
6/11/72	0.38	737	3490	4680	1.21	1.70
6/18/72	0.29	737	3510	4290	1.38	1.91
6/25/72	0.25	737	3990	4060	1.84	2.03
7/2/72	0.23	777	3950	3850	1.98	2.06
7/9/72	0.24	1255	3340	3190	1.64	1.65
7/16/72	0.18	1300	3090	2810	1.79	1.84
7/23/72	0.10	1300	2160	2140	2.03	2.35

TABLE 8

Pilot Plant Data, Southwest Philadelphia, Pennsylvania

Week of	Recycle fraction (R/Q)	Clarifier overflow rate (gal/day/ft²)	Mixed-liquor suspended solids (mg/l)		Recycle suspended solids (%)	
			Observed	Calculated	Observed	Calculated
12/26/71	0.55	562	6690	7520	2.37	2.12
1/2/72	0.29	737	5450	5710	2.26	2.54
1/9/72	0.29	737	4890	5710	2.01	2.54
1/16/72	0.29	737	5110	5710	2.30	2.54
1/23/72	0.29	737	5320	5710	2.28	2.54
1/30/72	0.17	830	4460	4550	2.63	3.13
2/6/72	0.15	857	4370	4280	3.34	3.28
2/13/72	0.26	630	3810	6000	3.68	2.91
2/20/72	0.19	650	4300	5350	3.04	3.35
2/27/72	0.13	865	4110	4040	3.13	3.51
3/5/72	0.10	1075	3750	3260	3.50	3.59
3/12/72	0.10	1085	3370	3250	3.37	3.57

Data courtesy of Union Carbide Corporation.

TABLE 9

Pilot Plant Data, Kansas City, Kansas

Week of	Recycle fraction (R/Q)	Clarifier overflow rate (gal/day/ft^2)	Mixed-liquor suspended solids (mg/l)		Recycle suspended solids (%)	
			Observed	Calculated	Observed	Calculated
6/3/73	0.3	347	5500	7100	3.1	3.08
6/10/73	0.3	347	6290	7100	3.41	3.08
6/17/73	0.3	347	5930	7100	2.84	3.08
6/24/73	0.3	460	5490	6350	2.50	2.75
7/1/73	0.34	460	5255	6650	2.31	2.62
7/8/73	0.39	465	6730	6930	2.44	2.47
7/15/73	0.36	465	6475	6750	2.92	2.55
7/22/73	0.31	465	6240	6410	2.97	2.71
7/29/73	0.31	465	6160	6410	2.85	2.71
8/5/73	0.31	465	6380	6410	2.84	2.71
8/12/73	0.23	615	5860	5100	3.23	2.73
8/19/73	0.23	633	5445	5030	3.06	2.69

Data courtesy of Union Carbide Corporation.

TABLE 10

Pilot Plant Data, Detroit, Michigan

Week of	Recycle fraction (R/Q)	Clarifier overflow rate (gal/day/ft²)	Mixed-liquor suspended solids (mg/l)		Recycle suspended solids (%)	
			Observed	Calculated	Observed	Calculated
5/27/73	0.23	780	4120	3900	2.75	2.10
6/3/73	0.24	770	3790	4000	2.21	2.07
6/10/73	0.24	750	3930	4050	2.15	2.08
6/17/73	0.24	770	3660	4000	2.17	2.07
6/24/73	0.19	915	3980	3400	2.76	2.12
7/1/73	0.19	925	2900	3350	1.86	2.10
7/8/73	0.21	850	3200	3650	1.93	2.10
7/15/73	0.21	830	3890	3650	2.18	2.12
7/22/73	0.21	845	3770	3650	2.17	2.11
7/29/73	0.20	1830	3590	2720	2.25	1.63
8/5/73	0.22	1620	3230	2970	1.79	1.65
8/12/73	0.21	820	4030	3700	2.62	2.12
8/19/73	0.18	1020	3790	3150	3.37	2.08
8/26/73	0.23	780	4180	3950	1.95	2.1
9/2/73	0.25	730	3940	4200	1.36	2.07
9/9/73	0.27	635	4760	4500	2.17	2.13
9/16/73	0.25	730	3960	4200	2.40	2.07
9/23/73	0.27	680	3980	4400	2.41	2.08
9/30/73	0.26	710	5726	4300	2.58	2.08
10/7/73	0.32	670	4870	4800	2.38	1.98
10/14/73	0.28	680	4800	4500	2.23	2.05
10/21/73	0.25	730	5440	4200	2.38	2.09
10/28/73	0.32	570	5210	5050	2.23	2.08
11/4/73	0.30	600	4600	4800	2.01	2.09
11/11/73	0.34	540	7130	5250	2.86	2.07
11/18/73	0.31	580	13360	4900	5.05	2.10
11/25/73	0.30	610	6510	4800	3.24	2.08
12/2/73	0.29	620	6000	4700	3.01	2.08
12/9/73	0.38	480	5080	5700	2.14	2.09
12/16/73	0.35	520	4660	5400	1.78	2.09

Data courtesy of Union Carbide Corporation.

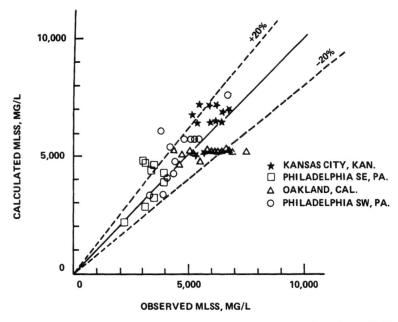

FIGURE 12. Comparison of calculated and observed pilot plant MLSS concentrations.

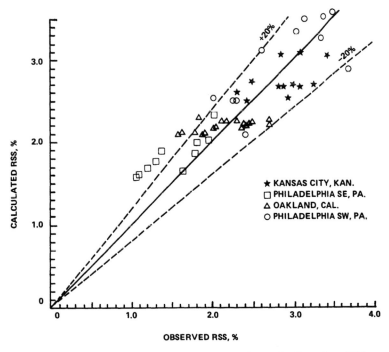

FIGURE 13. Comparison of calculated and observed pilot plant RSS concentrations.

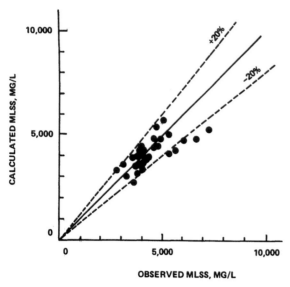

FIGURE 14. Comparison of calculated and observed MLSS concentrations for Detroit, Mich.

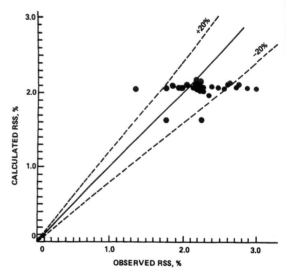

FIGURE 15. Comparison of calculated and observed RSS concentrations for Detroit, Mich.

C. Full-scale Plant Data

The performance data presented in this section have been collected at operating full-scale high-purity oxygen-activated sludge facilities. Unlike the pilot plant units, which are small-scale, highly instrumented and easy to keep calibrated because of their size, the full scale facilities oftentimes lack flow-measuring devices critical to effective clarifier performance evaluation. Additionally, the full-scale equipment is more difficult to calibrate and much more difficult to maintain in calibrated condition. The above operational difficulties, though, usually do not hinder the plant operator's ability to keep the facility on line, performing the pollution control function for which it was designed. These practical realities, however, do affect the amount of quantitative full-scale plant data available for evaluation. Therefore, of the many full-scale operating high purity oxygen activated sludge facilities to date, sufficient data are available from only a few to allow extensive analysis: (1) the American Cyanamide Lederle Laboratories Plant in Pearl River, N.Y., (2) the Speedway Wastewater Treatment Plant in Speedway, Ind., (3) the Chesapeake Corporation, West Point, Va., and (4) Detroit, Mich.

The Lederle Laboratories facility is a pharmaceutical manufacturing plant producing an average wastewater flow of 1.02 mgd containing primary effluent BOD_5 and SS concentrations during the evaluated period of approximately 1150 and 1050 mg/l, respectively. The data used for this evaluation were collected over a period of 9 months; however, because of scheduled summer time maintenance, only 5 months which encompassed spring and fall operation during 1973 have been included here.

The effluent from the Lederle aeration tanks is fed to three identical circular clarifiers, each measuring 40 ft in diameter. The bottoms of these clarifiers are pitched slightly (approximately 5°) toward the center, and sludge is moved to the center withdrawal point via conventional rotating scrapers. These clarifiers are the center feed-type with peripheral takeoff and have sidewater depths of 8 and 10 ft at the wall and center, respectively.

The Speedway, Ind. plant was one of the first full-scale municipal high purity oxygen activated sludge facilities operating in the U.S. It receives typical domestic municipal wastewater with primary clarification. After secondary treatment, the mixed liquor is fed to three identical circular clarifiers. Each clarifier is the centerfeed, peripheral takeoff-type, and each is 65 ft in diameter. The sidewater depth measures 10 ft at the perimeter and increases to 12.7 ft at the center. The sludge pickup mechanism is the rapid sludge return (RSR) type. The clarifier performance data at this plant are from two different operating periods — one is from the third and fourth quarter of 1972 when the oxygen-activated sludge system received raw degritted wastewater. The other is from mid 1973 through January 1974 when the oxygen activated sludge system received primary settled wastewater.

The full-scale high purity oxygen activated

sludge plant at the Chesapeake Corporation, West Point, Va. treats primary settled wastewater coming from pulp washing, liquor condensates, paper mill operation, and black liquor recovery. The plant was designed for a flow of 2.1 mgd with 470 mg/l BOD_5 influent to the three train three-stage oxygen-activated sludge system. The flow and BOD_5 during the evaluation period of September 1975 were 3.2 MGD and 380 mg/l, respectively. The two secondary clarifiers are circular peripheral feed and peripheral overflow type with sludge return by means of a differential hydraulic head RSR mechanism. Each clarifier is 136 ft in diameter with a 14-ft sidewater depth and a sloping bottom.

The Detroit, Mich. full-scale high purity oxygen activated sludge system was designed to treat 300 mgd of primary settled wastewater. The oxygen activated sludge facility is a five-stage design with submerged turbine, rotating sparger-type dissolution equipment. High-purity oxygen is supplied by a cryogenic oxygen generation plant capable of producing 180 TPD of 98% oxygen. After secondary treatment, the mixed liquor flows into six peripheral feed, peripheral takeoff, 200-ft diameter circular clarifiers. The total surface area of these six final clarifiers is 188,500 ft^2 providing an overflow rate of 1600 gal/day/ft^2 at the design flow of 300 MGD. The sludge is removed from the clarifier with a conventional rapid sludge removal system.

Figures 16 through 19 exhibit the measured relationship between the solids concentration and the initial settling velocity for these four full-scale operating facilities during the respective periods of evaluation. The magnitudes of the a and n settling constants (Equation 10) are

	a	n
Lederle	13×10^{-3}	1.24
Speedway (1972)	9×10^{-4}	1.87
(1973–1974)	1.17×10^{-3}	1.54
Chesapeake	14×10^{-3}	1.01
Detroit	6.52×10^{-5}	2.26

As can be noted in Figure 16, the Lederle sludge characteristically settles very well at high solids concentrations resulting in settling constant magnitudes that fall outside the "typical" oxygen sludge settling band at high sludge concentrations. This oxygen-activated sludge settling band is discussed further in Section VI. The above apparent anomaly is due to an operating procedure at the Lederle facility. Normal plant

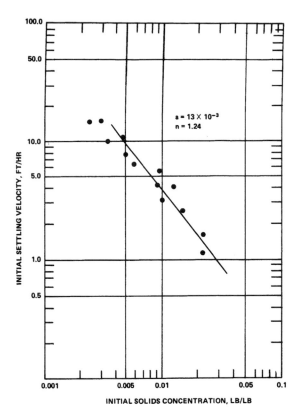

FIGURE 16. Full-scale plant solids-velocity relationship, Lederle Laboratory, Pearl River, New York. (Courtesy of Union Carbide Corporation.)

operating practice stipulates that the waste secondary sludge be fed into the primary clarifloculator where lime is added as a flocculant. The combined primary-secondary sludge is then wasted from the clarifloculator. As could be expected, the lime carry over into the secondary tank subsequently improves the settling characteristics of the secondary sludge.

As was done with the pilot plant data, the appropriate full-scale data were substituted into Equation 20 and the calculated performance was compared to the observed figures. Tabulations of the calculated and observed performance along with the actual recycle ratios and clarifier overflow rates are shown in Tables 11, 12, 13, and 14. These comparative data are plotted in Figures 20 and 21 showing the observed and calculated RSS and MLSS, respectively. It is seen that the results for the full-scale plant data are more scattered than the pilot scale data shown in Figures 14 and 15. The aspect ratio, i.e., the side water depth to the diameter ratio, in the full-scale clarifiers is much smaller than that in the pilot-scale clarifiers. As the aspect ratio of the clarifier decreases, the

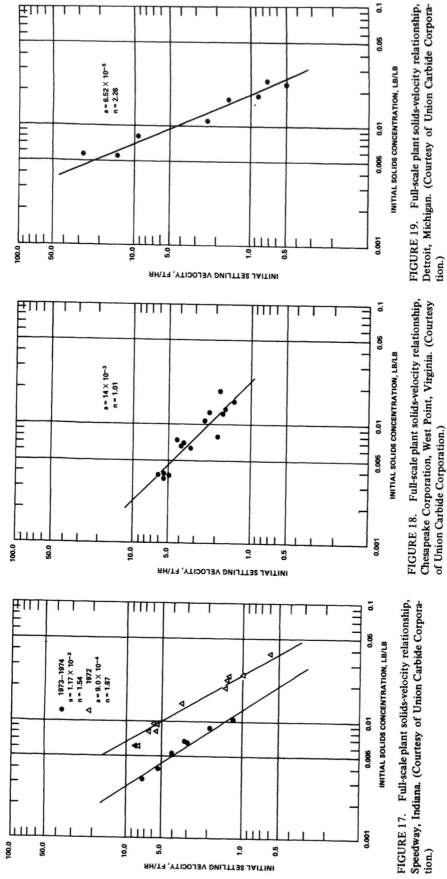

FIGURE 17. Full-scale plant solids-velocity relationship, Speedway, Indiana. (Courtesy of Union Carbide Corporation.)

FIGURE 18. Full-scale plant solids-velocity relationship, Chesapeake Corporation, West Point, Virginia. (Courtesy of Union Carbide Corporation.)

FIGURE 19. Full-scale plant solids-velocity relationship, Detroit, Michigan. (Courtesy of Union Carbide Corporation.)

TABLE 11

Data from Lederle Laboratories, Pearl River, New York

Month	Recycle fraction (R/Q)	Clarifier overflow rate (gal/day/ft²)	MLSS (mg/l)		Recycle suspended solids (%)	
			Observed	Calculated	Observed	Calculated
1	0.707	310	16070	17400	4.01	4.20
2	0.755	280	15120	18610	5.69	4.33
3	0.580	281	14180	19580	5.82	6.21
4	0.755	280	15100	18610	5.69	4.33
5	1.099	240	16000	18950	3.90	3.62

TABLE 12

Pilot Plant Data 1973–1974, Speedway, Indiana

Recycle fraction* (R/Q)	Clarifier overflow rate (gal/day/ft²)	MLSS (mg/l)		Recycle suspended solids (%)	
		Observed	Calculated	Observed	Calculated
0.43	520	4900	6230	1.5	2.07
0.40	560	6780	5910	1.6	2.07
0.56	400	6300	7430	1.6	2.07
0.54	410	7960	7310	1.5	2.08
0.60	370	7830	7810	1.3	2.08
0.53	420	7990	7200	1.3	2.08
0.37	600	7550	5620	1.7	2.08
0.63	415	11200	14800	2.9	3.84
0.60	500	13700	13400	3.4	3.56

*Not normally monitored at this facility; therefore, this number has been estimated from the pumping characteristics of the recycle pump.

Data courtesy of Union Carbide Corporation.

TABLE 13

Data from Chesapeake Corporation, West Point, Virginia

Date	Recycle fraction (R/Q)	Clarifier overflow rate (gal/day/ft²)	MLSS (mg/l)		Recycle suspended solids (%)	
			Observed	Calculated	Observed	Calculated
9/22/75	0.29	660	3000	2880	1.18	1.28
9/23/75	0.29	660	3100	2880	1.28	1.28
9/24/75	0.29	660	3190	2880	1.30	1.28
9/25/75	0.29	660	3170	2880	1.29	1.28

TABLE 14

Pilot Plant Data, Detroit, Michigan

Date	Recycle fraction (R/Q)	Clarifier (gal/day/ft²)	MLSS (mg/l)		Recycle suspended solids (%)	
			Observed	Calculated	Observed	Calculated
10/7/74	0.56	1330	4360	5350	0.85	1.49
10/14/74	0.56	1330	5090	5350	1.13	1.49
5/75	0.60	2040	3810	4500	1.05	1.20
6/75	0.35	2256	3800	3760	0.95	1.45

Data courtesy of Union Carbide Corporation.

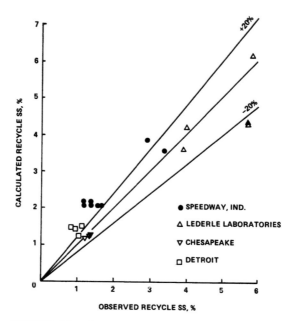

FIGURE 20. Comparison of calculated and observed RSS data from full-scale plants.

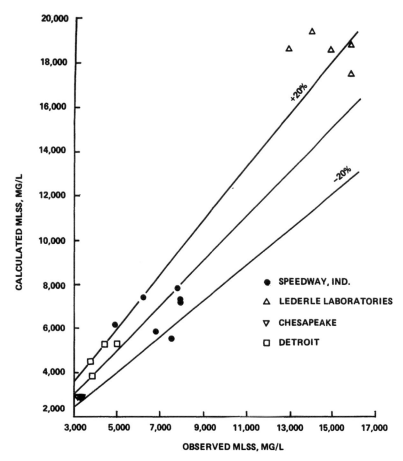

FIGURE 21. Comparison of calculated and observed MLSS data from full-scale plants.

assumptions made in the derivation of the solids flux model, namely, that solids are uniformly applied over the entire cross-sectional area of the settling tank and that removal of sludge produces uniform downward velocities throughout the cross-section of the tank, are evidently not as well satisfied. Nevertheless, the data did show that in spite of these limitations the application of this technique to large full-scale clarifier design and evaluation is justified.

VI. FACTORS INFLUENCING SLUDGE SETTLEABILITY

The value of any analytical design procedure is primarily related to the precision associated with obtaining the experimental parameter necessary to apply the procedure. It has already been shown that an extremely important parameter in the design is the settling characteristics of the biomass.

The accurate determination of these characteristics is a prerequisite for the application of this design approach. As previously mentioned, great care must be taken to eliminate the many sources of error in the batch settling test procedure. As also shown, general empirical relationships for determining settling constants must be judiciously applied to insure good quantitative clarifier designs.

In addition to solids concentration, sludge settleability has been shown to be a function of numerous operating conditions. Temperature, pH, type of influent wastewater, system loading, and dissolved oxygen concentration are the predominant factors affecting sludge settling characteristics. Quantification of many of these dependencies, however, has not been developed. For example, the effect of temperature on the settling properties of oxygen and air pilot plant sludges have been quite noticeable in some cases, while essentially insignificant in others. Oxygen pilot studies conducted jointly by the Environmental

Protection Agency and Union Carbide Corporation at the Blue Plains facility in Washington, D.C. indicated a decrease in settleability of both air and oxygen sludges with decreasing temperature.[22] However, at Louisville, Ky.[23] and Middlesex County, N.J.,[24] where oxygen pilot plants operated over wide ranges of temperatures, no detectable effect of temperature on settleability was observed.

The most significant parameter, in addition to the solids concentration, that affects the settleability of sludge is the D.O. concentration of the mixed liquor. Extensive factorial laboratory experiments carried out to determine the effects of environmental conditions on sludge characteristics have shown that a high D.O. environment results in much better settleability than low D.O. conditions when all other parameters are held constant. A detailed description of the carefully controlled factorial test format has been presented by Drnevich and Gay,[25] and the sludge settleability results from these tests are presented in Table 15. These data indicate that the settleability of the high D.O. sludge was consistently higher than the low D.O. sludge. The pH dependency noted earlier is also evident.

Other investigators have also noted the favorable effects of high dissolved oxygen concentrations on the settling characteristics of activated sludge. Jewell et al.[26] concluded from their study on brewery wastewater that at high food to biomass ratios, low sludge volume indices could only be achieved when using pure oxygen and high D.O. concentration. The effect of mixed-liquor oxygen concentration on sludge settleability is a phenomenon that has only recently received adequate attention from most investigators. The literature is replete with settling information developed during pilot plant studies wherein the D.O. concentration was an uncontrolled operational parameter. As Drnevich and Stuck[27] recently reported, small-scale air pilot facilities usually operate, unless carefully controlled, at high dissolved oxygen concentrations. Because of the dependency of settleability on D.O., the use of these data for full-scale plant design will usually result in significant design errors.

Another common practice reported by a number of investigators is the preaeration of the sludge prior to conducting the settling tests. This preaeration is routinely done to either keep the sludge "fresh" or to keep it completely mixed. However, preaeration at customary laboratory rates causes significant D.O. elevation. Subjecting sludge grown at low dissolved oxygen levels to a high dissolved oxygen concentration environment, even if for just a relatively short period of a few hours, will significantly improve its settleability. To qualitatively show the approximate magnitude of this improvement, Union Carbide Corporation[35] performed on-site settling tests on sludge from five different air-activated sludge treatment plants. Fresh sludges from these plants were taken and settling velocities were determined at various solids concentrations according to the procedures outlined previously. The same sludges were then

TABLE 15

Initial Settling Rate (ft/hr)

		pH = 6.5				pH = 7.2			
		0.5 ppm D.O.		5.0 ppm D.O.		0.5 ppm D.O.		5.0 ppm D.O.	
		ISV	MLSS	ISV	MLSS	ISV	MLSS	ISV	MLSS
Air	Week 1	4.85	5720	10.2	4955	7.85	5425	5.87	5875
	Week 2	2.75	4395	8.0	4335	7.4	4965	5.83	5120
	Week 3	2.77	4690	9.9	4355	2.90	5070	6.34	5200
Oxygen	Week 1	5.5	5765	9.57	5195	6.6	5140	4.59	5515
	Week 2	4.2	4875	8.36	4810	4.55	4755	7.25	4655
	Week 3	2.07	4995	8.9	4755	4.7	3615	6.59	4710

Note: ISV, initial settling velocity (ft/hr); MLSS, suspended solids concentration (mg/l).

Data courtesy of Union Carbide Corporation.

aerated in classical laboratory equipment for from 3 to 8 hr and the settling velocities redetermined. The results are presented in Table 16.

Inspection of these data reveals that with one exception the settling velocities after aeration were considerably higher. The one exception (Marlborough West, Massachusetts) was a sludge developed initially at a relatively high D.O. level of 3.6 mg/l. As expected, reaeration of this sludge to a D.O. of 6 mg/l resulted in only a very small increase in settling velocity.

Numerous side-by-side oxygen and air pilot plant programs have been run during the past 10 years. Because the significance of the dissolved oxygen level effect on settleability has not been widely appreciated, many of the settling data developed during these comparisons are not valid since the low D.O. air systems without D.O. control and with excess aeration capacity in pilot-scale units were operated at high D.O. concentrations. In spite of this operational problem,

however, enough low D.O. vs. high D.O. comparative data have been taken, in addition to those already presented, to allow a qualitative determination of the effect of D.O. on sludge settleability. These comparative data are presented in Figure 22 along with the unaerated settling velocities determined from the four low D.O. air plants referred to above. The data scatter within the two "bands" indicated in this figure can be attributed to the influence of parameters other than D.O. affecting sludge settleability that have already been discussed. The major difference between the "bands" themselves is only a function of the difference in the dissolved oxygen environment to which the sludge has been exposed. Inspection of these data allows the following conclusions:

1. The settling characteristics of activated sludges developed from various types of feed wastewater can differ significantly.
2. The band formed by the settling velocities of the high D.O. concentration activated

TABLE 16

Comparison of Aerated and Unaerated Air Settling Data

	D.O. before aeration (mg/l)	Unaerated ISR (ft/hr)	MLSS (mg/l)	D.O. after aeration (mg/l)	Aerated ISR (ft/hr)
Brockton, Massachusetts	1.5	4.9	2156	9.0	7.0
		5.0	2010		6.2
		2.1	4020		3.2
		0.95	6030		1.1
		0.49	8040		0.56
Hartford, Connecticut	1.2	7.7	2482	7.0	11.9
		7.8	2224		12.4
		3.2	4448		6.7
		1.4	6672		2.8
		0.7	8896		1.05
Marlborough, Massachusetts (East)	1.2	7.7	1502	8.5	8.4
		1.75	3484		2.1
Marlborough, Massachusetts (West)	3.6	4.9	4070	6.0	4.0
		8.1	2213		8.4
		3.4	4426		5.2
		1.0	6639		1.3
		0.7	8852		0.8
Pensacola, Florida	1.0	6.7	1650	6.0	8.1

Data courtesy of Union Carbide Corporation.

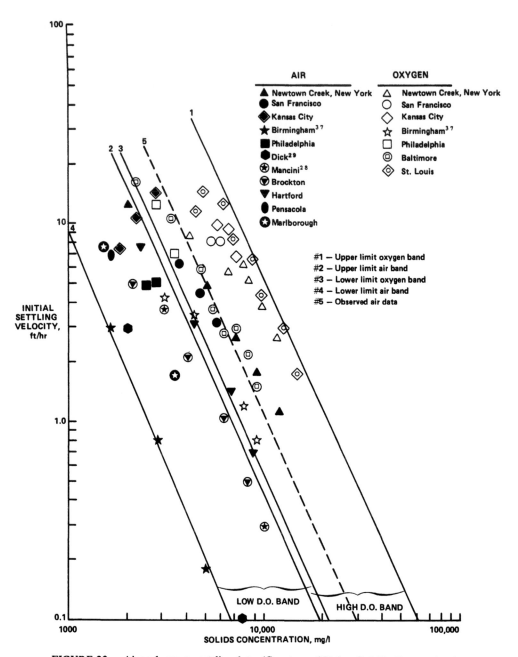

FIGURE 22.　Air and oxygen settling data. (Courtesy of Union Carbide Corporation.)

sludges lies much higher than the band formed by the settling velocities of the low D.O. concentration activated sludges.

3. The two bands slightly overlap, i.e., the highest reported low D.O. band settling velocities are slightly higher than the lowest high D.O. band settling velocities.

4. Notwithstanding point 3 above, whenever the low D.O. settling velocities are on the high or low side of the low D.O. band, the corresponding high D.O. settling velocities for the same wastewater are also on the high or low portion of the high D.O. band. A significant difference in relative settling velocity magnitude is always maintained.

5. Some air plant sludge settling velocities have been measured which are above the "air band" indicated in Figure 22. These data can be as high as shown by line 5 and are usually obtained at underloaded plants operated at relatively high D.O. levels.

Although difficult, if not impossible, to develop a general, quantitative mathematical correlation for all settleability dependencies, the designer must account for each of them. The most accurate method, of course, is to conduct pilot-scale studies to determine those functionalities. Since in many cases this is not possible, another viable alternative is to establish settling characteristics from previously collected data. Both of these methods, however, must be applied judiciously and with a cognizance of the effects of the above discussed parameters. Care must always be exercised to insure that the data collected during pilot plant studies or the characteristics developed from other investigations are valid for the design being considered. To assist the designer, therefore, the following checklist can be used to carefully compare the actual data with what will be anticipated in the full-scale plant facility under design. Significant variations between the pilot plant data and anticipated plant operation will invalidate the use of those data for design purposes.

1. Influent wastewater similarity?
2. Mixed-liquor solids concentration similarity?
3. Dissolved oxygen concentration similarity?
4. pH similarity?
5. System loading (F/M) similarity?
6. Operating temperature similarity?
7. Biomass characteristics (VSS/TSS, specie types, etc.) similarity?

The preceding discussion presents a quantitative procedure for the design and evaluation of oxygen treatment system final clarifiers. Although data have not been presented, this procedure is valid for the design of any gravity settling tank in which the settling solids and tank geometry satisfy the assumptions outlined earlier in the theoretical development of the model. Therefore, it is anticipated that this design procedure (or appropriately modified versions of it) could be used for air activated sludge settling tank design as well as gravity thickener design.

VII. NOMENCLATURE

a	Initial settling velocity constant (ft/hr)
A_x	Cross-sectional area for thickening of clarifier (ft^2)
c_i	Concentration of suspended solids (lb/lb)
c_1	Solids concentration at the minimum flux rate G_1 (lb/lb)
c_u	Concentration of suspended solids in the underflow (lb/lb)
D.O.	Dissolved oxygen level in mixed liquor (mg/l)
F/M	Food to biomass ratio (lb of BOD$_5$ applied/day/lb of MLVSS under aeration)
MLSS	Mixed-liquor suspended solids concentration (mg/l)
MLVSS	Mixed-liquor volatile suspended solids concentration (mg/l)
n	Sludge settling exponent
OR	Clarifier overflow rate (GPD/ft^2)
Q	Influent wastewater flow (mg)
R	Recycle flow (MGD)
RSS	Return sludge suspended solids concentration (mg/l)
RVSS	Return sludge volatile suspended solids concentration (mg/l)
SRT	Sludge retention time (days)
SS$_E$	Effluent suspended solids concentration (mg/l)
t	Time
V	Volume of aeration tank (mg)
v_i	Zone or initial settling velocity of suspended solids (ft/hr)
VSS/TSS	Fraction of suspended solids under aeration which are volatile
W	Waste sludge flow (MGD)
G_t	Total solids flux (lb/ft^2/day)
G_g	Solids flux due to gravity settling (lb/ft^2/day)
G_u	Solids flux due to removal of thickened sludge (lb/ft^2/day)
u	Average downward velocity caused by sludge removal (gal/ft^2/day)
c_o	Same as MLSS (lb/lb)
G_1	Limiting solids handling capacity (lb/ft^2/day)
r	Recycle sludge ratio (R/Q)
ρ	Density of sludge (lb/ft^3)
c_m	Minimum value of c_o (lb/lb)
Q_u	Total underflow sludge withdrawal rate (gal/day)

REFERENCES

1. Great Lakes-Upper Mississippi River Board of State Sanitary Engineers: (Ten States Standard), Recommended Standards for Sewage Works, 1968.
2. Kalbskopf, K.-H., Clarifiers for activated sludge plant, in New Developments in Wastewater Technology, *Wien. Mitt. Wasser Abwasser Gewaesser*, No. 15, 1974.
3. McKinney, R. E., Mathematics of complete-mixing activated sludge, *J. Sanit. Eng. Div. ASCE*, 88, SA3, 87, 1962.
4. Lawrence, A. W. and McCarty, P. L., Unified basis for biological treatment design and operation, *J. Sanit. Eng. Div. ASCE*, 96, SA3, 757, 1970.
5. Vosloo, P. B. B., Some factors relating to the design of activated sludge plants, *Water Pollut. Control*, 69(5), 486, 1970.
6. Dick, R. I. and Vesilind, P. A., The sludge volume index — what is it?, *J. Water Pollut. Control Fed.*, 41(7), 1285, 1969.
7. Kynch, G. J., A theory of sedimentation, *Trans. Faraday Soc.*, 48, 166, 1952.
8. Yoshioka, N., Hotta, Y., Tanaka, S., Naito, S., and Tsugoni, S., Continuous thickening of homogeneous flocculated slurries, *Chem. Eng.*, Tokyo, 21, 66, 1957.
9. Shannon, P. T., Stroupe, E., and Tory, E. M., Batch and continuous thickening, *Ind. Eng. Chem. Fundam.*, 2, 203, 1963.

10. Dick, R. I., Role of activated sludge final settling tanks, *J. Sanit. Eng. Div. ASCE,* 96, SA2, 423, 1970.

11. Dick, R. I. and Young, K. W., Analysis of Thickening Performance of Final Settling Tanks, presented at the 27th Annual Meeting, Purdue Industrial Waste Conference, Lafayette, Ind., May 2–4, 1972.

12. Richardson, J. F. and Zaki, W. N., Sedimentation and fluidization, Part I, *Trans. Inst. Chem. Eng.,* 32, 35, 1954.

13. Vesilind, P. A., The Influence of Stirring in the Thickening of Biological Sludge, Ph.D. thesis, University of North Carolina, Chapel Hill, N.C., 1968.

14. Javaheri, A. R. and Dick, R. I., Aggregate size variations during thickening of activated sludge, *J. Water Pollut. Control Fed.,* Part 2, 41(5), R197, 1969.

15. Dick, R. I., Sludge treatment, in *Manual of Physical and Chemical Processes,* Weber, W. J., Jr., Ed., John Wiley & Sons, N.Y., 1972.

16. Albertsson, J. G., McWhirter, J. R., Robinson, E. K., and Vahldieck, N. P., Investigation of the Use of High Purity Oxygen Aeration in the Conventional Activated Sludge Process, FWQA Department of the Interior Program No. 17050 DNW, Contract No. 14-12-465, May 1970.

17. Union Carbide Corporation, Pilot Plant Report for the Kaw Point Plant, Kansas City, Kan., Engineering Memorandum No. 5485, 1973.

18. Union Carbide Corporation, Pilot Plant Report for the SW Water Pollution Control Plant, Philadelphia, Pa., Engineering Memorandum No. 5347, 1972.

19. Union Carbide Corporation, Pilot Plant Report for the SE Water Pollution Control Plant, Philadelphia, Pa., Engineering Memorandum No. 5347, 1972.

20. Union Carbide Corporation, Pilot Plant Report for the Water Pollution Control Plant, Oakland, Cal., Engineering Memorandum No. 5242, 1972.

21. Young, K. W., Union Carbide Corporation, Internal Memorandum on UNOX Sludge Sedimentation, 1971.

22. Stamberg, J. B., EPA Research and Development Activities with Oxygen Aeration, presented at the Technology Transfer Design Seminar for Municipal Wastewater Treatment Facilities, N.Y., N.Y., February 29 and March 1–2, 1972.

23. Union Carbide Corporation, Internal Memorandum (Pilot Plant Report) on the Performance of UNOX Pilot Plant Operation at Fort Southworth Sewage Treatment Plant, Louisville, Ky., 1971.

24. Union Carbide Corporation, Internal Memorandum (Pilot Plant Report) on the Performance of UNOX Pilot Plant Operation at Middlesex County Sewage Authority, N.Y., 1971.

25. Drnevich, R. F. and Gay, D. W., Sludge Production Rates in Activated Sludge Systems, presented at the 28th Annual Meeting, Purdue Industrial Waste Conference, Lafayette, Ind., May 2–4, 1972.

26. Jewell, W. J. and Eckenfelder, W. W., Jr., The Use of Pure Oxygen for the Biological Treatment of Brewery Wastewaters, presented at the 26th Annual Meeting, Purdue Industrial Waste Conference, Lafayette, Ind., May 1971.

27. Drnevich, R. F. and Stuck, J. D., Error Sources in the Operation of Bench and Pilot Scale Systems used to Evaluate the Activated Sludge Process, presented at the 30th Annual Meeting, Purdue Industrial Waste Conference, Lafayette, Ind., May 6–8, 1975.

28. Mancini, J. L., Gravity Clarifier and Thickener Design, Proc. Purdue Industrial Waste Conference, Lafayette, Ind., 267, 1962.

29. Dick, R. I. and Schroepfer, G. J., personal communications, 1972.

30. Vesilind, P. A., *Treatment and Disposal of Wastewater Sludges,* Ann Arbor Science Publishers, Ann Arbor, Mich., 1974.

31. Dick, R. I. and Ewing, B. B., Evaluation of activated sludge thickening theories, *J. Sanit. Eng. Div. ASCE,* 93, SA4, 9, 1967.

32. Reed, S. C. and Murphy, R. S., Low temperature activated sludge settling, *J. Sanit. Eng. Div. ASCE,* 95, SA4, 1969.

33. Dick, R. I., Discussion of low temperature activated sludge settling by S. C. Reed and R. S. Murphy, *J. Sanit. Eng. Div. ASCE,* 96, SA2, 638, 1970.

34. Dick, R. I., Thickening, Seminar on Process Design in Water Quality Engineering, Vanderbilt University, Nashville, Tenn., November 9–13, 1970.

35. Chapman, T. C., Matsch, L. C., and Zander, E. H., Effect of high dissolved oxygen concentration in activated sludge systems, *J. Water Pollut. Control Fed.,* 48, 2486, 1976.

36. Dick, R. I., Folk lore in the design of final settling tanks, *J. Water Pollut. Control Fed.,* 48, 643, 1976.

37. Investigation of the Performance of the Modified (UNOX) and Conventional Activated Sludge Process under Comparable Conditions of Loading, Progress Report for the Period December 6, 1972 to August 15, 1973, Water Pollution Res. Rep. No. 441R, Water Pollution Research Laboratory, Stevenage Herts, Great Britain.

Chapter 8

OXYGENATION TANK - CLARIFIER DESIGN INTEGRATION

M. R. Lutz and S. E. King

TABLE OF CONTENTS

I. INTRODUCTION

A system is defined as a regularly interacting or interdependent group of items forming or operating as a unified whole. The two major "items" in a secondary treatment system are the oxygenation tank and the final clarifier. Both the oxygenation tank and clarifier are in themselves discrete systems, and they must be designed to perform specific functions according to applicable design criteria. They also interact in the secondary treatment system and, therefore, can only be designed properly if the interdependencies and interactions are taken into consideration. If the designs of the two systems are not properly integrated, overall total system performance, operating flexibility, and cost will not be optimum.

Optimum oxygenation tank-clarifier integration is achieved when the system provides: (1) the best practicable effluent quality with (2) maximum system operating flexibility at (3) minimum cost. In this chapter the methodology of proper oxygenation tank-clarifier integration will be discussed. Design models presented in earlier chapters will be utilized, guidelines will be established, and a method by which proper design integration can be achieved will be illustrated.

II. MATHEMATICAL MODELS

The reactor, clarifier, and return sludge system perform as a unit and, therefore, must be designed as an integral system to ensure maintenance of treatment efficiency for the entire range of flow and load conditions over which the plant must effectively operate.

The parameters which can readily be examined for the expected plant flow range include recycle suspended solids concentration, mixed-liquor suspended solids concentration, and sludge blanket levels. These parameters and the resulting plant performance are functions of clarifier geometry, oxygenation tank volume, and sludge recycle flow rate.

Utilizing the solids flux theory as developed in Chapter 7, Volume I, the recycle suspended solids concentration (RSS) as a function of clarifier overflow rate (OR) and the ratio of the return sludge flow rate to influent flow rate (R/Q) can be calculated for a circular clarifier. This is done from the uniform solids flux model, assuming typical operation where the wasting flow rate is much less than the recycle flow rate:

$$RSS = \frac{K}{[(OR)\,(R/Q)]^{1/n}} \tag{1}$$

where

$$K = \left(\frac{n}{n-1}\right) [(180)(a)(n-1)]^{1/n} (10^6) \qquad (1a)$$

Also, as discussed in Chapter 7, Volume I, the settling constants (a and n) must first be determined from pilot plant settling data or estimated based on settling data from other facilities treating a wastewater with similar characteristics. Settling constants are determined according to the settling velocity equation:

$$v_i = a \ (c_i)^{-n} \qquad (2)$$

Once the settling constants are obtained, the recycle suspended-solids concentration can be calculated for various clarifier overflow rates (OR) and recycle ratios (R/Q) using Equation 1.

The RSS concentration is inversely proportional to the product of OR × R/Q as shown in Equation 1. If a constant recycle ratio is maintained, the product of OR × R/Q will increase as the clarifier overflow rate increases, resulting in a corresponding decrease in RSS concentration. At constant recycle flow rate, however, the product of OR × R/Q is constant and, therefore, the RSS concentration remains constant.

Figure 1 illustrates the basic functional relationship between OR, R/Q, and RSS concentration. For comparative purposes, the RSS concentration is plotted as a function of OR at: (1) constant recycle flow rate and (2) constant recycle ratio. As can be seen, at a constant recycle ratio, the RSS concentration decreases as the OR increases, whereas at a constant recycle rate, the RSS concentration remains constant.

Certain limitations must be applied to the recycle suspended solids concentrations predicated by Equation 1 because the settling velocity correlation, upon which the model was developed, does not apply over the full range of concentrations. At very low concentrations discrete particle settling occurs and a constant settling velocity is approached which is independent of concentration. As the solids concentration increases, on the other hand, a maximum concentration value is reached, and the settling velocity decreases to essentially zero. Therefore, if the product of OR × R/Q is extremely small or large, RSS values will be calculated which are not realistic. Use of predicted values of RSS outside the range of 3,000 to 40,000 mg/l is not recommended.

Corresponding mixed-liquor suspended solids concentrations (MLSS) can be determined by mass balance for the values selected for OR and R/Q. Assuming the influent solids are negligible compared to RSS:

$$MLSS = \frac{(R/Q)(RSS)}{1 + R/Q} = \frac{(R/Q)(K)}{(1 + R/Q)[(OR)(R/Q)]^{1/n}} \qquad (3)$$

If a comparatively large quantity of nondegradable influent suspended solids (SS_A) are present, Equation 3 can be modified as follows:

$$MLSS = \frac{(SS_A)(Q) + (RSS)(R)}{Q + R}$$
$$= \frac{SS_A + \dfrac{(R/Q)K}{[(OR)(R/Q)]^{1/n}}}{(1 + R/Q)} \qquad (3a)$$

Figure 2 presents the relationship between OR, R/Q, and MLSS concentration for a typical waste where SS_A is much less than RSS. The MLSS

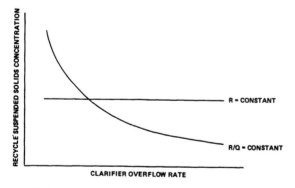

FIGURE 1. Recycle suspended solids concentration as a function of clarifier overflow rate and recycle rate. (Courtesy of Union Carbide Corporation.)

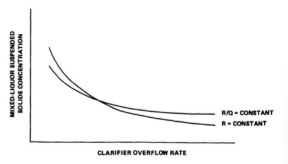

FIGURE 2. Mixed-liquor suspended solids concentration as a function of clarifier overflow rate and recycle rate. (Courtesy of Union Carbide Corporation.)

concentration is plotted as a function of OR at both constant recycle rate and recycle ratio, as is the RSS concentration in Figure 1. As depicted in Figure 2, the MLSS concentration decreases for both constant recycle ratio and constant recycle rate as the OR increases. The variation in MLSS concentration is not as great, however, with a constant recycle ratio. Therefore, a constant recycle ratio is more effective than a constant recycle rate in maintaining steady operating conditions (F/M) within the oxygenation tank for a given clarifier size and range of overflow rates.

The secondary clarifier sludge blanket level is also a key operating parameter which must be controlled to: (1) provide a reasonable clarifier sludge blanket retention time, (2) prevent excessive solids loss over the effluent weir during peak flow periods, and (3) prevent recycle and wasting of an extremely dilute sludge stream during low flow periods brought about through an inventory shift from the clarifier to the oxygenation tank.

The sludge blanket retention time for a circular clarifier is a function of the average sludge blanket height, the clarifier cross sectional area, and the recycle flow rate.

$$t_{S.B.} = \frac{(H_B)(A_X)(7.48)(24)}{(R)(10^6)} \qquad (4)$$

$$t_{S.B.} = \frac{(H_B)(180)}{(R/Q)(OR)} \qquad (4a)$$

where

$t_{S.B.}$ = sludge blanket retention time (hours)
H_B = blanket height (feet)
A_X = clarifier thickening cross-section area (square feet)
R = recycle flow (MGD)
OR = clarifier overflow rate (gal/day/square feet)

A circular clarifier, operating at, for example, an overflow rate of 600-gpd/ft^2 and a recycle ratio of 30% would provide approximately a 1-hr sludge blanket retention time per foot of blanket. For typical operating conditions, the blanket level would be controlled between 2 and 3 ft, thus providing a 2- to 3-hr sludge blanket retention time.

During peak flows, the clarifier sludge blanket level will increase; therefore, the average blanket level must be kept low enough, and the clarifier must have sufficient depth or storage capacity to prevent solids loss over the effluent weir. Conversely, during low flow periods, the clarifier blanket level will decrease, and the average blanket level must be high enough to prevent channeling or complete loss of the blanket and resulting recycle and wasting of an extremely dilute stream. Typically, the clarifier blanket level should not drop below 0.75 ft, nor should it rise to within 4 ft below the clarifier overflow weir. The clarifier sludge blanket level can be calculated as follows:

$$H_B = \frac{SS_{BO}}{SS_B}(H_{BO}) + \frac{(OR_O)(t_{QO})}{(180)(SS_B)}(MLSS_O - MLSS) \quad (5)$$

where

SS_B = the sludge blanket suspended solids concentration (mg/l)
t_{QO} = hydraulic retention time in the aeration tank at average wastewater flow (hours)

Subscript O refers to conditions existing at average wastewater flow. The first term in Equation 5 takes into account the changes which result from the variation in the clarifier blanket solids concentration. These changes can be equated to variations in the RSS and MLSS concentrations which are functions of OR and R/Q. The second term is the change in the clarifier blanket level which results from the shift of biomass to or from the clarifier during peak or low flow periods. During peak flows, mixed-liquor solids are displaced or washed into the clarifier at a rate faster than they can be returned, resulting in an increase in blanket level and decrease in MLSS concentration. During low flow periods, the MLSS concentration increases, and the clarifier blanket level decreases as the mass of solids being returned from the blanket is greater than that leaving the oxygenation tank. The RSS concentration also increases during low flow periods if R/Q is constant, further decreasing the clarifier sludge blanket level.

The average clarifier sludge blanket suspended-solids concentration (SS_B) will vary depending on several factors including sludge settling characteristics, the type of clarifier pickup or scraper mechanism, clarifier geometry, and sludge blanket height. The average sludge blanket suspended solids concentration can be expressed as:

$$SS_B = \frac{RSS + MLSS}{A_B} \qquad (6)$$

where A_B is a factor which accounts for any difference between the average blanket and recycle suspended solids concentrations. If for example, SS_B is the average between the MLSS and RSS concentrations, then

$$A_B = 2$$

If the average blanket suspended solids concentration is low, the blanket volume per unit mass will be high, and expansion during peak flows and subsequently clarifier depth becomes more critical.

Utilizing Equations 1, 3, and 6 and assuming the wasting rate is much less than the recycle rate, Equation 5 can be rearranged as follows:

$$H_B = \left[\frac{(OR)(R/Q)}{(OR_O)(R/Q_O)}\right]^{1/n} \left(\frac{1 + R/Q}{1 + R/Q_O}\right)$$
$$\left(\frac{1 + 2\,R/Q_O}{1 + 2\,R/Q}\right) H_{BO} + \left[\frac{(OR_O)(t_{QO})}{180}\right] \qquad (7)$$
$$\left[\frac{(A_B)(R/Q)}{1 + 2\,R/Q}\right] \left\{ \frac{R/Q_O\,(1 + .R/Q)}{R/Q\,(1 + R/Q_O)} \right.$$
$$\left. \left[\frac{(OR)(R/Q)}{(OR_O)(R/Q_O)}\right]^{1/n} -1 \right\}$$

By choosing an A_B (2 is a realistic number) and combining this with values for R/Q, OR, and t_Q, the sludge blanket level can be calculated using Equation 7. By applying this calculation to the entire range of flow rates expected, the clarifier depth or storage capacity required for design can be calculated for any oxygenation tank-clarifier system.

Figure 3 illustrates typical blanket level changes, again using R and R/Q constant and, for comparison, two values of H_{BO}. As shown, the blanket level fluctuations are greatest with R/Q constant as compared to constant recycle rate. Also, the average blanket level (H_{BO}) greatly affects the magnitude of the blanket level fluctuations as clarifier overflow rate changes.

Utilizing these mathematical models and the appropriate process design calculations, a general procedure for oxygenation tank-clarifier design can be developed as presented in the following section.

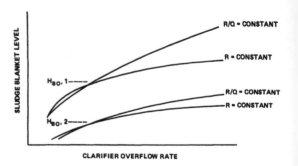

FIGURE 3. Clarifier sludge blanket level as a function of clarifier overflow rate and recycle rate. (Courtesy of Union Carbide Corporation.)

III. OXYGENATION TANK-CLARIFIER INTEGRATION

A. Definition of Process Constraints

Proper secondary treatment system operation can only be achieved if the basic process design variables remain within acceptable limits. In determining these design limits, operation at all conditions — design, peak, and minimum — must be considered. By properly sizing the oxygenation tank and clarifier, a system that will operate within these limits at the design point and allow the maintenance of high operating efficiency at off-design conditions can be developed. Several of the process design variables important in achieving proper system operation are: (1) biomass loading (F/M), (2) MLVSS concentration and recycle rate, (3) clarifier overflow rate and clarifier depth, and (4) oxygenation tank retention time. These parameters and their interrelationships will now be discussed.

1. Biomass Loading

The proper biomass loading for a given system will depend upon the wastewater degradability and the required substrate removal, the sludge production desired, and the effect of biomass loading on the settling characteristics of the biological sludge. Details of the kinetic model used in relating biomass loading and substrate removal, as well as a method for determining waste sludge production, are presented in Chapter 5, Volume I. As discussed in Chapters 5 and 7, Volume I, limits must also be set on the biomass loading in order to maintain a nondispersed, well-flocculated sludge.

For municipal wastewaters, typical design point biomass loadings for an oxygen-activated sludge system are 0.5 to 0.9 lb BOD_5/lb MLVSS/day. If the average biomass loading is maintained within this range and the oxygenation tank and clarifier are properly integrated, the loadings experienced during minimum or peak flow periods will not result in a decrease in system performance.

Biomass loading is calculated as follows:

$$F/M = \frac{(BOD_5)(Q)}{(MLVSS)(V)} \qquad (8)$$

$$= \frac{(BOD_5)(24)}{(MLVSS)(t_Q)} \qquad (8a)$$

$$= \frac{(BOD_5)(24)}{\left(\dfrac{VSS}{TSS}\right)\left\{\dfrac{(R/Q)(K)}{(1+R/Q)\,[(OR)(R/Q)]}-1/n\right\}(t_Q)} \qquad (8b)$$

As shown by Equation 8b, the biomass loading is a function of the influent BOD_5 concentration, the biomass settling characteristics, the recycle ratio, the clarifier overflow rate, and the oxygenation tank retention time.

The influent BOD_5 loading range and sludge settling constants, once determined, can be considered as fixed parameters that the design engineer can judiciously select but cannot control (excluding provisions for equalization or chemical addition). The design engineer does have control, however, in sizing the clarifiers, recycle pumps, and oxygenation tank; this control is the key to proper oxygenation tank-clarifier integration to insure continuous operation within a desired F/M range.

2. MLSS Concentration and Recycle Rate

The MLSS (or MLVSS) concentration that can be maintained in an activated sludge system is described by Equation 3. As shown, the MLSS concentration depends upon the settling characteristics of the biomass, the clarifier overflow rate, and the recycle ratio. The maximum acceptable design MLSS concentration is dictated in most cases by system economics (i.e., large clarifiers) rather than process considerations. Low design MLSS concentration will reduce the clarifier area but increase oxygenation tank volume also resulting in overall unfavorable economics. In addition, low MLSS concentrations may result in dispersed floc and poor effluent suspended solids levels. Design MLSS and MLVSS concentrations typically range from 5,000 to 8,000 and 3,500 to 6,000 mg/l, respectively, in oxygen-activated sludge systems.

The MLSS concentration can be controlled by varying the clarifier recycle ratio and Figure 4 illustrates the effect of recycle ratio on MLSS and RSS concentrations at a constant clarifier overflow rate. As can be seen, the RSS (and waste sludge) concentration decreases continuously as the recycle ratio increases. The MLSS concentration increases, however, until R/Q = n − 1 (n is a constant determined by the settling characteristics of the biomass; see Equation 2). If the recycle ratio for example is increased to greater than 100%, the MLSS concentration actually decreases and for this reason the design point recycle flow range is typically established at 25 to 50% of design flow in oxygen-activated sludge systems. In addition, it is recommended that maximum recycle pumping capacity of up to 50% of the sustained peak flow be installed. This allows adequate flexibility to increase the MLSS concentration during peak load periods and results in high RSS (and waste sludge) concentrations during average or low flow conditions.

3. Clarifier Overflow Rate and Depth

The importance of clarifier design is often overlooked to some extent in the design of activated sludge systems. As pointed out in Chapter 7, Volume I, the clarifier actually performs two functions, clarification and thickening, and generally it is the thickening function that is

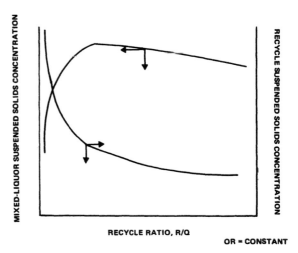

FIGURE 4. MLSS and RSS concentration as a function of clarifier recycle rate. (Courtesy of Union Carbide Corporation.)

limiting in the process design. If the clarifier overflow rate is too great, the unit, with its high upflow velocities and potential hydraulic short circuiting, may not allow proper flocculation, settling, and sludge blanket formation; high effluent solids and poor overall system performance may result. If the design clarifier overflow rate is too low, costs will be high, and it may become difficult to control the required low sludge blanket levels resulting in excessive clarifier sludge retention times, septic or rising sludge, and poor effluent quality. Typical design clarifier overflow rates for oxygen-activated sludge systems are 500 to 1000 gal/ft^2/day.

As previously discussed, clarifier depth is also an important factor, and the depth must be great enough to allow for the shift of biomass from the oxygenation tank to the clarifier and clarifier sludge blanket expansion during peak flow periods. Therefore, the clarifier side water depth should be at least 10 ft for typical applications.

4. Oxygenation Tank Retention Time

A minimum retention time must be maintained in the oxygenation tank if the microorganisms are to successfully solubilize, absorb, and oxidize the influent BOD$_5$ during peak flow periods. Typically, the product of retention time and MLVSS concentration (the denominator in the biomass loading term) is more important than retention time alone in determining BOD$_5$ removal efficiency; however, some minimum physical contact time between the mixed-liquor and wastewater must be provided if system performance is to be maintained at a high level.

A minimum oxygenation tank retention time must also be maintained in order to provide sufficient tank volume for the dissolution equipment to achieve the required oxygen transfer at maximum efficiency. This limitation can be quite important, especially in the treatment of high strength wastewaters that exhibit a high oxygen uptake rate per unit volume. In a secondary treatment system employing high-purity oxygen, this factor is minimized, however, due to the high driving force and reduced oxygen dissolution power requirement. Oxygen transfer efficiency can also be lost if the oxygenation tank volume is allowed to become too great because the system power requirement will then be controlled by mixing rather than mass transfer considerations, resulting in decreased efficiency and higher

operating costs. The minimum oxygenation tank volume will vary for each particular location and must be based on a detailed review of the above parameters.

B. Application of Reactor Clarifier Mathematical Models

The interrelationships between biomass loading, retention time, and clarifier overflow rate are illustrated in Figure 5. By establishing limits such as those described above for these parameters, an integrated range of acceptable design conditions can be determined. The range is shaded for the example presented in Figure 5 and as illustrated, if a design biomass loading of 0.7 lb BOD$_5$/lb MLVSS/day is chosen, retention times of 1.5 to 2.2 hr and corresponding overflow rates of 500 to 1,000 gal/ft^2/day may be used.

If the unit sizes are physically practical, an economically as well as operationally optimum design can be selected by calculating total system capital and operating costs over the range of acceptable oxygenation tank and clarifier sizes. The optimum oxygenation tank clarifier design that provides minimum total cost will vary depending on location, and many cost factors must be considered for each specific case including land, equipment, construction material, power, and sludge disposal.

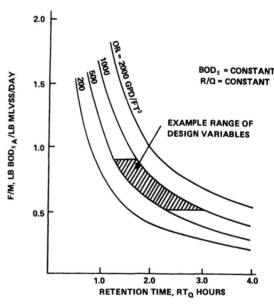

FIGURE 5. Relationship between biomass loading, retention time, and clarifier overflow rate. (Courtesy of Union Carbide Corporation.)

Through the procedure outlined above — (1) selecting limits on process design variables, (2) relating these variables to equipment requirements through basic functional relationships, and (3) selecting the economic optimum that satisfies the process requirements — proper oxygenation tank-clarifier integration can be achieved for any wastewater treatment system in any location.

IV. NOMENCLATURE

a	Initial settling velocity constant
A_B	Factor which accounts for any difference between the average clarifier blanket and recycle suspended solids concentrations
A_X	Cross-sectional area for thickening of clarifier (ft^2)
BOD_5	Five-day biochemical oxygen demand (mg/l)
c_i	Concentration of suspended solids (lb/lb)
F/M	Food to biomass ratio, lb BOD_5 applied/lbMLUSS/day
H_B	Clarifier sludge blanket height (ft)
H_{BO}	Clarifier sludge blanket height at average wastewater flow (ft)
K	Clarifier underflow concentration constant
MLSS	Mixed-liquor suspended solids concentration (mg/l)
$MLSS_O$	Mixed-liquor suspended solids concentration at average wastewater flow (mg/l)
MLVSS	Mixed-liquor volatile suspended solids concentration (mg/l)
n	Sludge settling exponent
OR	Clarifier overflow rate (GPD/ft^2)
OR_O	Clarifier overflow rate at average wastewater flow (GPD/ft^2)
Q	Influent wastewater flow (millions of gallons per day)
R	Recycle flow (millions of gallons per day)
RSS	Return suspended solids concentration (mg/l)
SS_A	Suspended solids applied in influent wastewater (mg/l)
SS_B	Sludge blanket suspended solids concentration (mg/l)
SS_{BO}	Sludge blanket suspended solids concentration at average wastewater flow (mg/l)
t_{SB}	Sludge blanket retention time (hr)
t_Q	Hydraulic retention time in aeration tank based on wastewater flow (hr)
t_{QO}	Hydraulic retention time in aeration tank at average wastewater flow (hr)
TSS	Total suspended solids concentration (mg/l)
V	Volume of aeration tank (million gal)
v_i	Zone or initial settling velocity of suspended solids (ft/hr)
VSS	Volatile suspended solids concentration (mg/l)

Chapter 9

OXYGENATION SYSTEM MASS TRANSFER DESIGN CONSIDERATIONS

J. R. McWhirter and N. P. Vahldieck

TABLE OF CONTENTS

I. INTRODUCTION

The unit operation of gas absorption, which includes the absorption of gases from gaseous mixtures into liquids and the desorption of dissolved gases from liquids into gases, has widespread application in the field of environmental engineering. In water treatment, gas absorption mass transfer operations are used for the removal or addition of carbon dioxide; the removal of hydrogen sulfide, methane, and other volatile organic compounds causing tastes and odors; and for the supply of ozone or chlorine for disinfection. In wastewater treatment operations, gas absorption mass transfer operations are used to supply the oxygen required by biochemical oxidation treatment processes, either from air or from purified oxygen; to supply ozone or chlorine for disinfection; and to strip ammonia from wastewater with air.

The objectives of this chapter are to present a comprehensive mathematical model for the multicomponent interphase mass transfer process which occurs in multi-stage oxygenation systems; to verify validity of the model by comparing predicted performance results with field experience on operating oxygen activated sludge systems; and to describe its application to the process design of oxygenation systems for secondary wastewater treatment.

II. OXYGEN MASS TRANSFER FUNDAMENTALS

A. Air Aeration System Design

The design of oxygenation systems using atmospheric air as the source of oxygen gas is typically a simple and relatively straightforward mass transfer operation involving only the transfer of oxygen gas from the air to the wastewater being aerated. This is a consequence of the fact that wastewater is normally in equilibrium with the nitrogen in atmospheric air and that only a small percentage of the oxygen in the air is absorbed into the wastewater in the aeration process. Also, since relatively large quantities of "excess air" are employed in the aeration process, the desorption of carbon dioxide into the aeration gas does not appreciably influence the overall composition of the aeration gas. Thus, under these conditions, the mass transfer process practically involves only the transfer of oxygen from the gas phase to the liquid phase with the oxygen content of the aeration gas phase remaining essentially constant and equal to that of atmospheric air ($Y_{O_2} = 0.21$). This enables the rate of interphase oxygen transfer to be described by the following simple relationship:

$$GTR_{O_2} = \alpha K_L a_{O_2} (\beta \cdot HC_{O_2} \cdot P_e \cdot Y_{O_2} - X_{O_2}) \quad (1)$$

where GTR_{O_2} = rate of transfer of oxygen from the aeration gas to the liquid phase (lb O_2/hr); $K_L a_{O_2}$ = overall liquid phase oxygen mass transfer coefficient for the gas-liquid contacting device (lb O_2/hr/mg/l); α = alpha value for the system ($K_L a_{O_2}$ in actual mixed-liquor/$K_L a_{O_2}$ in tap water); β = beta value for the system (HC_{O_2} in actual mixed-liquor/HC_{O_2} in tap water); HC_{O_2} = Henry's law constant for oxygen in tap water = 44.4 mg/l/atm at 20°C; P_e = effective pressure for the gas-liquid contacting device under the operating conditions in the system (atm); Y_{O_2} = mole fraction of oxygen

in the aeration gas = 0.21 for atmospheric air; and X_{O_2} = dissolved oxygen concentration in the mixed-liquor (mg/l). The evaluation of the oxygen transfer capability or rating capacity of air aeration devices is typically conducted in "pure" tap water at a zero dissolved oxygen level. An extensive discussion of the test methods and calculational procedures used in these evaluations is contained in Chapter 1, Volume II. The oxygen transfer rate under these conditions is referred to as the "standard transfer rate." The standard transfer rate is useful in comparing the relative oxygen transfer capability of air aeration devices under a carefully controlled set of standard conditions. As discussed in Chapter 1, Volume II, however, the test procedures and methods must be very carefully controlled and implemented in order to obtain meaningful results.

Substituting into Equation 1 the specific factors at standard conditions:

$$STR_{O_2} = K_L a_{O_2} [(44.4)(1.0)(0.21) - 0.0] = K_L a_{O_2} (9.3) \quad (2)$$

since α, β = 1.0 for tap water; X_{O_2} = 0.0 for standard conditions; P_e = 1.0 atm; Y_{O_2} = 0.21; STR_{O_2} = oxygen transfer rate at standard conditions of tap water at 20°C and zero dissolved oxygen at 1 atm pressure (lb O_2/hr) and

$$STE_{O_2} = \frac{STR_{O_2}}{shp} \quad (2a)$$

where STE_{O_2} = standard transfer efficiency and shp = shaft horsepower input.

A typical value of the standard transfer efficiency for an optimally designed, low speed surface aerator is about 3.5 lb of oxygen transferred per hour per shaft horsepower for aerators in the range of 10 to 75 hp. For an aerator with a shaft horsepower input of 20 hp, for example

$$STR_{O_2} = (3.5)(20) = K_L a_{O_2} (9.3) \quad (3)$$

$$K_L a_{O_2} = \frac{70.0}{9.3} = 7.53 \frac{lb \; O_2}{(hr)(mg/l)} \quad (4)$$

The value of $K_L a_{O_2}$ represents the intrinsic oxygen mass transfer capability of the aeration device and can be used to calculate the oxygen transfer rate under any specific actual process operating conditions. For operation under typical air-activated sludge mixed-liquor conditions, the following oxygen transfer rate for the above surface aerator example can be calculated:

$$GTR_{O_2} = (0.85)(7.53)[(0.95)(44.4)(1.0)(0.21) - 2.0] \qquad (5)$$

$$GTR_{O_2} = 43.9 \text{ lb } O_2/\text{hr} \qquad (6)$$

where $\alpha = 0.85$ for activated sludge mixed-liquor; $K_L a_{O_2} = 7.53 \text{ (hr)} \frac{\text{lb } O_2}{\text{(hr)(mg/l)}}$; $\beta = 0.95$ for activated sludge mixed-liquor; $HC_{O_2} = 44.4$ mg/l/atm at 20°C; $P_e = 1.0$ atm; $Y_{O_2} = 0.21$; and $X_{O_2} = 2.0$ mg/l in the mixed-liquor.

The transfer efficiency under actual activated sludge mixed-liquor aeration conditions and a mixed-liquor dissolved oxygen concentration of 2.0 mg/l is then

$$ATE_{O_2} = \frac{43.9}{20} = 2.20 \frac{\text{lb } O_2}{\text{shp hr}} \qquad (7)$$

Correcting for typical motor efficiency and gear box efficiency of 90 and 95% respectively:

$$ATE_{O_2} = (2.20)(0.90)(0.95) = 1.88 \frac{\text{lb } O_2}{\text{whp hr}} \qquad (8)$$

where whp = wire input motor Hp.

Thus, the true aerator oxygen transfer efficiency per unit of actual input horsepower is about 50% of that which is typically quoted for aerator efficiencies under standard transfer conditions. It is, of course, the actual input motor horsepower requirement of the aeration device which is important in achieving proper sizing of the aeration equipment under activated sludge process operating conditions.

B. Oxygenation System Design

As discussed in Chapter 3, Volume I, the additional constraints and requirements of an economical high-purity oxygenation system result in the need for a considerably more complex analysis and overall system design procedure. The necessity of achieving a high percentage oxygen absorption and utilization efficiency results in a multicomponent interphase mass transfer process involving the transfer of significant quantities of gases other than oxygen. The multicomponent process is a result of wastewater being in equilibrium with atmospheric air and the significant amounts of carbon dioxide generated in the bio-oxidation process. The transfer of other gaseous components, including nitrogen, carbon dioxide, argon, and water vapor, must be properly taken into account in addition to the transfer of oxygen.

C. Multicomponent Mass Transfer Fundamentals

It has been shown by Toor,[3-5] Toor, Seshadri, and Arnold,[6] and Stewart and Prober,[7] that the rate of interphase mass transfer from a gas to a liquid in a multicomponent system can be described by the vector-matrix generalization of the conventional rate equation used to describe the transfer rate in a binary or two component system. Therefore, taking the vector-matrix generalization of the simple binary relationship as expressed in Equation 1, for an $n + 1$ component system we have:

$$(GTR_i) = \alpha [K_L a_i] (\beta P_e [X_i^*] - [X_i]) \qquad (9)$$

where

$$(GTR_i) = \begin{bmatrix} GTR_1 \\ GTR_2 \\ \cdot \\ \cdot \\ \cdot \\ GTR_n \end{bmatrix}$$

$$(\beta P_e [X_i^*] - [X_i]) = \begin{bmatrix} \beta P_e (X_1^*) - (X_1) \\ \beta P_e (X_2^*) - (X_2) \\ \cdot \\ \cdot \\ \cdot \\ \beta P_e (X_n^*) - (X_n) \end{bmatrix}$$

$$(X_i^*) = [HC_i] (Y_i)$$

$$[HC_i] = \begin{bmatrix} HC_{11} & HC_{12} \ldots HC_{1n} \\ HC_{21} \ldots HC_{2n} \\ \cdot \\ \cdot \\ HC_{n1} \ldots HC_{nn} \end{bmatrix}$$

$$(Y_i) = \begin{bmatrix} Y_1 \\ Y_2 \\ \cdot \\ \cdot \\ Y_n \end{bmatrix}$$

$$(X_i) = \begin{bmatrix} X_1 \\ X_2 \\ \cdot \\ \cdot \\ \cdot \\ X_n \end{bmatrix}$$

$$[K_L a_i] = \begin{bmatrix} K_L a_{11} & K_L a_{12} & \cdots & K_L a_{1n} \\ K_L a_{21} & \cdots & & K_L a_{2n} \\ \cdot \\ \cdot \\ \cdot \\ K_L a_{nl} & \cdots & & K_L a_{nn} \end{bmatrix}$$

(GTR$_i$) is the column vector of individual component mass transfer rates or fluxes. [$K_L a_i$] is the overall liquid phase mass transfer coefficient matrix (an n × n square matrix) for the mixture of n + 1 components. (X$_i$) and (Y$_i$) are the column vectors of individual component bulk liquid and gas phase compositions, respectively. The term ($\beta P_e (X^*) - (X_i)$) is the column vector of overall liquid phase mass transfer driving forces for the individual components. [HC$_i$] is the equilbrium constant coefficient matrix for the n + 1 component mixture and, like [$K_L a_i$], is a square matrix with n × n elements.

By way of example, consider a four component system, which requires a system of 3 × 3 matrices to describe the multicomponent mass transfer process. For three of the four components we have:

$$\begin{aligned} GTR_1 =\ & \alpha K_L a_{11} (\beta P_e [X_1^*] - X_1) + \\ & \alpha K_L a_{12} (\beta P_e [X_2^*] - X_2) + \\ & \alpha K_L a_{13} (\beta P_e [X_3^*] - X_3) \end{aligned} \tag{10}$$

$$\begin{aligned} GTR_2 =\ & \alpha K_L a_{21} (\beta P_e [X_1^*] - X_1) + \\ & \alpha K_L a_{22} (\beta P_e [X_2^*] - X_2) + \\ & \alpha K_L a_{23} (\beta P_e [X_3^*] - X_3) \end{aligned} \tag{11}$$

$$\begin{aligned} GTR_3 =\ & \alpha K_L a_{31} (\beta P_e [X_1^*] - X_1) + \\ & \alpha K_L a_{32} (\beta P_e [X_2^*] - X_2) + \\ & \alpha K_L a_{33} (\beta P_e [X_3^*] - X_3) \end{aligned} \tag{12}$$

In Equations 10 to 12, it is seen that the net rate of transfer or flux of each component is dependent upon the transfer rate or flux of all components as determined by the interaction terms in each equation. These interactive terms can be determined, at least for mixtures of ideal gases and liquids, from the basic equations of molecular diffusion in multicomponent mixtures.

Similarly, the equilibrium relationships in Equations 10 to 12 are as follows:

$$X_1^* = HC_{11} Y_1 + HC_{12} Y_2 + HC_{13} Y_3 \tag{13}$$

$$X_2^* = HC_{21} Y_1 + HC_{22} Y_2 + HC_{23} Y_3 \tag{14}$$

$$X_3^* = HC_{31} Y_1 + HC_{32} Y_2 + HC_{33} Y_3 \tag{15}$$

The equilibrium relationships in Equations 13 to 15 also contain interactive terms, which means that the equilibrium concentration of any component can be influenced by the concentration of the other components of the mixture. If the gas phase is an ideal gas mixture and the liquid phase concentrations are dilute, however, the interactive terms, HC_{12}, HC_{13}, HC_{21}, HC_{23}, HC_{31}, and HC_{32}, will be zero. Under these conditions:

$$X_1^* = HC_{11} Y_1 \tag{16}$$

$$X_2^* = HC_{22} Y_2 \tag{17}$$

$$X_3^* = HC_{33} Y_3 \tag{18}$$

Equations 9 to 15 can be used to describe any general multicomponent gas absorption mass transfer process between a gas and liquid phase. The general process is quite complex and requires determination of the interactive terms in both the overall liquid phase mass transfer coefficient matrix [$K_L a_i$] and the equilibrium constant matrix [HC$_i$]. A number of important conditions exist, however, in the case of the oxygenation of mixed-liquor which allow the description of the multicomponent mass transfer process to be considerably simplified:

1. The transfer of all solute gases is liquid phase mass transfer controlled and, therefore, the interfacial liquid compositions can be assumed to be in equilibrium with the bulk gas phase.
2. Multicomponent diffusional interactions in both the gas and liquid phases can be ignored. Gas-phase interactions can be ignored because the interphase transfer of all solute gases is liquid phase mass transfer controlled. Liquid-phase diffusional interactions can be ignored because all solute gases are present in a dilute aqueous solution and their binary diffusion coefficients in water are approximately equal.

3. The gas phase closely follows the ideal gas law.
4. Henry's law can be used to independently describe the equilibrium distribution of all solute gases. This means that the dissolution of the solute gases behaves as an ideal liquid solution because all dissolved gases are present in a very dilute solution.

Under these conditions, the interactive (off-diagonal) terms in both the mass transfer coefficient matrix $[K_L a_i]$ and the equilibrium constant matrix $[HC_i]$ are all equal to zero. Equations 10 to 12 then simplify to the following relationships for a four component mixture:

$$GTR_1 = \alpha \, K_L a_{11} \, (\beta \, P_e \, [X_1^*] - X_1) \qquad (19)$$

$$GTR_2 = \alpha \, K_L a_{22} \, (\beta \, P_e \, [X_2^*] - X_2) \qquad (20)$$

$$GTR_3 = \alpha \, K_L a_{33} \, (\beta \, P_e \, [X_3^*] - X_3) \qquad (21)$$

where $X^* = HC_{11} Y_1$; $X^* = HC_{22} Y_2$; and $X^* = HC_{33} Y_3$ from Equations 16 to 18.

Thus, the rate of transfer of all nonionized component gases in multicomponent oxygenation systems can be accurately described by a set of independent uncoupled equations of the following form:

$$GTR_i = \alpha \, K_L a_i \, (\beta \, HC_i \cdot P_e \cdot Y_i - X_i) \qquad (22)$$

where GTR_i = rate of transfer of component i from the gas phase to the liquid phase (lb/hr); HC_i = Henry's law constant for component i (mg/l/atm); $K_L a_i$ = overall liquid phase mass transfer coefficient for the gas-liquid contacting device in the system (lb/hr/mg/l); P_e = effective pressure for the gas-liquid contacting device under the operating conditions existing in the system (atm); X_i = dissolved concentration of component i in the mixed-liquor (mg/l); Y_i = mole fraction of component i in the gas phase; α = alpha value for the system ($K_L a_i$ actual/$K_L a_i$ tap water); and β = beta value for the system (HC_i actual/HC_i tap water). The product ($\alpha \cdot K_L a_i$) in Equation 22 represents the mass transfer capability of the gas-liquid contacting device for component i. Both α and $K_L a_i$ can vary with changes in operating conditions for the system. The term ($\beta \cdot HC_i \cdot P_e \cdot Y_i - X_i$) represents the driving force available for mass transfer of component i from the gas to the liq-

uid phase. P_e can also vary with changes in system operating conditions. Equation 22 indicates that the rate of interphase mass transfer is equal to the product of the intrinsic mass transfer capability of the contacting device for component i and the driving force available for mass transfer for component i. The product ($\beta \cdot HC_i \cdot P_e \cdot Y_i$) represents the concentration of component i at equilibrium with a gas of composition Y_i, assuming that nonionized gases in solution obey Henry's law. If the equilibrium concentration of component i exceeds the actual concentration of component i in the liquid (X_i), the driving force is positive and component i will be transferred from the gas phase to the liquid phase. If the equilibrium concentration of component i is less than the actual concentration of component i in the liquid phase, the driving force is negative, and component i will be transferred from the liquid phase to the gas phase. For gases which partially ionize in solution, such as ammonia, carbon dioxide, and hydrogen sulfide, Henry's law applies only to the nonionized portion of the dissolved gas.

For such gases, X_i in Equation 1 is assumed to be the sum of the concentrations of all the nonionized forms of component i, and HC_i refers to the sum of the concentrations of all of the nonionized forms of component i. With these modified definitions of X_i and HC_i, Equation 22 also describes the interphase multicomponent mass transfer for ionized components in oxygenation systems.

The design of gas-liquid contacting equipment for staged oxygenation of wastewater treatment systems is considerably more difficult than the sizing of aeration equipment for conventional wastewater treatment systems, even though both are fundamentally sized by employing Equation 22. The value of α in Equation 22 depends on the characteristics of the mixed-liquor, which can vary as treatment progresses, as well as with the type of gas-liquid contacting equipment, the geometric configuration of the aeration tanks, and the operating conditions for the system. The value of $K_L a_i$ is also a function of the type of gas-liquid contacting equipment, the geometric configuration of the aeration tanks, and the operating conditions, as well as the characteristics of component i in the wastewater. The value of β is a characteristic of the mixed-liquor which can vary as treatment progresses. HC_i is a characteristic of component i and a function of temperature. The system temperature can change from stage to stage in some instances. For example, the temperature can increase from stage to stage in thermophilic aerobic digestion with purified oxygen as well as with high strength wastes.

The value of P_s depends on the type of gas-liquid contacting equipment employed, the geometric configuration of the aeration tanks, and the operating conditions. P_s, the effective pressure, is fixed by the operating pressure and the effective depth of the gas-liquid contacting device. The effective depth is a specific characteristic of the gas-liquid contacting device which should be experimentally determined at the same time as the mass transfer coefficient, K_La_i. The effective depth is in turn a function of the type of contacting device, the geometric configuration of the aeration tanks, and the operating conditions.

When sizing air aeration equipment for conventional wastewater treatment systems, the value for Y_{O_2} is assumed to remain constant. If a value for GTR_{O_2} is known or assumed, Equation 22 is readily solved for X_{O_2}, if $K_La_{O_2}$ is known, or for a value of $K_La_{O_2}$ corresponding to a specific value of X_{O_2}. The transfer of other gaseous components is usually ignored in sizing air aeration equipment for conventional wastewater treatment systems. However, for staged oxygenation systems with high utilization of oxygen, Y_{O_2} varies considerably from stage to stage because of the influence of other gaseous constituents in the system. The value of Y_{O_2} in each stage depends on the composition, flow rate, and temperature of each gas and water stream entering the stage; the rate of consumption or formation of any component undergoing chemical reaction within the stage; and the rate and direction of mass transfer between the gas and liquid phases for each gaseous component. Thus, Equation 22 must be satisfied for each of the gaseous components of the system, instead of just for oxygen.

A rigorous mathematical model of the multicomponent mass transfer process is needed to accurately describe the mass transfer and related water chemistry of the multi-stage system. This is required so that pilot plant results can be extrapolated to various design conditions in order to optimize the process design of acti-

vated sludge carbonaceous and nitrification systems, aerobic digestion, and ozonation systems employing purified oxygen. The mathematical model can also be used by process designers to estimate the distribution of volatile hydrocarbon contaminants between gas and liquid phases in staged oxygenation systems. The model can further be used to estimate the degree of removal of ammonia, hydrogen sulfide, or carbon dioxide attainable with oxygen or air stripping; the chemical requirements for pH control (if pH control is desired, or necessary); or the solubility limits for sparingly soluble salts.

III. AQUEOUS-PHASE IONIC EQUILIBRIUM AND RELATED WATER CHEMISTRY

Oxygen, ozone, nitrogen, argon, methane, and other volatile organic compounds are not ionized in aqueous solutions. They directly obey Henry's law which states that at equilibrium, their concentration in solution is directly proportional to their partial pressure in the gas phase in contact with the solution. Carbon dioxide, hydrogen sulfide, ammonia, and chlorine ionize in water, and Henry's law applies only to the concentrations of their nonionized forms in solution. Calculations of the rate of interphase mass transfer of ionized gases, therefore, requires the determination of the fractions of the dissolved gases which are in the ionized and nonionized form. Ionization reactions also influence other parameters of significance in secondary wastewater treatment, such as the pH and the solubility of certain sparingly soluble salts. This section contains a brief summary of the important ionic equilibria and related water chemistry pertinent to oxygenation in secondary wastewater treatment.

A. Hydrogen Ion Equilibria

For dilute aqueous solutions of intermediate acidity, i.e., for a range of pH from 3 to 11, the pH value can be adequately defined as the negative logarithm of the hydrogen ion activity:

$$pH = -\log \alpha_H = -\log m_H \gamma \tag{23}$$

where α_H = conventional hydrogen ion activity; m_H = molality of hydrogen; and γ = activity coefficient. For most natural waters and wastewaters at temperatures below 30°C, the difference between molality and molarity is small, and the activity coefficient has a value close to one. For such waters, the pH value can be adequately defined as the negative logarithm of the hydrogen ion concentration in gram moles/liter:

$$pH = -\log (H^+) = \log \frac{1}{(H^+)} \qquad (24)$$

where (H^+) = hydrogen ion concentration (gram moles/liter). The parentheses in the above equation and in all the equations in this section are used to indicate concentrations in gram moles/liter (lb mol/1000 lb H_2O). H^+ is used as a symbol for the hydrated hydrogen ion rather than H_3O+ as a matter of convenience.

The pH of water is ordinarily determined by measuring the potential difference between glass and reference electrodes immersed in the water. The potential must be measured with a sensitive electrometer that does not permit a significant flow of current. Modern pH meters can measure pH in the field or in the laboratory to the nearest 0.01 pH unit. However, because the pH is defined as the logarithm of a concentration, measurement to the second decimal place may still be imprecise compared to the usual measurement of concentrations of other solute species.

Published water analyses frequently contain pH values whose significance can be misinterpreted. The pH of water represents the interrelated result of a large number of chemical equilibria. The equilibria can be altered by the sampling procedure. A pH measurement taken at the moment of sampling will usually represent the original equilibrium conditions satisfactorily. If the water is put into a sample bottle and the pH is not determined until the sample is taken out for analysis at some later time, the measured pH may have no relation to the original value. The solution may be influenced by gains or losses of carbon dioxide or the oxidation of components such as ferrous iron. Further, the pH measured at one temperature has no quantitative meaning relative to the pH at another temperature. A laboratory determination of pH can be considered as applicable only to the solution in the sample bottle at the time the determination is made. Therefore, accurate measurement of pH values in the field should be standard practice.

Water is weakly and reversibly ionized as indicated by the following equilibrium relationship:

$$H_2O \rightleftharpoons H^+ + OH^- \qquad (25)$$

The extent of the ionization of water is expressed by the equation:

$$(H^+)(OH^-) = K(H_2O) \cong 10^{-14} \text{ at } 25°C \qquad (26)$$

$K(H_2O)$, the ionization constant for water, is a function of temperature as shown in Figure 1.

B. Carbon Dioxide Ionic Equilibria

Carbon dioxide dissolved in water undergoes a series of reversible chemical reactions which can be represented by the following equilibrium reactions:

$$CO_2 + H_2O \rightleftharpoons H_2CO_3 \qquad (27)$$

$$H_2CO_3 \rightleftharpoons H^+ + HCO_3^- \qquad (28)$$

$$HCO_3^- \rightleftharpoons H^+ + CO_3^{--} \qquad (29)$$

The extent or degree of completion of each of the reactions is expressed by the following equations:

$$\frac{(H_2CO_3)}{(CO_2)} = K_0(CO_2) \qquad (30)$$

$$\frac{(H^+)(HCO_3^-)}{(H_2CO_3)} = K_1(CO_2) \qquad (31)$$

$$\frac{(H^+)(CO_3^{--})}{(HCO_3^-)} = K_2(CO_2) \qquad (32)$$

$K_0(CO_2)$, $K_1(CO_2)$, and $K_2(CO_2)$ are called the equilibrium constants for carbon dioxide and are shown in Figure 2 as a function of temperature. Because $K_0(CO_2)$ is 4 to 5 orders of magnitude greater than $K_1(CO_2)$, Reaction 27 is generally assumed to be complete for purposes of estimating carbon dioxide ionization equilibrium constants. All of the dissolved carbon dioxide which is not ionized is assumed to be in the form of carbonic acid, H_2CO_3, when calculating the ionization equilibria. Literature values for $K_1(CO_2)$ and $K_2(CO_2)$ are calculated by making the practical assumption that all of the aqueous carbon dioxide is in the form of carbonic acid. However, at usual wastewater temperatures, the value of $K_0(CO_2)$ will be between 0.03 and 0.06. Almost all of the nonionized car-

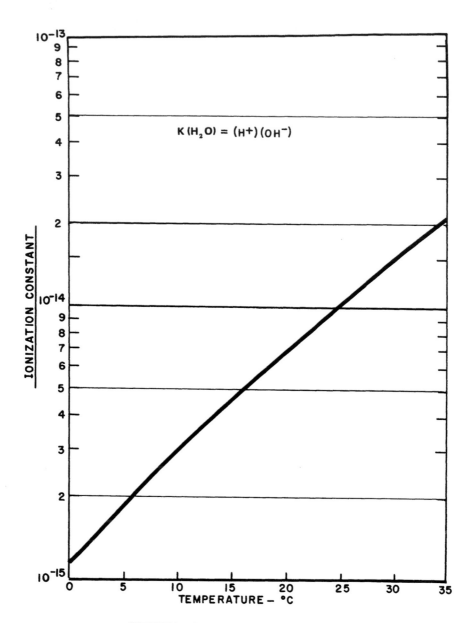

$$K(H_2O) = (H^+)(OH^-)$$

FIGURE 1. Ionization constant for water.

bon dioxide is, therefore, in the form of dissolved gas rather than carbonic acid. Contrary to the usual assumption, interphase mass transfer calculations are consequently based on the usual assumption that all of the nonionized carbon dioxide is in the form of dissolved gas.

Calculation of the fraction of the dissolved carbon dioxide in the nonionized form is required before Henry's law can be used to relate the gas and liquid phase concentrations at equilibrium. This can be accomplished from the equilibrium relationships contained in Equations 30 to 32. The fraction of dissolved carbon

dioxide in the nonionized form is defined as follows:

$$XF(CO_2) = \frac{(H_2CO_3)}{(H_2CO_3) + (HCO_3^-) + (CO_3^{--})} =$$

$$\frac{1}{1 + \dfrac{(HCO_3^-)}{(H_2CO_3)} + \dfrac{(CO_3^{--})}{(H_2CO_3)}} \quad (33)$$

Equation 33 assumes that, from an ionization equilibria standpoint, all of the nonionized dissolved carbon dioxide is in the form of carbonic

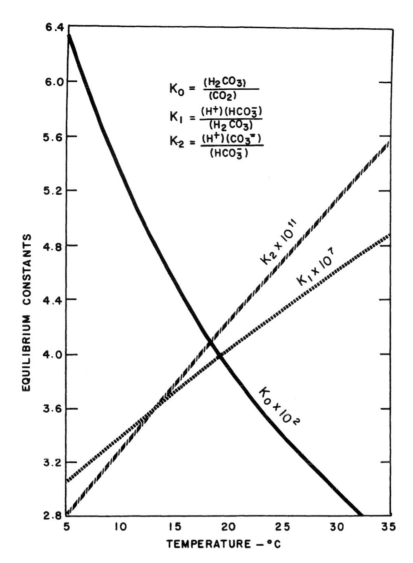

$$K_0 = \frac{(H_2CO_3)}{(CO_2)}$$

$$K_1 = \frac{(H^+)(HCO_3^-)}{(H_2CO_3)}$$

$$K_2 = \frac{(H^+)(CO_3^=)}{(HCO_3^-)}$$

FIGURE 2.   Carbon dioxide equilibrium constants.

acid, H_2CO_3. Substituting from Equations 31 and 32 into Equation 33:

$$XF(CO_2) = \frac{1}{1 + \dfrac{K_1(CO_2)}{(H^+)} + \dfrac{K_1(CO_2)K_2(CO_2)}{(H^+)^2}} \qquad (34)$$

Henry's law can be expressed as:

$$(H_2CO_3) = HC_{CO_2} \cdot Y_{CO_2} \cdot P_e \qquad (35)$$

The Henry's law constant for dissolved carbon dioxide gas as a function of temperature is shown in Figure 3.

C. Hydrogen Sulfide, Ammonia, and Chlorine Equilibria

Hydrogen sulfide dissolved in water can also undergo a series of reversible chemical reactions which can be represented by the following equilibrium relationships:

$$H_2S \rightleftharpoons H^+ + HS^- \qquad (36)$$

$$HS^- \rightleftharpoons H^+ + S^{--} \qquad (37)$$

The degree of completion of each of these reactions can be expressed by the following equations:

243

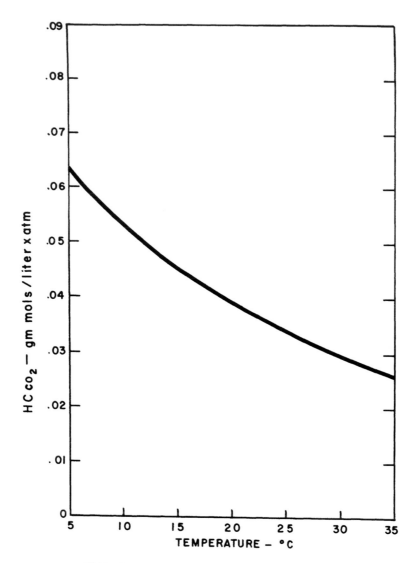

FIGURE 3.   Henry's law constant for carbon dioxide.

$$\frac{(H^+)\,(HS^-)}{(H_2S)} = K_1\,(H_2S) \qquad\qquad (38)$$

$$\frac{(H^+)\,(S^{--})}{(HS^-)} = K_2\,(H_2S) \qquad\qquad (39)$$

Ammonia dissolved in water also reacts incompletely as indicated by the following equilibrium relationship:

$$NH_3 + NH_4^+ \rightleftarrows H_2O + OH^- \qquad\qquad (40)$$

The degree of completion of the reaction can be expressed by the following equation:

$$\frac{(NH_4^+)\,(OH^-)}{(NH_3)} = K\,(NH_3) \qquad\qquad (41)$$

Ammonia dissolved in water does not initially form NH_4OH molecules in a manner analogous to the formation of H_2CO_3 molecules from CO_2. Each ammonia molecule may be loosely bound to a number of water molecules.

Calculation of the fraction of hydrogen sulfide and ammonia which is in the nonionized form can be calculated from the above relationships, as was previously done in Equations 33 to 35 for carbon dioxide. Therefore, we have for the nonionized fraction of hydrogen sulfide:

$$XF(H_2S) = \frac{(H_2S)}{(H_2S) + (HS^-) + (S^{--})}$$

$$= \frac{1}{1 + \frac{(HS^-)}{(H_2S)} + \frac{(S^{--})}{(H_2S)}} \tag{42}$$

Substituting from Equations 38 and 39:

$$SF(H_2S) = \frac{1}{1 + \frac{K_1(H_2S)}{(H^+)} + \frac{K_1(H_2S)K_2(H_2S)}{(H^+)^2}} \tag{43}$$

In a similar manner, for the nonionized fraction of ammonia:

$$SF(NH_3) = \frac{(NH_3)}{(NH_3) + (NH_4^+)} = \frac{1}{1 + \frac{(NH_4^+)}{(NH_3)}} \tag{44}$$

Substituting from Equations 21 and 41:

$$XF(NH_3) = \frac{1}{1 + \frac{(H^+)K(NH_3)}{K(H_2O)}} \tag{45}$$

Chlorine reacts with water in a similar series of reversible chemical reactions:

$$Cl_2 + H_2O \rightleftharpoons HOCl + H^+ + Cl^- \tag{46}$$

$$HOCl \rightleftharpoons H^+ + OCl^- \tag{47}$$

The extent of the reactions can be expressed by the following equations:

$$\frac{(HOCl)(H^+)(Cl^-)}{(Cl_2)} = K_0(Cl_2) \tag{48}$$

$$\frac{(H^+)(OCl^-)}{(HOCl)} = K_1(Cl_2) \tag{49}$$

Both HOCl and OCl$^-$ react with organic material in solution or suspension in water, so a disinfection reaction model must be employed simultaneously when estimating chlorine ionization equilibria.

D. Phosphoric Acid Equilibria

Dissolved minerals, such as phosphates, silicates, and borates, affect the hydrogen ion equilibria of water. Orthophosphoric acid, dissolved in water, undergoes a series of reversible ionization reactions which can be represented by the following relationships:

$$H_3PO_4 \rightleftharpoons H^+ + H_2PO_4^- \tag{50}$$

$$H_2PO_4 \rightleftharpoons H^+ + HPO_4^{--} \tag{51}$$

$$HPO_4^{--} \rightleftharpoons H^+ + PO_4^{---} \tag{52}$$

The degree of completion of these ionization reactions can be expressed by the following equations:

$$\frac{(H^+)(H_2PO_4^-)}{(H_3PO_4)} = K_1(PO_4) \tag{53}$$

$$\frac{(H^+)(HPO_4^{--})}{(H_2PO_4^-)} = K_2(PO_4) \tag{54}$$

$$\frac{(H^+)(PO_4^{---})}{(HPO_4^{--})} = K_3(PO_4) \tag{55}$$

E. Alkalinity

The alkalinity (A) of water is defined as its capacity for reacting with hydrogen ions to a pH corresponding to stoichiometric formation of carbonic acid (or carbon dioxide and water). The alkalinity can be readily determined by a simple titration with 0.02 N sulfuric acid to the methyl purple end point (pH = 4.5). The alkalinity is then determined as follows:

Alkalinity (milligrams per liter as $CaCO_3$) =

$$\frac{\text{milliliters acid}}{\text{milliliters sample}} \times 1000 \tag{56}$$

Defining the alkalinity in terms of equivalents per liter,

A = alkalinity (equivalents per liter) =

$$\frac{\dfrac{\text{milliliters acid}}{\text{milliliters sample}} \times 1000}{50,000} \tag{57}$$

The alkalinity of water is due to the combined presence of many ionic species including HCO_3^-, CO_3^{--}, OH^-, $HSiO_3^-$, $H_2BO_3^-$, $H_2PO_4^-$, HPO_4^{--}, HS^-, and NH_3. Expressing the definition for alkalinity in the form of an equation

$$A = \text{alkalinity (equivalents/l)} = (HCO_3^-) + 2(CO_3^{--}) +$$

$$(OH^-) + (HSiO_3^-) + (H_2BO_3^-) + (H_2PO_4^-) +$$

$$2(HPO_4^{--}) + (HS^-) + (NH_3) - (H^+) \tag{58}$$

Note that (CO_3^{--}) and (HPO_4^{--}) are multiplied

by two in defining alkalinity as gram equivalents/liter. Because of their valence, the equivalent weights of these ions are one half of their molecular weights. (NH₃) refers to nonionized ammonia; the ammonium ion does not contribute to the alkalinity. From a practical standpoint, the borate and silicate ion contributions to the alkalinity can be ignored with little resulting error.

A buffer is a substance which resists a change in the hydrogen ion concentration, or pH value, of an aqueous system upon addition of either acidic or basic materials. In oxygen-activated sludge secondary wastewater treatment systems, the alkalinity of the wastewater buffers it against depression of its pH caused by the carbon dioxide generated and dissolved during treatment. The higher the alkalinity of the wastewater, the less will be the depression of the pH during treatment. The alkalinity of water is not altered by the gain or loss of gaseous carbon dioxide because equivalent amounts of hydrogen and bicarbonate or carbonate ions are always added to or subtracted from the water simultaneously.

In most wastewaters, the HCO_3^-, CO_3^{--}, and OH^- ions account for the bulk of the titrated alkalinity. If it is assumed that the alkalinity is due to only these ions, the alkalinity as defined in Equation 58 simplifies to:

$$A = (HCO_3^-) + 2(CO_3^{--}) + (OH^-) - (H^+) \qquad (59)$$

The concentrations of each of these components of the alkalinity can be expressed from the equilibrium relationships given above in Equations 26, 31, and 32:

$$(OH^-) = \frac{K(H_2O)}{(H^+)} \qquad (60)$$

$$(HCO_3^-) = K_1(CO_2) \cdot \frac{(H_2CO_3)}{(H^+)} \qquad (61)$$

$$(CO_3^{--}) = K_2(CO_2) \cdot \frac{(HCO_3^-)}{(H^+)} =$$

$$K_1(CO_2) \cdot K_2(CO_2) \cdot \frac{(H_2CO_3)}{(H^+)^2} \qquad (62)$$

Substituting Equations 60, 61, and 62 into 59 results in:

$$A = K_1(CO_2) \cdot \frac{(H_2CO_3)}{(H^+)} + 2K_1(CO_2)K_2(CO_2) \cdot \frac{(H_2CO_3)}{(H^+)^2} + \frac{K(H_2O)}{(H^+)} - (H^+) \qquad (63)$$

Substituting for the concentration of carbonic acid, (H_2CO_3), from Equation 35, for a system at equilibrium with respect to the dissolved carbon dioxide concentration, Equation 63 becomes:

$$A = \left[\frac{K_1(CO_2)}{(H^+)} + \frac{2K_1(CO_2)K_2(CO_2)}{(H^+)^2} \right] \cdot HC_{CO_2} \cdot Y_{CO_2} \cdot P_e + \frac{K(H_2O)}{(H^+)} - (H^+) \qquad (64)$$

This expression relates the alkalinity, A, and the hydrogen ion concentration, (H^+), to the mole fraction of carbon dioxide in the gas in equilibrium with the water at a given system temperature and pressure. Figure 4 shows the equilibrium relationship between the alkalinity, pH, and percent carbon dioxide in the gas phase for a gas-liquid system at 25°C and 1.0 atm total pressure. (Note that the alkalinity has been converted to milligrams per liter as $CaCO_3$ and the hydrogen ion concentration to pH values in Figure 4.) With CO_2 concentrations ranging from about 1 to 10% and aqueous-phase alkalinites ranging from about 70 to 1000 mg/l as $CaCO_3$, the pH of the liquid varies from only 6 to 8. This is well within the acceptable range for peak efficiency operation of the activated sludge process.

The concentration of NH_3 in wastewater also affects its alkalinity (see Equation 58). Nitrogen is present in wastewater as dissolved free nitrogen, organic nitrogen, free or unreacted ammonia, ammonium ions, nitrite ions, and nitrate ions. In the biological nitrification process, some of the organic nitrogen is converted to ammonia, and some of the ammonium ions are oxidized to nitrite and nitrate ions. These reactions, therefore, affect the alkalinity of the wastewater. In order to show the relationship between nitrogen and ammonia removal and alkalinity change, let:

$$F = \frac{NH_3 \text{ formed}}{\text{organic N removed}} \qquad (65)$$

with $NH_3 \cdot N$ = sum of free ammonia plus am-

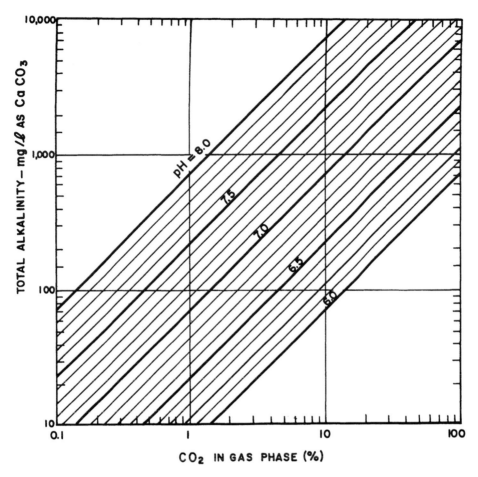

FIGURE 4. Equilibrium relationship between pH, alkalinity, and percent CO_2 in the gas phase for water at 25°C and at 1.0 atm. (Courtesy of Union Carbide Corporation.)

monium nitrogen (parts per million N) (this is the so-called "ammonia nitrogen" usually measured analytically); Δ = concentration of a constituent in the effluent minus the concentration in the influent; ΔA = change in alkalinity (milligrams per liter as $CaCO_3$); and NH_3 = free ammonia (milligrams per liter N). From the definition of alkalinity:

$$\Delta A = \frac{50.0}{14.01} \, (NH_3 \text{ formed} - NH_3 \text{ lost but not oxidized}$$

$$- NH_3 \text{ oxidized to } NO_2^-) \qquad (66)$$

where NH_3 formed = $-F \cdot \Delta$ organic N; NH_3 lost but not oxidized = NH_3 (in) $-F \cdot \Delta$ organic N $- NH_3$ (out); and NH_3 oxidized = $\Delta (NO_2^- + NO_3^-)$

Substituting into Equation 66:

$$\Delta A = 3.57 \, [\Delta \, NH_3 - \Delta \, (NO_2^- + NO_3^-)] \qquad (67)$$

Since $\Delta NH_3 = \Delta NH_3 \cdot N$ (assuming NH_4^+ concentration is constant)

$$\Delta A = 3.57 \, [\Delta \, NH_3 \cdot N - \Delta \, (NO_2^- + NO_3^-)] \qquad (68)$$

where ΔA = milligrams per liter as $CaCO_3$; $\Delta NH_3 \cdot n$ = milligrams per liter $(NH_3 + NH_4^+)$ as N; and $\Delta (NO_2^- + NO_3^-)$ = milligrams per liter $(NO_2^- + NO_3^-)$ as N. The alkalinity change with nitrification and its resultant effect on mixed-liquor pH can be a very important process design consideration in some cases as discussed in Chapter 5, Volume II.

F. Sparingly Soluble Salts

Increasing the pH of the mixed-liquor of an oxygen-activated sludge system by adding alkaline reagents can cause the precipitation of certain slightly soluble salts. The equilibrium relationships between several important slightly soluble salts and their ions are given below:

$$Ca^{++} + CO_3^{--} \rightleftarrows CaCO_3 \qquad (69)$$

$$Mg^{++} + CO_3^{--} \rightleftarrows MgCO_3 \cdot 2H_2O \qquad (70)$$

$$Mg^{++} + 20H^- \rightleftarrows Mg(OH)_2 \qquad (71)$$

$$10Ca^{++} + 6PO_4^{---} + 20H^- \rightleftarrows Ca_{10}(PO_4)_6(OH)_2 \quad (72)$$

The products of the concentrations of the reacting species at saturation are known as the solubility products of the salts:

$$K_{CaCO_3} = (Ca^{++})(CO_3^{--}) \qquad (73)$$

$$K_{MgCO_3 \cdot 2H_2O} = (Mg^{++})(CO_3^{--}) \qquad (74)$$

$$K_{Mg(OH)_2} = (Mg^{++})(OH^-)^2 \qquad (75)$$

$$K_{Ca_{10}(PO_4)_6(OH)_2} = (Ca^{++})^{10}(PO_4^{---})^6(OH^-)^2 \quad (76)$$

These equations define the conditions of saturation for the respective salt solutions. If the solubility products are exceeded, the water is supersaturated with respect to the salt, and precipitation can occur. At the pH levels to which mixed-liquor might be increased (approximately 7-8), it is very unlikely that the solubilities of $Mg(OH)_2$ or $MgCO_3 \cdot 2H_2O$ would ever be exceeded. Under conditions of high calcium ion and phosphate ion concentrations, raising the pH of mixed-liquor could cause it to become saturated with respect to hydroxylapatite, $Ca_{10}(PO_4)_6(OH)_2$. However, the rate of precipitation of hydroxylapatite is usually very slow. The effect of the precipitation of hydroxylapatite on the pH of mixed-liquor can usually be ignored. The most important precipitation reaction which must be considered in the calculation of the effects of carbon dioxide equilibria on the pH of mixed-liquor is Equation 69, the precipitation of calcium carbonate.

Unless an alkaline reagent is added, the solubility of calcium carbonate in the mixed-liquor of an oxygen-activated sludge system is always greater than its solubility in the raw wastewater. The reason for this is that the carbon dioxide generated and maintained in solution by the closed system tends to decrease the pH of the mixed-liquor. As shown by Equations 29 and 73, the increased hydrogen ion concentration decreases the concentration of carbonate ion, while a decrease in the concentration of carbon-ate ion increases the concentration of calcium ion at calcium carbonate saturation.

IV. GENERALIZED MATHEMATICAL MODEL OF MULTI-STAGE OXYGENATION SYSTEM

In order to satisfy all of the needs of process designers, a generalized mathematical model of the multi-stage oxygenation system must be capable of handling (a) all of the major gaseous constituents likely to be encountered in the oxygenation of wastewater, such as oxygen, ozone, nitrogen, argon, ammonia, hydrogen sulfide, carbon dioxide, chlorine, methane, and other volatile organic compounds; (b) cocurrent or countercurrent gas and mixed-liquor flow with single or multiple gas and liquid feed streams of any composition, flow rate, or temperature to any stage; (c) consumption or formation of any system constituent in any stage if the constituent undergoes chemical change within the stage; (d) heat loss from or generation in any stage; (e) buffering effects of ammonia and inorganic phosphorus; (f) pH control by the addition of acid or alkali; and (g) solubility of sparingly soluble salts.

In the generalized mathematical model to be developed in this section, the gas-liquid devices are divided into two classes. Liquid dispersion devices, such as surface aerators and sprays, disperse liquid into a continuous gas phase. Gas dispersion devices, such as bubble diffusers and impeller-sparger systems, inject gas bubbles into a continuous liquid phase. The six process variations contained in the generalized model being developed are defined and illustrated in Figures 5 to 10. The symbols used in the figures are defined in the section on nomenclature.

Figure 5 is a schematic flow diagram for the nth stage of a multi-stage oxygenation system with a liquid dispersion gas-liquid contacting device. Figure 6 is a schematic flow diagram for the nth stage of a multi-stage oxygenation system with a gas dispersion gas-liquid contacting device. It may sometimes be desirable to have both types of gas-liquid contacting devices in a single stage. Figure 7 is a schematic flow diagram for the nth stage of a multi-stage oxygenation system with both liquid and gas dispersion gas-liquid contacting devices. Figures 8, 9, and 10 are schematic flow diagrams for air

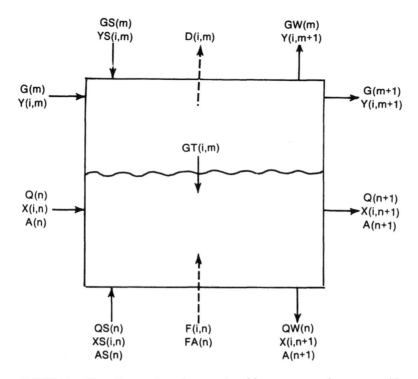

FIGURE 5. Flow diagram for n^{th} stage of multi-stage oxygenation system with liquid dispersion gas-liquid contacting device. (Courtesy of Union Carbide Corporation.)

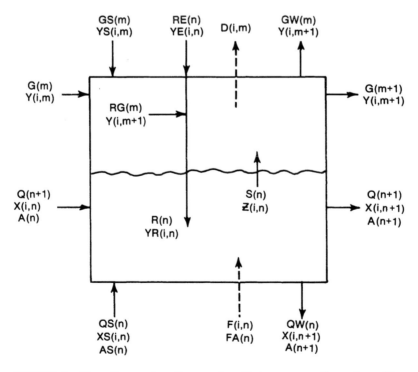

FIGURE 6. Flow diagram for n^{th} stage of multi-stage oxygenation system with gas dispersion gas-liquid contacting device. (Courtesy of Union Carbide Corporation.)

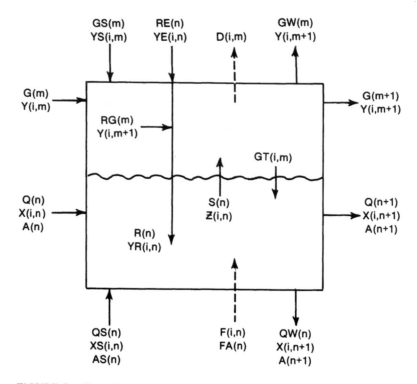

FIGURE 7. Flow diagram for n^{th} stage of multi-stage oxygenation system with both liquid and gas dispersion gas-liquid contacting devices. (Courtesy of Union Carbide Corporation.)

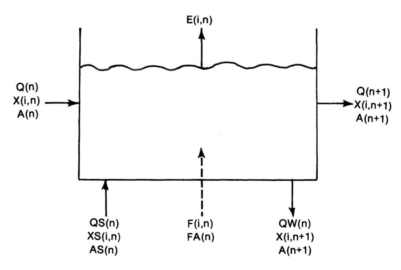

FIGURE 8. Flow diagram for air stripping n^{th} stage of multi-stage oxygenation system with liquid dispersion gas-liquid contacting device. (Courtesy of Union Carbide Corporation.)

stripping the nth stage of a multi-stage oxygenation system with liquid dispersion, gas dispersion, or both liquid and gas dispersion gas-liquid contacting devices, respectively. Conventional air aeration in secondary waste-water treatment systems can be simulated with the model by considering all stages to be air stripped.

Table 1 lists the unknown variables for each of the process variations described in Figures 5

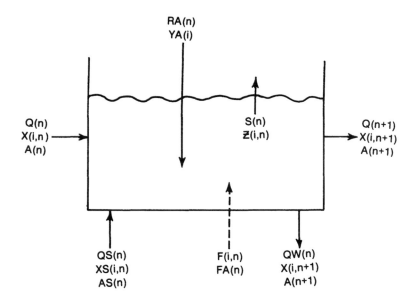

FIGURE 9. Flow diagram for air stripping of nth stage of multi-stage oxygenation system with gas dispersion gas-liquid contacting device. (Courtesy of Union Carbide Corporation.)

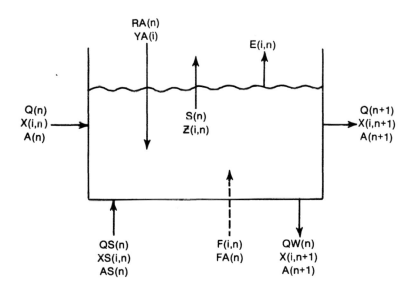

FIGURE 10.. Flow diagram for air stripping n^{th} stage of multi-stage oxygenation system with both liquid and gas dispersion gas-liquid contacting devices. (Courtesy of Union Carbide Corporation.)

to 10. The mathematical model assumes that all variables other than those listed in Table 1 are known. The known values are obtained from design specifications, from calculations for the previous stage, from a biochemical oxidation model of the process, from a mixing model of the stage, or from a disinfection model, if ap-

plicable. Only the gas-liquid mass transfer and water chemistry aspects of the generalized mathematical model are discussed in this chapter.

The different sets of equations which comprise the various mathematical models will now be developed. For the process variations de-

TABLE 1

Summary of Unknown Variables for the Six Different Generalized Flow Model Variations Depicted in Figures 5 to 10

n^{th} Stage of multi-stage oxygenation system			Air stripping n^{th} stage of multi-stage oxygenation system		
Figure 5: liquid dispersion gas-liquid contacting device	Figure 6: gas dispersion gas-liquid contacting device	Figure 7: both liquid and gas dispersion gas-liquid contacting devices	Figure 8: liquid dispersion gas-liquid contacting device	Figure 9: gas dispersion gas-liquid contacting device	Figure 10: both liquid and gas dispersion gas-liquid contacting devices
$Q(n+1)$	$Q(n+1)$	$Q(n+1)$	$Q(n+1)$	$Q(n+1)$	$Q(n+1)$
$A(n+1)$	$A(n+1)$	$A(n+1)$	$A(n+1)$	$A(n+1)$	$A(n+1)$
$T(n+1)$	$T(n+1)$	$T(n+1)$	$T(n+1)$	$T(n+1)$	$T(n+1)$
$X(i,n+1)$	$X(i,n+1)$	$X(i,n+1)$	$X(i,n+1)$	$X(i,n+1)$	$X(i,n+1)$
$XF(i,n+1)$	$XF(i,n+1)$	$XF(i,n+1)$	$XF(i,n+1)$	$XF(i,n+1)$	$XF(i,n+1)$
$H(n+1)$	$H(n+1)$	$H(n+1)$	$H(n+1)$	$H(n+1)$	$H(n+1)$
$G(m+1)$	$G(m+1)$	$G(m+1)$	$E(i,n)$	$S(n)$	$S(n)$
$Y(i,m+1)$	$Y(i,m+1)$	$Y(i,m+1)$		$Z(i,n)$	$Z(i,n)$
$GT(i,m)$	$S(n)$	$S(n)$			$E(i,n)$
	$Z(i,n)$	$Z(i,n)$			
	$YR(i,n)$	$YR(i,n)$			
		$GT(i,m)$			
Total number of unknown variables					
$2+4I$	$3+5I$	$3+6I$	$1+3I$	$2+3I$	$2+4I$

scribed in Figures 5 to 7, the relationships between the gas and liquid stage numbers are given by:

$$m = n \qquad (77)$$

for cocurrent gas and liquid flow, and

$$m = N - n + 1 \qquad (78)$$

for countercurrent gas and liquid flow. The first three calculation steps for each stage consist of calculating the effluent liquid flow rate, alkalinity, and temperature from the following mass balance, alkalinity balance, and heat balance relationships:

$$Q(n+1) = Q(n) + QS(n) - QW(n) \qquad (79)$$

$$[Q(n+1) + QW(n)] \cdot T(n+1) = Q(n) \cdot T(n) + QS(n) \cdot AS(n) + FA(n) \cdot 10^{-3} \qquad (80)$$

$$[Q(n+1) + QW(n)] \cdot A(n+1) = Q(n) \cdot A(n) + QS(n) \cdot TS(n) + NHR(n) - NHL(n) \qquad (81)$$

The mass balance equation (79) assumes that

the quantity of water lost by evaporation caused by gas or air stripping has a negligible effect on the liquid flow rate. The heat balance equation (81) assumes that in multi-stage oxygenation systems, the heat content of the gas streams has a negligible effect on the temperature of the liquid streams. The specific heat of all water streams are assumed to be constant and equal to 1.0 Btu/lb/°F, and the gas phase is assumed to be at the temperature of the water phase and saturated with water vapor. After the temperature in the stage has been calculated, values of parameters that are functions of temperature, such as ionization constants, gas solubilities, solubility products, and mass transfer coefficients are calculated. Henry's law for gas solubility applies only to nonionized gases and to the nonionized portion of the ionized gases, as discussed previously. The following equations summarize the nonionized fraction of the various gases included in the model. For all nonionized gases:

$$XF(i,n+1) = 1.0 \qquad (82)$$

For ionized gases, the nonionized fraction is given by:

$$XF(NH_3, n+1) = \cfrac{1}{1 + \cfrac{H(n+1) \cdot KN(n+1)}{KW(n+1)}} \tag{82a}$$

$$XF(H_2S, n+1) = \cfrac{1}{1 + \cfrac{KS1(n+1)}{H(n+1)} + \cfrac{KS1(n+1) \cdot KS2(n+1)}{H(n+1)^2}} \tag{82b}$$

$$XF(CO_2, n+1) = \cfrac{1}{1 + \cfrac{KC1(n+1)}{H(n+1)} + \cfrac{KC1(n+1) \cdot KC2(n+1)}{H(n+1)^2}} \tag{82c}$$

The necessary electroneutrality of the system, or charge balance, is expressed by the following equation:

$$A(n+1) - \frac{XF(NH_3, n+1 \cdot X(NH_3, n+1)}{W(NH_3) \cdot 10^3} + H(n+1) = \frac{KW(n+1)}{H(n+1)} + \frac{KS1(n+1) \cdot XF(H_2S, n+1) \cdot X(H_2S, n+1)}{H(n+1) \cdot W(H_2S) \cdot 10^3}$$

$$+ \frac{2 \cdot KS1(n+1) \cdot KS2(n+1) \cdot XF(H_2S, n+1) \cdot X(H_2S, n+1)}{H(n+1)^2 \cdot W(H_2S) \cdot 10^3} + \frac{KC1(n+1) \cdot XF(CO_2, n+1) \cdot X(CO_2 n+1)}{H(n+1) \cdot W(CO_2) \cdot 10^3}$$

$$+ \frac{2 \cdot KC1(n+1) \cdot KC2(n+1) \cdot XF(CO_2, n+1) \cdot X(CO_2, n+1)}{H(n+1)^2 \cdot W(CO_2) \cdot 10^3}$$

$$+ \frac{X(\text{ortho } P, n+1)}{W(\text{ortho } P) \cdot 10^3} \cdot \left[\cfrac{1}{\cfrac{H(n+1)}{KP1(n+1)} + 1 + \cfrac{KP2(n+1)}{H(n+1)} + \cfrac{KP2(n+1) \cdot KP3(n+1)}{H(n+1)^2}} + \right.$$

$$\cfrac{2}{\cfrac{H(n+1)^2}{KP1(n+1) \cdot KP2(n+1)} + \cfrac{H(n+1)}{KP2(n+1)} + 1 + \cfrac{KP3(n+1)}{H(n+1)}} +$$

$$\left. \cfrac{3}{\cfrac{H(n+1)^3}{KP1(n+1) \cdot KP2(n+1) \cdot KP3(n+1)} + \cfrac{H(n+1)^2}{KP2(n+1) \cdot KP3(n+1)} + \cfrac{H(n+1)}{KP3(n+1)} + 1} \right] \tag{83}$$

The following equations summarize the mass balances for each balances for each individual component i. For the process variation described in Figure 5:

$$G(m) \cdot Y(i,m) - G(m+1) \cdot Y(i,m+1) + GS(i,m) - GW(m) \cdot Y(i,m+1) - D(i,m) +$$

$$\frac{Q(n) \cdot X(i,n) - Q(n+1) \cdot X(i,n+1) + QS(n) \cdot XS(i,n) - QW(n) \cdot X(i,n+1)}{W(i)} + F(i,n) = 0 \tag{84}$$

For the process variation described in Figures 6 and 7:

$$G(m) \cdot Y(i,m) - G(m+1) \cdot Y(i,m+1) + GS(i,m) \cdot YS(i,m) - GW(m) \cdot Y(i,m+1)$$

$$- D(i,m) + \frac{Q(n) \cdot X(i,n) - Q(n+1) \cdot X(i,n+1) + QS(n) \cdot XS(i,n) - QW(n) \cdot X(i,n+1)}{W(i)}$$

$$+ F(i,n) + RE(n) + RE(n) \cdot YE(i,n) = 0 \tag{84a}$$

For the process variations described in Figures 8, 9, and 10, respectively:

$$\frac{Q(n) \cdot X(i,n) - Q(n+1) \cdot X(i,n+1) + QS(n) \cdot XS(i,n) - QW(n) \cdot X(i,n+1)}{W(i)}$$

$$+ F(i,n) - E(i,n) = 0 \tag{84b}$$

$$\frac{Q(n) \cdot X(i,n) - Q(n+1) \cdot X(i,n+1) + QS(n) \cdot XS(i,n) - QW(n) \cdot X(i,n+1)}{W(i)}$$

$$+ F(i,n) + RA(n) \cdot YA(i) - S(n) \cdot Z(i,n) = 0 \qquad (84c)$$

$$\frac{Q(n) \cdot X(i,n) - Q(n+1) \cdot X(i,n+1) + QS(n) \cdot XS(i,n) - QW(n) \cdot X(i,n+1)}{W(i)}$$

$$+ F(i,n) - E(i,n) + RA(n) \cdot YA(i) - S(n) \cdot Z(i,n) = 0 \qquad (84d)$$

The gas-space mass balances for component i for the process variations described in Figures 5, 6, and 7, respectively are as follows:

$$G(m) \cdot Y(i,m) - G(m+1) \cdot Y(i,m+1) + GS(m) \cdot YS(i,m) - GW(m) \cdot Y(i,m+1) - D(i,m)$$

$$- GT(i,m) = 0 \qquad (85)$$

$$G(m) \cdot Y(i,m) - G(m+1) \cdot Y(i,m+1) + GS(m) \cdot YS(i,m) - GW(m) \cdot Y(i,m+1) - D(i,m)$$

$$- RG(m) \cdot Y(i,m+1) + S(n) \cdot Z(i,n) = 0 \qquad (85a)$$

$$G(m) \cdot Y(i,m) - G(m+1) \cdot Y(i,m+1) + GS(m) \cdot YS(i,m) - GW(m) \cdot Y(i,m+1) - D(i,m)$$

$$- GT(i,m) - RG(m) \cdot Y(i,m+1) + S(n) \cdot Z(i,n) = 0 \qquad (85b)$$

The mass transfer calculations are based on the assumption that complete mixing occurs in both the gas and liquid phases in the individual stages. Alpha values for a specific gas-liquid contacting device in a given stage are assumed to be the same for all gases. Beta values in a given stage are also assumed to be the same for all gases. The following equations are modifications of Equation 22 which describe the transfer rate of component i between the gas and liquid phases in multi-stage oxygenation systems by liquid and gas dispersion gas-liquid contacting devices, respectively:

$$GT(i,m) = AL(n) \cdot \frac{KLAL(i,n)}{W(i)} \cdot (B[n] \cdot HC[i,n+1] \cdot PL[n] \cdot Y[i,m+1] - XF[i,n+1] \cdot X[i,n+1]) \qquad (86)$$

$$R(n) \cdot YR(i,n) - S(n) \cdot Z(i,n) = AG(n) \cdot \frac{KLAG(i,n)}{W(i)} \cdot \left(B[n] \cdot HC[i,n+1] \cdot PG[n] \cdot \left[\frac{YR(i,n)}{2} \right. \right.$$

$$\left. \left. + \frac{Z(i,n)}{2} \right] - XF[i,n+1] \cdot X[i,n+1] \right) \qquad (86a)$$

Equation 22 can also be modified to describe the rate of transfer of component i between the air and water phases while air stripping with liquid and gas dispersion gas-liquid contacting devices:

$$-E(i,n) = AL(n) \cdot \frac{KLAL(i,n)}{W(i)} \cdot (B[n] \cdot HC[i,n+1] \cdot PL[n] \cdot YA[i] - XF[i,n+1] \cdot X[i,n+1]) \qquad (86b)$$

$$RA(n) \cdot YA(i) - S(n) \cdot Z(i,n) = AG(n) \cdot \frac{KLAG(i,n)}{W(i)} \cdot \left(B[n] \cdot HC[n+1] \cdot PG[n] \cdot \left[\frac{YA(i)}{2} + \right. \right.$$

$$\left. \left. \frac{Z(i,n)}{2} \right] - XF[i,n+1] \cdot X[i,n+1] \right) \qquad (86c)$$

By definition, the mole fractions must sum up to unity:

$$\Sigma\, Z(i,n+1) = 1 \qquad (87)$$

$$\Sigma\, Y(i,m+1) = 1 \qquad (88)$$

The mass balance in a gas dispersion mass transfer device is given by:

$$RG(n)\cdot Y(i,m+1) + RE(n)\cdot YE(i,n) = R(n)\cdot YR(i,n) \qquad (89)$$

The solubility limit of sparingly soluble carbonates and phosphates can be calculated from their solubility products. As an example, the solubility limit of calcium ion, $X(Ca,n+1)$, is given by:

$$\frac{X(Ca,n+1)}{W(Ca)\cdot 10^3} \cdot \frac{\dfrac{X(CO_2,n+1)}{W(CO_2)\cdot 10^3}}{\dfrac{H(n+1)^2}{KC1(n+1)\cdot KC2(n+1)} + \dfrac{H(n+1)}{KC2(n+1)} + 1} = K(n+1) \qquad (90)$$

Values for the unknown variables listed in Table 1 are obtained by solving the simultaneous sets of equations listed in Table 2.

V. EXPERIMENTAL VERIFICATION OF THE MATHEMATICAL MODEL

Figure 11 compares the calculated and experimental values for Y_{O_2} in the gas stages of two full-scale oxygen-activated sludge systems.[1,2] Figure 12 compares the caculated and experimental pH values in the liquid stages of the same oxygen-activated sludge systems. Agreement between the calculated and the measured values is quite good. The slight deviations between the calculated and measured values are believed to be caused more by inaccuracies in the input parameters and the experimental data rather than by discrepancies or deficiencies in the basic mathematical model itself.

The beneficial effect of increased gas staging in oxygenation systems is also seen in Figure 11. It is readily observed by comparing the overall average oxygen gas concentrations in the six-vs. the three-stage system that the average oxygen gas concentration is significantly higher in the six-stage system, even though the exhaust gas oxygen concentration is about the same in both cases. Most of the oxygen transfer also occurs in the first few stages, where the difference in oxygen gas purities between the two systems is the greatest. The higher average oxygen gas concentration in the six-stage system would

TABLE 2

Summary of Sets of Simultaneous Equations for the Six Different Generalized Mathematical Models for the Flow Model Variations Depicted in Figures 5 to 10

n^{th} Stage of multi-stage oxygenation system			Air stripping n^{th} stage of mutli-stage oxygenation system		
Figure 5: liquid dispersion gas-liquid contacting device	Figure 6: gas dispersion gas-liquid contacting device	Figure 7: both liquid and gas dispersion gas-liquid contacting devices	Figure 8: liquid dispersion gas-liquid contacting device	Figure 9: gas dispersion gas-liquid contacting device	Figure 10: both liquid and gas dispersion gas-liquid contacting devices
(82)*	(82)*	(82)*	(82)*	(82)*	(82)*
(83)	(83)	(83)	(83)	(83)	(83)
(84)*	(84a)*	(84a)*	(84b)*	(84c)*	(84d)*
(85)*	(85a)*	(85b)*	(86b)*	(86c)*	(86b)*
(86)*	(86a)*	(86)		(87)	(86c)*
(88)	(87)	(86a)*			(87)
	(88)	(87)			
	(89)*	(88)			
		(89)*			

* One equation required for each component considered.

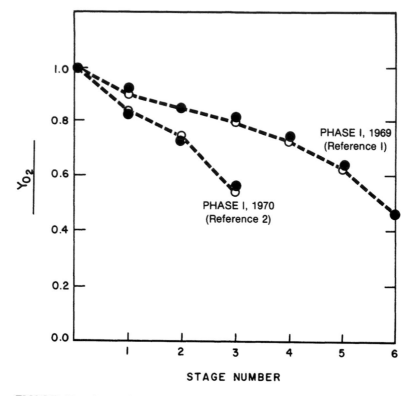

● MEASURED VALUES
○ CALCULATED VALUES

PHASE I, 1969
(Reference I)

PHASE I, 1970
(Reference 2)

Y_{O_2}

STAGE NUMBER

FIGURE 11. Comparison of measured and calculated values for Y_{O_2} in the UNOX® System at Batavia, N. Y. (Courtesy of Union Carbide Corporation.)

translate into a reduced oxygen dissolution power requirement for the same overall oxygenation system capacity and oxygen utilization efficiency.

VI. VARIOUS APPLICATIONS OF THE MATHEMATICAL MODEL

All of the variables contained in the equations listed in the process variations defined in Table 2, except for the variables listed in Table 1, are assumed to be known. Their values must be supplied for each stage and are obtained from (a) specified process design conditions; (b) calculations for the preceding stage; (c) a biochemical oxidation model for the process; (d) known characteristics of the specified gas-liquid contacting devices at the selected stage geometry and operating conditions; (e) a mixing model for the stage; or (f) an ozone or chlorine disinfection model, if applicable. Appropriate models for the biochemical oxidation reactions,

mixing, and disinfection processes must also be developed.

The calculation procedures described earlier considered the gas-liquid contacting characteristics of the mass transfer device to be known and solved for the dissolved oxygen concentration in the liquid. An alternative use of the model is to specify a value for the dissolved oxygen concentration in the liquid, $X(O_2, n + 1)$, and calculate either $KLAL(O_2, n + 1)$ or $KLAG(O_2, n + 1)$. The same sets of simultaneous equations are solved in either case.

VII. COMPARISON OF AIR AND HIGH-PURITY OXYGENATION SYSTEM POWER REQUIREMENTS

Once the individual stage gas-phase compositions and the required mass transfer coefficients are calculated to supply the necessary dissolved oxygen, as determined from the appropriate biochemical oxidation process

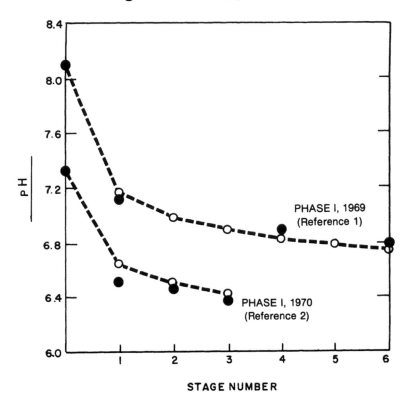

FIGURE 12. Comparison of measured and calculated pH values in the UNOX®
System at Batavia, N. Y. (Courtesy of Union Carbide Corporation.)

model, it is a relatively straightforward procedure to size the aerators for the individual
stages. Each stage is, of course, operating at a
different gas-phase oxygen purity and overall
oxygen transfer rate throughout the multi-stage
contacting system. A meaningful comparison
with air aerated systems, however, can be obtained by calculating the oxygen transfer rate
weighted average oxygen purity for the overall
multi-stage oxygenation system. This
"weighted average" oxygen purity can then be
used to compare the energy dissolution efficiency of oxygenation systems as compared
with air systems operating with the same basic
gas-liquid contacting device such as a surface
aerator.

The oxygen transfer rate weighted average
oxygen purity is defined as follows:

$$Y_{O_2(WA)} = \frac{\sum\limits_{i=1}^{N} \left(OTR_i\right)\left(Y_{O_{2_i}}\right)}{\sum\limits_{i=1}^{N} \left(OTR_i\right)} \quad (91)$$

where OTR_i = oxygen transfer rate of Stage i,
lb O_2/hr. $Y_{O_{2_i}}$ = oxygen gas-phase purity of
Stage i (mole fraction); N = number of stages
in multi-stage oxygenation system; and $Y_{O_{2(WA)}}$
= oxygen transfer rate weighted average oxygen purity of multi-stage oxygenation system
(mole fraction). The oxygen transfer rate
weighted average oxygen purity of a multi-stage
oxygenation system can then be related directly
to the atmospheric air oxygen composition of
0.21 mol fraction oxygen in comparing the relative dissolution energy-transfer efficiency of
the two systems. For a typical three- or four-
stage UNOX® System design, this weighted
average oxygen purity is equal to 0.71 mol fract
oxygen. This purity will now be used to compare the intrinsic relative energy-transfer efficiency of air and oxygen systems by using the
same 20 hp surface aerator example that was
discussed in the first section of this chapter.

Adapting Equation 1 to the high-purity oxygenation system weighted average oxygen purity and using the same α and β factors:

$$GTR_{O_2} = \alpha K_L a_{O_2} \left[\beta \cdot HC_{O_2} \cdot P_e \cdot Y_{O_2 (WA)} - X_{O_2} \right] \quad (92)$$

Substituting the same parameters as used in Equation 5 for the 20 hp surface aerator example, except using a mixed-liquor D.O. of 6.0 mg/l instead of 2.0 mg/l, we obtain:

$\alpha = 0.85$

$K_L a_{O_2} = 7.53 \dfrac{lb\ O_2}{(hr)\ (mg/l)}$

$\beta = 0.95$

$HC_{O_2} = 44.4$ mg/l/atm at $20°C$

$X_{O_2} = 6.0$ mg/l

$$GTR_{O_2} = (0.85)\ (7.53)\ [(0.95)\ (44.4)\ (1.0)\ (0.71) - 6.0)]$$

$$(93)$$

$$GTR_{O_2} = 153.3\ lb\ O_2/hr \quad (94)$$

Thus, it is seen that the 20 hp surface aerator which transferred only 43.9 lb O_2/hr at a mixed-liquor D.O. level of 2.0 mg/l in the air system, can transfer on an overall weighted average basis 153.3 lb O_2/hr in a multi-stage oxygenation system at a D.O. level of 6 mg/l. This corresponds to an energy transfer efficiency of:

$$ATE_{O_2} = \frac{153.3}{20} = 7.67 \frac{lbs\ O_2}{shp\ hr} \quad (95)$$

Correcting for typical motor efficiency and gear box efficiency losses of 90 and 95% respectively, as in the air case, we have:

$$ATE_{O_2} = (7.67)\ (0.90)\ (0.95) = 6.6 \frac{lb\ O_2}{whp\ hr} \quad (96)$$

where whp = wire input motor hp. This represents an energy-transfer efficiency which is 3.5 times greater than that of the corresponding air system using exactly the same aerator at a mixed-liquor D.O. of 2.0 instead of 6.0 mg/l for the multi-stage oxygenation system.

Of course, for a meaningful overall power comparison, the energy required for oxygen generation must be added to the dissolution energy for the oxygenation system. The energy required for oxygen generation is 0.17 kWh/lb O_2 for cryogenic generation and 0.22 kWh/lb O_2

for PSA generation, as shown in Figure 18 in Chapter 3, Volume I. Adding the dissolution energy requirement from Equation 96 (0.11 kWh/lb O_2), there is a total energy requirement of 0.28 kWh/lb O_2 for cryogenic generation and 0.33 kWh/lb O_2 for PSA generation. This corresponds to energy-transfer efficiencies of 2.63 and 2.27 lb O_2/hp hr, respectively, for the total energy requirement of the oxygen system. This compares to the energy-transfer efficiency of 1.88 lb O_2/hp hr for the same aeration device in air aeration service. This represents a 40% increase in overall energy-transfer efficiency for an oxygen system with a cryogenic oxygen generator and a 21% increase in overall energy-transfer efficiency for an oxygen system with a PSA oxygen generator.

NOMENCLATURE

A(n)   Alkalinity of Q(n), lb equivalents/10^3 lb H_2O (g equivalents/l).

AG(n)   Alpha value for gas dispersion gas liquid contacting device in water stage n, $K_L a$ (actual)/$K_L a$ (tap water).

AL(n)   Alpha value for liquid dispersion gas liquid contacting device in water stage n, $K_L a$ (actual)/$K_L a$ (tap water).

AS(n)   Alkalinity of QS(n), lb equivalents/10^3 lb H_2O (g equivalents/l).

B(n)   Beta value for water stage n, gas solubility (actual)/gas solubility (tap water).

D(i,m)   Rate of decomposition of component i in gas stage m, lb mol/hr.

E(i,m)   Rate of air stripping of component i from water stage n with liquid dispersion gas transfer device, lb mol/hr.

F(i,n)   Rate of formation of component i in water stage, lb mol/hr.

FA(n)   Alkalinity formed in or added to the water stage n, lb equivalents/hr.

G(m)   Gas flow to gas stage m from gas stage m-1, lb mol dry gas/hr.

GS(m)	Supplemental gas flow to gas stage m, lb mol dry gas/hr.
GT(i,m)	Rate of transfer of component i from gas stage m to water stage n by liquid dispersion gas transfer device, lb mol dry gas/hr.
GW(m)	Withdrawal gas flow from gas stage m, lb mol dry gas/hr.
H(n)	Hydrogen ion concentration in Q(n), lb mol/10^3 lb H_2O (g mol/l).
i	Individual gaseous component.
I	Number of gaseous components considered.
HC(i,n)	Henry's law constant for component i in Q(n), ppm/atm (mg/l/atm).
K(n)	Solubility product for $CaCO_3$ in Q(n).
KC1(n)	First ionization constant for CO_2 in Q(n).
KC2(n)	Second ionization constant for CO_2 in Q(n).
KLAG(i,n)	K_La (tap water) of gas dispersion gas-liquid contacting device in water stage n for component i, lb/hr/ppm (lb/hr/mg/l).
KLAL(i,n)	K_La (tap water) of liquid dispersion gas-liquid contacting device in water stage n for component i, lb/hr/ppm (lb/hr/mg/l).
KN(n)	Ionization constant for NH_3 in Q(n).
KP1(n)	First ionization constant for H_3PO_4 in Q(n).
KP2(n)	Second ionization constant for H_3PO_4 in Q(n).
KP3(n)	Third ionization constant for H_3PO_4 in Q(n).
KS1(n)	First ionization constant for H_2S in Q(n).
KS2(n)	Second ionization constant for H_2S in Q(n).
KW(n)	Ionization constant for H_2O in Q(n).
m	Gas stage number.
n	Water stage number.
N	Number of stages.
NHL(n)	Net heat loss from water stage n and gas stage m to the surroundings, millions of Btu/hr.
NHR(n)	Net heat released in water stage n and gas stage m as a result of chemical or biochemical reaction, million Btu/hr.
PG(n)	Effective pressure for gas dispersion gas-liquid contacting device in water stage n, atm.
PL(n)	Effective pressure for liquid dispersion gas-liquid contacting device in water stage n, atm.
Q(n)	Water flow to water stage n from water stage n-1, millions of lb/hr.
QS(n)	Supplemental water flow to water stage n, millions of lb/hr.
QW(n)	Withdrawal water flow from water stage n, millions of lb/hr.
R(n)	Gas flow rate to gas dispersion/gas-liquid contacting device in water stage n, lb mol dry gas/hr.
RA(n)	Air flow to gas dispersion gas-liquid contacting device air stripping water stage n, lb mol dry air/hr.
RE(n)	Part of R(n) supplied from an external source, lb mol dry gas/hr.
RG(m)	Part of R(n) supplied from gas stage m, lb mol dry gas/hr.
S(n)	Gas surfacing rate from gas dispersion gas-liquid contacting device in water stage n, lb mol dry gas/hr.
T(n)	Temperature of Q(n), °F.
TS(n)	Temperature of QS(n), °F.
W(i)	Molecular weight of component i.
X(i,n)	Sum of the concentrations of all forms of component i in QS(n), ppm (mg/l).
XF(i,n)	Fraction of X(i,n) not ionized, dimensionless.
XS(i,n)	Sum of the concentrations of all forms of component i in QS(n), ppm (mg/l).
Y(i,m)	Mole fraction of component i in G(m), dry basis.
YA(i)	Mole fraction of component i in dry air.
YE(i,n)	Mole fraction of component i in RE(n), dry basis.
YR(i,n)	Mole fraction of component i in R(n), dry basis.
YS(i,n)	Mole fraction of component i in GS(m), dry basis.
Z(i,n)	Mole fraction of component i in S(n), dry basis.

GTR_{O_2}	Oxygen transfer rate from gas to liquid phase, lb O_2/hr.	whp	Wire input motor horsepower, hp.
$K_L a_{O_2}$	Overall gas-phase oxygen mass transfer coefficient, lb O_2/hr/mg/l.	(GTR_i)	Column vector of individual mass transfer rates in multicomponent gas mixture, lb/hr.
α	Alpha value for the system, $K_L a_{O2}$ (in actual mixed-liquor)/$K_L a_{O2}$ (in tap water).	$[K_L a_i]$	Gas-phase mass transfer coefficient matrix for multicomponent mixture, lb/hr/mg/l.
β	Beta value for the system, HC_{O2} (in actual mixed-liquor)/HC_{O2} (in tap water).	(X^*_i)	Column reactor of equilibrium liquid phase compositions in equilibrium with gas phase in multicomponent mixture, mg/l.
HC_{O_2}	Henry's law constant for oxygen, mg/l/atm.	(X_i)	Column vector of liquid-phase compositions in multicomponent liquid mixture, mg/l.
$P.$	Effective pressure for gas-liquid contacting device, atm.		
Y_{O_2}	Mole fraction oxygen in gas phase aeration gas.	$[HC_i]$	Equilibrium constant coefficient matrix for multicomponent gas-liquid mixture, mg/l/atm.
X_{O_2}	Dissolved oxygen concentration in mixed-liquor, mg/l.	(Y_i)	Column vector of gas-phase compositions in multicomponent gas mixture, mol fraction.
STR_{O_2}	Oxygen transfer rate at "standard conditions" of tap water at 20°C and zero dissolved oxygen concentration and 1 atm pressure, lb O_2/hr.	$Y_{O_{2(WA)}}$	Oxygen transfer rate weighted average oxygen purity for multistage oxygenation system, mol fraction.
ATE_{O_2}	Actual oxygen transfer efficiency under activated sludge mixed-liquor aeration conditions, lb O_2/hp hr.	OTR_i	Oxygen transfer rate of stage i of multi-stage oxygenation system, lb O_2/hr.
shp	Aerator shaft horsepower input, hp.	$Y_{O_{2_i}}$	Oxygen gas-phase purity of stage i of multi-stage oxygenation system, mol fraction.

REFERENCES

1. **Albertsson, J. G., McWhirter, J. R., Robinson, E. K., and Vahldieck, N. P.,** Investigation of the Use of High Purity Oxygen Aeration in the Conventional Activated Sludge Process, Report to Federal Water Quality Administration on Program No. 17050 DNW, Contract No. 14-12-465, May 1970.
2. **McDowell, M. A., Vahldieck, N. P., Wilcox, E. A., and Young, K. W.,** Continued Evaluation of Oxygen Use in the Conventional Activated Sludge Process, Report to Environmental Protection Agency on Project No. 17050 DNW, Contract No. 14-12-867, September 1971.
3. **Toor, H. L.,** Solution of the linearized equations of multicomponent mass transfer. I, *AIChe. J.*, 10(4), 448, 1964.
4. **Toor, H. L.,** Solution of the linearized equations of multicomponent mass transfer. II. Matrix Methods, *AIChe. J.*, 10(4), 460, 1964.
5. **Toor, H. L.,** Prediction of efficiencies and mass transfer on a stage with multicomponent systems, *AIChe. J.*, 10(4), 545, 1964.
6. **Toor, H. L., Seshadri, C. V., and Arnold, K. R.,** Diffusion and mass transfer in multicomponent mixtures of ideal gases, *AIChe. J.*, 11(4), 746, 1965.
7. **Stewart, W. E. And Prober, R.,** Matrix calculations of multicomponent mass transfer in isothermal systems, *Ind. Eng. Chem. Fundam.*, 3, 224, 1964.

Index

INDEX

A

Acclimation, see Biomass, acclimation

Acetylene
  hazards of, II:203, 204, 223, 234
  Removal, II:232

Active mass fraction, I:125

Adenosine triphosphate concentration, II:64, 119

Adsorbents, molecular sieve, II:214—215, 243—251

Adsorbent vessels, PSA system, II:254

Adsorption isotherms, II:243, 249

Adsorption traps, oxygen production systems, II:203, 220, 232, 234—235

Aeration, see Air-aerated sludge systems; Oxygenation systems

Aeration tanks, UNOX® systems, see UNOX® Systems, aeration tanks

Aeration test methods, II:30—36

Aerator performance testing, II:30—31

Aerators, surface, see Surface aerators

Aerobic autotrophic bacteria in nitrification systems, II:140—171

Aerobic sludge digestion
  biological reactions, II:119—120
  design and operation, II:135—136
  field testing, II:129—135
  floc metabolism, II:77—79
  general discussion, II:118, 136—137
  kinetics, II:121—122
  laboratory testing, II:123
  mathematical model, II:122—123
  mesophilic, II:118, 126
    design and operation, II:135
  pasteurization, II:128—129, 136
  pilot plant studies, II:123—128
  stoichiochemistry and thermochemistry, II:120—121
  theory, II:119—123
  thermophilic, II:118—137
    design and operation, II:135—136

Aftercooler, cryogenic oxygen production systems, II:221, 232, 236

Air, oxygen separation from, process, see Air separation technology

Air-aerated sludge systems, compared to oxgenated sludge systems
  aerobic digestion of sludge, II:118, 121, 123—128, 136
  bio-precipitation process, compared to oxygenation systems, I:18
  dissolved-oxygen control systems, I:30
  mass transfer, see also Mass transfer, I:236—237
    compared to oxygenation systems, I:32—35, 256—258
    tests, I:236—237
    compared to UNOX® system, I:44—45
  mixed liquor dissolved oxygen level effects compared to oxygenation systems, I:50—60
  nitrification systems, II:151
  oxygenation systems, compared to, I:28—29
  oxygen dissolution energy requirements for, I:27—28

ozonation of wastewater, II:179—182

process design considerations, compared to UNOX® systems, I:46—47

sludge dewatering, II:84—87, 89—91, 100—102, 106—108

sludge production and settling characteristics, compared to oxygenation systems, I:53—60, 220—223; II:71—73, 76—80

submerged turbine oxygenators, II:24

Air compressors, see Compressors

Airco, Incorporated's forced free-fall oxygenation (F³O) system, I:13—14

Air flotation thickeners, see Dissolved air-pressure flotation thickeners

Air separation technology
  composition of air, II:223, 225
  distillation of air, II:223—231
  Linde's system, II:219—220
  oxygen removed from, I:27—28
  process flowsheet, II:231—237

Air stripping stage, multi-stage oxygenation system, flow diagrams, I:251

Alkalinity
  mixed liquor, effect on, I:33
  nitrification systems, II:148—151
  water, definition and description of, I:245—247

Alpha factor in wastewater, II:8—9

Aluminosilicates, II:246—250

Ammonia, II:120—122, 125, 132, 178
  equilibria, water, I:244—245
  nitrogen oxidation, see Nitrification systems
  removal from wastewater, see Nitrification systems

Anaerobic digestion of sludge, II:113, 118, 128

Anaerobic heterotrophic bacteria in nitrification systems, II:140—171

Anionic polymers, in sludge dewatering, II:100

Aqueous-phase ionic equilibria of water, I:240—248

Arbitrary flow reactor, see Nonideal flow reactors

Argon, boiling point, II:223, 225

Arrhenius equation, II:121

Atmospheric vaporizers, II:241

ATP, see Adenosine triphosphate

Attainable concentrations, organic combustibles in reactors, II:195—196

Autooxidation of sludge, II:118

Autothermal thermophilic aerobic digestion process, II:118—137

Autotrophic bacteria in nitrification systems, II:140—171

B

Backbone structure, biofloc particles, I:56

Backmix reactor, see Complete-mix reactor

Backup systems
  liquid oxygen, see Liquid oxygen backup systems
  safety, II:205—207
  UNOX® Systems, II:52, 54

ozonation systems, II:178—179, 183, 186—187
pressure swing oxygen production systems, II:251—256
supports, UNOX® reactors, II:49—50
UNOX® Systems, see UNOX® Systems, equipment
wastewater disinfection systems, II:178—179, 183
Expansion turbine, cryogenic oxygen production systems, II:222, 237—238
External weirs, UNOX® Systems, II:38

F

Fecal streptococci in sludge, II:128—135
Fecal coliforms in sludge, II:128—135, 178
Feed air compressor package, PSA systems, II:252
Feed characteristics and fluctuations, II:211
Feed gas, effect on oxygenation systems, I:33—35, 41—43
Feed gas oxygen mole fraction, I:133—136, 255
Feed solids concentrations, sludge dewatering systems, II:86—113
Fermenter system, see Laboratory fermenter system
Ferric chloride, in sludge conditioning, II:91—92, 100, 102, 110
Field testing
  aerobic sludge digestion, II:129—135
  bio-precipitation process, I:26
  EPA studies, see Environmental Protection Agency, evaluations of oxygenation systems
  oxygenation systems, in current use, I:9—13
  pressure swing oxygen production systems, II:257—258
  sludge dewatering, II:85, 89—90, 94—97, 101—102, 109—110
  sludge production, I:57—60; II:76—77
  solid liquid separation and clarifier design, I:204—219
  UNOX® Systems, I:7, 152—169
Fill-and-draw studies, bio-precipitation process, I:15, 18
Film, gas, mass transfer theory, II:22
Film manufacturing and processing industry, wastewater treatment, UNOX® studies, I:152, 159, 193—194
Filter media, II:98, 100, 101, 108—109
Filter presses, II:108—113
Filters, types of, II:98, 220
Filtration, in sludge dewatering
  belt pressure, see Belt pressue filtration
  pressure, see Pressure filtration
  vacuum, see Vacuum filtration
First-stage pressure control, UNOX® Systems, II:54—56
Five day biochemical oxygen demand, defined, I:131
Flammability of gases, II:193—195
Floc, see also Biomass
  aerobic, II:77—80
  biopolymeric materials and, see Biopolymeric materials
  bio-precipitation treatment of, I:15—24
  oxygen utilization of, in bio-precipitation systems, I:18—22, 103, 107
  particles, size, shape, and activity, I:48—53, 56—57, 103, 107, 127—128; II:147
Flotation thickening of activated sludges, II:87—88
  compared to gravity method, II:89—91
  UNOX® Systems, II:89—91
Flow conditions, nonideal, see Nonideal flow

Flow measurement
  oxygen, UNOX® Systems, II:57—58
  sludge, see Mass transfer
Flux, see Solids flux
FMC open-tank oxygenation systems, I:83—94
Foam control, UNOX® reactors, II:48—49, 52
Food processing industry, wastewater treatment, UNOX® studies, I:152, 154, 158, 183—184, 191—192
Food to microorganism (F/M) ratio, see also Biomass
  nitrification systems, II:141—142, 155—170
  oxygen supply systems, II:210—211
  sludge production, I:102—103, 105—107; II:63—65, 71—72, 76—80
F³O System, I:13—14
Fractional distillation of air, II:223—225
Full-scale operations, see Field testing

G

Gas
  considerations, UNOX® reactors, II:37—40, 49
  phase, UNOX® Systems, II:27—28
Gas dispersion devices, see also Bubble diffusers; Submerged turbines
  mathematical model, multi-stage oxygenation system, use in, I:248—256
Gas-liquid contacting system, oxygen mass transfer performance, see also Contact stabilization systems, I:33—34, 239—240
  UNOX® Systems, see UNOX® Systems, gas-liquid contacting method
Gas-liquid contactor design, UNOX® System, see UNOX® System, gas-liquid contactor design
Gas sample ports, UNOX® Systems, II:49, 51, 59
Gassed operation, UNOX® Systems, II:26—28
Gas space oxygen material balances, I:124, 254
Gel traps, II:234—235
Generator characteristics, oxygen production
  cryogenic oxygen production, see also Cryogenic oxygen production, II:213—214
  pressure swing adsorption oxygen production, see also Pressure swing oxygen production, II:214—215, 250—258
  sizing considerations, II:215—216
Grab sample ports, UNOX® Systems, II:51
Graphical clarifier design procedure, I:200—202
Gravity thickening of activated sludge, II:86—87
  compared to flotation method, II:89—91
  UNOX® Systems, II:89—91

H

Heat balances, aerobic digestion systems, II:122—123, 125
Heat exchangers, see also Reversing heat exchanger, II:220, 232—236
Heats of combustion, organic compounds in sludge, II:120—121
Heat treatment, for dewatering sludge, II:92—97
  commercial systems, II:94—97

Heterotrophic bacteria in nitrification systems,
II:140—171
High-rate activated sludge system, compared to bio-
precipitation process, figure, I:20
High-speed direct-drive surface aerators, II:12
Hydraulics, II:10—11
  ozonation systems, II:183
  testing methods, II:31—32
  thickener units, II:86—87
  UNOX® Systems, II:16—21, 25—28, 31—32, 37,
    48—49
Hydrocarbons, see also Contaminants
  biodegradation, II:121
  monitoring, II:58, 204
  prechlorination and, II:152
  removal, II:51—52, 203, 232, 234—235, 251
  safety considerations, II:192—204, 223, 235
  stripping, II:49, 56, 198
Hydrogen ion equilibria of water, I:240—241
Hydrogen sulfide equilibria of water, I:122, 243—245

I

Ignition of combustibles, prevention of, II:199
Ilosvay test, II:204
Impeller design and geometries, UNOX® Systems, II:17,
  19, 25—27
Impeller-sparger systems, see Submerged turbines
Industrial applications of ozonation, II:184
Industrial wastewater treatment, UNOX® studies, I:152,
  154—159, 187—194
Influent distribution UNOX® reactors, II:37—38
Inhibitory substances, nitrification systems, II:152—156
Inorganic coagulants, in sludge conditioning, II:92—93
Instantaneous oxygen demand, I:121—122
Instrument air skid, II:254—256
Instrumentation, see also Equipment; specific instruments
    by name
  aeration systems, II:201
  aerobic digestion systems, II:136
  cryogenic oxygen production systems, II:220, 241—242
  ozonation systems, II:186—187
  pressure swing oxygen production systems, II:252
  UNOX® Systems, II:50, 53—58
Internal weirs, UNOX® Systems, II:38
Ionic equilibria of mixed liquor, I:33, 126, 240—248
Isotherms, adsorption, see Adsorption isotherms

J

Joints, UNOX® reactors, II:49
Joule-Thomson effect, cooling from, II:238

K

k, determination of, I:132—133

Kinetics
  aerobic sludge digestion, II:121—122
  Monod model, nitrification, II:142—143, 145, 163
  nitrification systems, II:141—155, 161—170
  oxygen activated sludge process, I:102—104, 125—133,
    145—146
  sludge production, II:61—65
Kjeldahl nitrogen, see Total Kjeldahl nitrogen
k_N factor, nitrification systems, II:143, 161—167

L

Laboratory fermenter system, UNOX® Systems,
  I:168—169
Laboratory test models
  aerobic sludge digestion, II:123, 128—135
  bio-precipitation process, I:15—21
    schematic diagram, laboratory apparatus, I:19
  nonideal flow reactors, II:42—45
  pathogens in sludge, II:128—137
  sludge dewatering, II:86, 111—112
  sludge production, I:58—60; II:65—68
  treatability studies, see Treatability studies, UNOX®
    Systems
Las Virgenes (Calif.) EPA project, evaluation of
    oxygenation systems, I:79—83
  schematic and flow diagrams, I:80
Lime, in sludge conditioning, II:91—92, 100, 110
Linde® molecular sieve adsorbents, II:243—244
Liquid dispersion devices, see also Surface aerators
  mathematical model, multi-stage oxygenation system,
    use in, I:248-256
Liquid oxygen backup systems, II:54, 56, 215—217, 223,
    238—241
  production of liquid oxygen, II:213, 220, 237, 241
  safety, II:205—207
  storage tanks, I:30; II:213, 239—241, 256
  UNOX® Systems, see UNOX® Systems, liquid oxygen
    backup
  vaporizers, I:30; II:240—241
Liquid/volume ratio, cryogenic oxygen production,
    II:228-231, 235
Lower column, see Double column separation of air
Lower explosion limit, II:192—200, 204
Low-speed surface aerators, II:12
Low temperature processing of air, see Cryogenic oxygen
    production
Lowther plate ozonator II:177
Lubrication system, cryogenic oxygen production,
    II:220—221

M

Macroscale mixing, activated sludge aeration basins,
  I:49—50, 130
Main condenser, air separation systems, II:236-237
Maintenance, UNOX® reactors, II:50
Marox system, I:13
Mass balances, aerobic digestion systems, II:122

N

Natural nitrogen cycle, II:140—141
Net yield, see also Yield
    nitrification systems, II:144—145, 163
    sludge production, II:65, 73—77
Newtown Creek (N.Y.) EPA project, evaluation of
    oxygenation systems, I:57—60; 71—79
    schematic diagram, I:73
Nitrate, II:120—122
    oxidation, see Nitrification
Nitrification systems
    biological processes, types, I:140—141; II:155—156
    design, see Design, nitrification systems
    general discussion, II:140—141
    nomenclature, list of abbreviations, II:171
    operation, II:161—170
    oxygen, advantages of, II:170
    pH effects, II:147—148
    sludge retention time, I:140—141
    studies of, I:68—69, 181
    UNOX® application studies, I:181; II:156—163
Nitrifier population dynamics, II:141—142
Nitrite oxidation, see Nitrification
Nitrobacter sp. in nitrification systems, I:140;
    II:140—141, 145—146, 151, 168
Nitrogen
    adsorption, see Adsorbents, molecular sieve
    aerobic digestion systems, II:122, 125, 132
    bio-precipitation process, effect on, I:20
    boiling point, II:223, 225—226, 236
    cycle, II:140—141
    Kjeldahl, see Total Kjeldahl nitrogen
    loading, II:94—95, 161, 164
    nitrification, see Nitrification systems
    production of, II:231
    regulations, reason for, II:140
    solubility in wastewater, I:33
    superheater, II:235—236
    transfer in wastewater, I:17
    wastewater, content of, effect on alkalinity, I:246—247
Nitrosomonas sp. in nitrification systems, I:140;
    II:140—141, 144—146, 151, 168
Nomenclature
    mass transfer design, I:258—260
    nitrification processes, II:171
    oxygentation system process design, I:146—148
    oxygenation tank-clarifier design integration, I:233
    solid-liquid separation and clarifier design, I:224
Nonbiodegradable volatile solids, II:61, 80, 119—120
Nonbackmix-flow reactor, see Plug-flow reactor systems
Nonideal flow reactors, II:41—48

O

OASES® system, I:13, 67
Odor problems, II:87, 95, 110, 184
Open-top treatability reactor, UNOX® Systems, I:168
Optimum oxygen utilization, I:133—136
    UNOX® Systems, I:35—37
Organic combustibles, see Combustible materials

Organic loading rate, calculation of, I:102—103,
    112—118, 230—231
Organic loadings, UNOX® Systems, I:51—53, 57—60,
    171
Organic shock, see Shock loadings
Organisms, nitrifying, II:140
Otto plate ozonator, II:176
Overflow rate, see Surface overflow rate
Oxidation of sludge, see Air aeration systems;
    Oxygenation systems
Oxygen
    aeration, related to generation, II:210
    air, separation from, process, see Air separation
        technology
    balances, aerobic digestion systems, II:122
    bio-precipitation process and, I:15—24, 26
    boiling point, II:223, 225—226, 236
    characteristics of pure, I:17
    combustion, requirements for, II:192
    cryogenic, see Cryogenic oxygen production
    demand, see Oxygen demand
    diffusion coefficient, I:106—107
    disposal of, II:207
    dissolution equipment, see Dissolution equipment
    dissolved, see Dissolved oxygen
    generator sizing considerations, II:215—216
    handling of, safety considerations, II:206—207
    leakage, II:39—40
    liquid, see Liquid oxygen backup systems
    monitoring, UNOX® System, II:58
    nitrification, advantages of, II:170
    pressure swing adsorption, see Pressure swing
        adsorption oxygen production
    production, see Supply (oxygen) considerations
    properties of, II:192
    requirements for nitrification, II:145, 168
    safety, see Safety considerations
    superheater, II:236
    supply systems, see Supply (oxygen) considerations
    uptake
        plug flow aeration tanks, I:26—27
        rate, determination of, I:123—124
        UNOX® Systems, I:34, 174—175
Oxygen-activated sludge systems, see Oxygenation systems
Oxygenation systems
    activity of sludge equations, II:77—80
    aerobic sludge digestion, see Aerobic sludge digestion
    air-aeration systems, compared to, I:28—29
    Airco, Incorporated, F³O System, I:13—14
    anaerobic sludge digestion, see Anaerobic digesters
    biological design influences, II:210—211
    bio-precipitation, see Bio-precipitation process
    clarifiers, see Clarifiers
    cleaning of systems, II:207
    conventional, bio-precipitation systems compared with,
        I:22—23, 26
    cryogenic oxygen production, see Cryogenic oxygen
        production
    design, see Design
    EPA evaluations, see Environmental Protection
        Agency, evaluations of oxygenation systems
    equipment, see Equipment
    feed gas, effects of, I:33—35, 41—43

Weirs, UNOX® Systems, II:38—40
Weir sight ports, UNOX® Systems, II:51
Wet air oxidation, in sludge conditioning, II:92—94
Winkler titration method, II:32—35

Y

Yield
   filter systems, II:98—100

nitrification systems, II:142—145
   sludge production systems, see Net yield
Yield coefficient, I:125

Z

Zeolites, II:246—250
Zimpro heat treatment systems, II:94—97

CRC PUBLICATIONS OF RELATED INTEREST

CRC HANDBOOKS:

CRC HANDBOOK OF CHROMATOGRAPHY
Edited by **Gunter Zweig, Ph.D.**, Chief, Chemistry Branch, Criteria and Evaluation Division., Office of Pesticide Programs, EPA and **Joseph Sherma, Ph.D.**, Chemistry Department, Lafayette College.
Volume I contains a listing of 549 tables expressing chromatographic data for substance identification.
Volume II provides an easy-to-read overview of theory and practices of various fields of chromatography.

CRC HANDBOOK OF TABLES FOR MATHEMATICS
Edited by **William H. Beyer, Ph.D.**, Professor of Mathematics and Head of Department of Mathematics and Statistics, University of Akron.
This new, revised edition offers concise information, free from redundancies.

CRC HANDBOOK OF TABLES FOR ORGANIC COMPOUND IDENTIFICATION
Edited by **Zvi Rappoport, Ph.D.**, Hebrew University of Jerusalem.
Compounds are divided in 26 groups with information on each compound including name, boiling or melting point, refractive index, density, and properties of up to eight derivatives.

CRC UNISCIENCE PUBLICATIONS:

FIXED BIOLOGICAL SURFACES — WASTEWATER TREATMENT
Edited by **Ronald L. Antoinie**, Manager of Technical Services, Autotrol Corp.
Author traces development and use of fixed film treatment systems, including the rotating contactor, discussed in detail.

INDUSTRIAL AIR POLLUTION CONTROL EQUIPMENT FOR PARTICULATES
Edited by **Louis Theodore, D. Eng. Sc.**, Manhattan College and **Anthony J. Buonicore, P. E.**, Entoleter, Inc., New Haven, Ct.
This book is directed toward the fundamental and design principles of industrial control equipment for particulate pollutants.

INDUSTRIAL CONTROL EQUIPMENT FOR GASEOUS POLLUTANTS, VOL I AND II
Edited by **Louis Theodore, D. Eng. Sc.**, Manhattan College, and **Anthony J. Buonicore, P. E.**, Entoleter, Inc., New Haven, Ct.
This reference is directed towards the fundamentals and design principles of industrial control equipment for gaseous pollutants with appropriate practical applications.

NOISE AND NOISE CONTROL
Edited by **M. J. Crocker, Ph.D.**, Purdue University, and **A. J. Price, Ph.D.**, University of British Columbia.
This reference offers a comprehensive evaluation of the various noise control principles and procedures in buildings, the community, industry, and transportation vehicles.

RECENT DEVELOPMENTS IN SEPARATION SCIENCE
Edited by **Norman N. Li, Sc.D.**, Corporate Research Laboratories, Exxon Research and Engineering Co.
This multivolume reference series discusses recent developments in science and technology of separation and purification.

TRACE ELEMENT MEASUREMENTS AT THE COAL-FIRED STEAM PLANT

Edited by **W. S. Lyon, Jr., M.S.,** Oak Ridge National Laboratory.

This is an examination of an exhaustive trace element balance made at the T. A. Allen Steam Plant in Memphis, Tenn. by an interdisciplinary group of scientists and technologists.

CRC CRITICAL REVIEWS:

CRC CRITICAL REVIEWS™ IN BIOCHEMISTRY

Edited by **Gerald D. Fasman, Ph.D.,** Brandeis University.

CRC CRITICAL REVIEWS™ IN MICROBIOLOGY

Edited by **Allen I. Laskin, Ph.D.,** Exxon Research and Engineering Co., and **Hubert Lechevalier, Ph.D.,** Rutgers University.